WATER DEFICITS
AND PLANT GROWTH

VOLUME IV

Soil Water Measurement, Plant Responses,
and Breeding for Drought Resistance

CONTRIBUTORS TO THIS VOLUME

W. G. ALLAWAY

S. G. BOATMAN

J. S. BOYER

F. J. BURROWS

B. R. BUTTERY

E. A. HURD

T. T. KOZLOWSKI

F. L. MILTHORPE

S. L. RAWLINS

JANET I. SPRENT

WATER DEFICITS
AND PLANT GROWTH

EDITED BY

T. T. KOZLOWSKI

DEPARTMENT OF FORESTRY
THE UNIVERSITY OF WISCONSIN
MADISON, WISCONSIN

VOLUME IV

Soil Water Measurement, Plant Responses,
and Breeding for Drought Resistance

1976

ACADEMIC PRESS   New York   San Francisco   London

A Subsidiary of Harcourt Brace Jovanovich, Publishers

ACADEMIC PRESS, INC.
111 Fifth Avenue, New York, New York 10003

United Kingdom Edition published by
ACADEMIC PRESS, INC. (LONDON) LTD.
24/28 Oval Road, London NW1

Library of Congress Cataloging in Publication Data

Kozlowski, Theodore Thomas, (date)
        Water deficits and plant growth.

        Includes bibliographies.
        CONTENTS: v. 1.   Development, control and measure-
ment.—v. 2.   Plant water consumption and response.—
v. 3. Plant response and control of water balance.—
v. 4.   Soil water measurement, plant responses, and
breeding for drought resistance.
        1.   Plants—Water requirements.   2.   Growth (Plants)
3.   Plant water relationships.   I.   Title.
QK870.K69   582'.01'335   68-14658
ISBN 0–12–424154–9

CONTENTS

1. MEASUREMENT OF WATER CONTENT AND THE STATE OF WATER IN SOILS
S. L. RAWLINS

2. STRUCTURE AND FUNCTIONING OF STOMATA
W. G. ALLAWAY AND F. L. MILTHORPE

3. STOMATAL CONDUCTANCE IN THE CONTROL OF GAS EXCHANGE
F. J. BURROWS AND F. L. MILTHORPE

4. WATER DEFICITS AND PHOTOSYNTHESIS
J. S. BOYER

5. WATER SUPPLY AND LEAF SHEDDING
T. T. KOZLOWSKI

6. WATER DEFICITS AND FLOW OF LATEX
B. R. BUTTERY AND S. G. BOATMAN

7. WATER DEFICITS AND NITROGEN-FIXING ROOT NODULES
JANET I. SPRENT

8. PLANT BREEDING FOR DROUGHT RESISTANCE
E. A. HURD

Contents

LIST OF CONTRIBUTORS

Numbers in parentheses indicate the pages on which the author's contributions begin.

W. G. ALLAWAY (57), School of Biological Sciences, University of Sydney, N. S. W., Australia

S. G. BOATMAN* (233), ICI Plant Protection Division, Jealott's Hill Research Station, Bracknell, Berks, England

J. S. BOYER (163), Department of Botany and Agronomy, University of Illinois, Urbana, Ilinois

F. J. BURROWS (103), School of Biological Sciences, Macquarie University, North Ryde, N. S. W., Australia

B. R. BUTTERY (233), Agriculture Canada Research Station, Harrow, Ontario, Canada

E. A. HURD (317), Research Branch, Canada Department of Agriculture, Swift Current, Saskatchewan, Canada

T. T. KOZLOWSKI (191), Department of Forestry, University of Wisconsin, Madison, Wisconsin

F. L. MILTHORPE (57, 103), School of Biological Sciences, Macquarie University, North Ryde, N. S. W., Australia

S. L. RAWLINS (1), U. S. Salinity Laboratory, USDA, ARS, Riverside, California

JANET I. SPRENT (291), Department of Biological Sciences, University of Dundee, Dundee, Scotland

* Present address: ICI United States Inc., Biological Research Center, P.O. Box 208, Goldsboro, North Carolina.

PREFACE

Following the enthusiastic response to the first three volumes of this work a number of investigators called to my attention the pressing need for reviewing some additional topics and their implicatons in crop yield, water use, and drought resistance. Hence, this volume was planned to supplement the earlier ones, with particular emphasis on subject matter areas in which there has been unabated interest, intense research activity, and marked progress.

The opening chapter is a modern treatment of measurement of soil water content and the state of water in soils. Particular emphasis is placed on methods developed from recent technological advances. The next two chapters deal at an advanced level with structure and functioning of stomata and stomatal conductance in control of gas exchange (measurement of stomatal and leaf conductance; internal and external control of stomatal aperture). Separate chapters follow on effects of water supply on photosynthesis (state of knowledge of how desiccation inhibits photosynthesis in terms of stomatal aperture, chloroplast activity, and respiration); leaf shedding (effects of summer drought, winter desiccation, flooding, diseases, and chemical desiccants); flow of latex (laticifer anatomy, rubber biosynthesis, and water relations of latex vessels); nitrogen-fixing root nodules (effects on nitrogen fixation activity, respiration, and enzyme localization). The final chapter is a comprehensive treatment of plant breeding for drought resistance, with emphasis on breeding and testing methods as well as on parameters of drought resistance and their application in breeding programs.

In planning this volume invitations to submit chapters were extended to researchers of demonstrated competence in the United States and abroad. I am indebted to each of them for their scholarly work, cooperation, and patience during the production phases. Mr. T. L. Noland and Mr. S. G. Pallardy assisted with preparation of the Subject Index.

T. T. Kozlowski

CONTENTS OF OTHER VOLUMES

CHAPTER 1

MEASUREMENT OF WATER CONTENT AND THE STATE OF WATER IN SOILS

S. L. Rawlins

U.S. SALINITY LABORATORY, USDA, ARS, RIVERSIDE, CALIFORNIA

I. INTRODUCTION

A full description of the soil-water system requires information on the relative quantities of soil, water, and dissolved constituents, as well as information on the state of water in the system. Any method or technique for measuring a specific property presupposes its careful definition. Though many proposals for terminology have been made, none will completely satisfy all users under all conditions. The definitions that follow, taken from a report by the Soil Physics Terminology Committee of Commission I of the International Society of Soil Science, ISSS (1975), are chosen because they are based on experimentally accessible parameters. The term water is reserved for the chemical constituent, H_2O. "Liquid phase" is used for aqueous solutions *in situ,* whereas the liquid phase is referred to as "soil solution" when considered separately from the soil.

1

A. Terms Relating to the Composition of the Soil

1. Densities

The composition of the soil can be fully described in terms of the densities of all constituents within the solid (s), liquid (l), and gas (a) phases. Densities, ρ, can specify the mass of a component present in a unit volume of either the appropriate phase or the bulk soil. To simplify the notation, if the density of a constituent present in a phase is given in terms of a unit volume of its own phase, no superscript is used. If its density is given in terms of a unit bulk soil volume, it is preceded by a superscript b. Densities of pure components in pure phases (viz., the density of pure water in the liquid phase) are preceded by a superscript 0.

a. Phase Density. The phase density of a constituent i (present in a phase α), ρ_i or $\rho_{i(\alpha)}$, is the mass of the constituent i in a unit bulk volume of phase α. Summing up the phase densities of all constituents of a phase gives the phase density, $\rho_\alpha = \Sigma_i \, \rho_{i(\alpha)}$.

b. Bulk Density. The bulk density of a constituent i (present in a phase α), $^b\rho_i$ or $^b\rho_{i(\alpha)}$, is the mass of constituent i (present in phase α) per unit bulk volume of soil. The phase density and bulk density of a constituent i present in a phase are related via the volume fraction of the phase, $\phi_\alpha = {}^b\rho_{i(\alpha)}/\rho_{i(\alpha)}$. Summing up the bulk densities of all constituents in all phases gives the bulk density of the soil,

$$^b\rho = \Sigma_\alpha \Sigma_i {}^b\rho_{i(\alpha)} = \Sigma_\alpha \phi_\alpha \rho_\alpha$$

Because the density of the gas phase usually is negligible, its bulk density in a dry soil equals the bulk density of the solid phase, $^b\rho = {}^b\rho_s$.

2. Water Content

Each of the several ways of expressing the liquid content of the soil has advantages under certain conditions.

a. Wetness. The water content or wetness, w, is the amount of water lost from the soil upon drying at 105°C, expressed as mass of water per unit mass of solid phase after drying.

If the water present in the vapor phase is disregarded, $w = {}^b\rho_w/{}^b\rho_s$.

b. Volume Fraction of Liquid. The volume fraction of liquid, θ, is the volume of the liquid phase per unit bulk of soil. Accordingly, $\theta = w({}^b\rho_s/\rho_w)$. (Usually, for lack of information, one is satisfied with the density of pure water, $^0\rho_w$, for this purpose.)

B. TERMS RELATING TO THE STATE OF WATER IN SOILS

At equilibrium, the constituent water of the liquid phase is subject to the action of the gravitational field, to the influence of dissolved salts and of the solid phase (including adsorbed ions) in its given geometry of packing, and to the action of the local pressure in the gas phase. Together these factors determine the value of the total potential, ψ_t, of the water in the soil relative to a chosen standard state. Selecting as standard a system S_0, that consists of a pool of pure (i.e., water not influenced by dissolved salts, or in other words, water whose osmotic pressure, π, is zero), free (i.e., water not influenced by the solid phase) water at temperature T_0, height h_0, and atmospheric pressure P_0, one defines:

a. Total Potential. The total potential, ψ_t, of the constituent water in the soil at temperature T_0, is the amount of useful work per unit mass of pure water that must be done by means of externally applied forces to transfer reversibly and isothermally an infinitesimal amount of water from the standard state S_0 to the soil liquid phase at the point under consideration.

It is convenient to divide the transfer process into several steps by introducing substandards according to

S_1: a pool of pure, free water as in S_0, but situated at the same height as the soil liquid phase under consideration, h_x, i.e., S_1 is at T_0, h_x, P_0.

S_2: a pool of free soil solution identical in composition with the soil liquid phase at the point under consideration, thus having an osmotic pressure, π, but otherwise identical with S_1, i.e., S_2 is at T_0, h_x, P_0.

Considering the transfer of water from standard state S_0, via the substandards S_1 and S_2, to the soil liquid phase leads to the following definitions of the component potentials of the constituent water.

b. Gravitational Potential. The gravitational potential, ψ_g, of the constituent water in soil at temperature T_0, is the amount of useful work per unit mass of pure water that must be done to transfer reversibly and isothermally an infinitesimal quantity of water from the standard S_0 to the substandard S_1, as defined above. This potential may be expressed as the difference in height between S_0 and S_1, $h = h_x - h_0$, according to $\psi_{gf} = g\,\Delta h$, in which g is the magnitude of the gravitational force per unit mass.

c. Osmotic Potential. The osmotic potential, ψ_o, of the constituent water in soil at temperature T_0, is the amount of useful work per unit mass of pure water that must be done to transfer reversibly and isothermally an infinitesimal quantity of water from the substandard S_1 to the substandard S_2, as defined above. This potential may be expressed in terms of the experimentally accessible osmotic pressure of the solution, π, according to

$$\psi_o = - \int_0^\pi \bar{V}_w \, dp$$

in which $\bar{V}_w$ is the partial specific volume of the constituent water in the soil solution.

d. Pressure Potential. The tensiometer-pressure potential, ψ_p, briefly referred to as the pressure potential of the constituent water (*in situ*), is the amount of useful work per unit mass of pure water that must be done to transfer reversibly and isothermally an infinitesimal quantity of water from the substandard S_2 to the soil liquid phase at the point under consideration. This potential may be expressed in terms of the experimentally accessible tensiometer pressure of the soil liquid phase *in situ*, p, according to

$$\psi_p = \int_0^p \bar{V}_w \, dp$$

Accordingly, the total potential may be found from the relation

$$\psi_t = \psi_g + \psi_o + \psi_p = g \, \Delta h - \int_0^\pi \bar{V}_w \, dp + \int_0^p \bar{V}_w \, dp$$

Although these three components completely define the energy status of the soil water in terms of easily measurable parameters, for some purposes it is convenient to divide the pressure potential into subcomponents. As defined above, pressure potential ranges from negative to positive values, depending on the relative magnitudes of overburden pressure, gas pressure, matrix configuration, and water content and configuration. Again using the tensiometer reading as the experimentally accessible measure of the pressure of the soil liquid phase *in situ*, p, one may introduce two components of the pressure potential according to $\psi_p = \psi_p{}^m + \psi_p{}^a$. One may then define the pneumatic and matric potentials.

e. Pneumatic Potential. The pneumatic potential, $\psi_p{}^a$, is the increment of ψ_p when an excess gas pressure $p_a = p_a - p_0$ is applied to a soil sample with given wetness and subject to a given envelope pressure. Insofar as the application of p_a does *not* influence the geometry of the liquid phase, this potential may be calculated according to

$$\psi_p{}^a = \int_0^{\Delta p_a} \bar{V}_w{}' dp$$

in which $\bar{V}_w'$ is the partial specific volume of water in the soil liquid phase *in situ*. In practice it is assumed to be equal to $\bar{V}_w$.

f. Matric Potential. The matric potential, $\psi_p{}^m$, is the value of ψ_p of a soil sample at given wetness and subject to a given envelope pressure but with $p_a = 0$.

For swelling soils, the matric potential is further subdivided into the "envelope-pressure" potential, and the wetness potential.

II. METHODS OF MEASURING SOIL-WATER CONTENT

Wetness is the fundamental measure from which all other expressions for water content can be derived by use of the phase and bulk densities. It can be measured either directly or indirectly.

Water content can be measured directly by determining the amount of water removed from a sample by evaporation, displacement, or chemical reaction (Gardner, 1965). Most of these measurements involve taking a soil sample to the laboratory. Because a measurement cannot be repeated at the location where a sample is removed, these methods do not permit continuous measurement at a given site. However, direct methods are generally well suited for detailed one-time surveys because their cost per sample is low.

Water content can be determined indirectly by measuring some physical or chemical property of the soil that varies with water content. Many indirect methods can be used to measure soil water content *in situ,* either continuously with permanently installed transducers or periodically by use of access tubes into which portable transducers can be lowered. Because *in situ* methods generally involve a relatively large investment per site, they are usually not well suited for one-time surveys requiring numerous samples. They are the only means, however, for following the water content changes with time at specific sites.

In this chapter, emphasis will be placed on new methods that have evolved from recent advances in technology. For more details on traditional methods, see the reviews by Gardner (1965) and Cope and Trickett (1965), as well as discussions in several recent textbooks on soil physics (Baver *et al.,* 1972; Hillel, 1971; Taylor and Ashcroft, 1972).

A. DIRECT METHODS

1. Thermogravimetric Methods

The thermogravimetric method is the standard to which all others are usually calibrated. It consists of weighing a moist soil sample in a tared container, drying it in an oven at 105°C to a constant weight, and reweighing it to determine the water lost. Soil samples are generally transported from the field to the laboratory in metal containers that can be placed directly into the oven. Any water condensing on their walls during transport is thus included in the first weighing. If the lids of the containers are not sealed hermetically, this procedure should be used only if the samples can be weighed soon after they are taken, or if water loss can be reduced by refrigeration until the first weighing. Otherwise, samples can be sealed in plastic bags for transport and any condensed water can easily be remixed with the sample by kneading the bag thoroughly before the sample is placed into containers for oven drying.

Gardner (1965) points out that the thermogravimetric method does not give an unequivocal dry state. The colloidal fraction of soils retains water in both the structural and adsorbed states. As the drying temperature increases, the material loses more and more water. The adsorption isotherms for most materials do not show which part of the water is adsorbed and which part is structural. But it is clear from the continuous variation in water content with increasing temperature that the temperature of drying must be specified precisely. Consequently, water content will not be measured precisely except where extraordinary attention is given to temperature measurement and control. Because oven air temperature can vary many degrees with time (and with position in convection ovens), precise soil water content measurements require that temperature be measured within the soil sample, rather than in the oven.

Another problem is defining the dry condition if the organic fraction of the soil is significant. Because organic materials oxidize at high temperatures, it is not practical to dry the samples containing significant amounts of organic matter to a constant weight. Therefore, the dry state for such samples must be regarded as an arbitrary value. Gardner (1965) suggested that since accuracy in water content determinations hinges so much upon a definable dry condition, it is often more appropriate to refer to reproducibility rather than to accuracy. Reproducibility in water content measurements can be achieved by treating every sample of a set to be compared in exactly the same way in terms of sample size, depth in the container, drying temperature, and drying time, or by following techniques that lead to equilibria that are as nearly independent as possible of such variables.

Although the thermogravimetric method usually yields the water content or wetness w on a mass basis, the volume fraction of liquid θ can be determined directly by simply measuring the water content in a given volume of soil. A volumetric soil sampler, such as that described by Richards and Stumpf (1960), can be used satisfactorily with many soils, but not with gravelly and stony soils. Because the scale of nonuniformity with such soils is large, it is necessary to take large sample volumes. One method for determining the volume of soil removed involves sampling with a shovel, placing a rubber or plastic membrane in the hole, and filling it with a measured quantity of water, sand, or some other material that readily packs to a constant bulk density.

Methods other than conventional oven drying are often chosen to decrease the drying time. Forced-draft ovens usually require a minimum of 10 to 12 hr and convection ovens can take twice as long to bring samples to constant weight. Some alternatives are radiation drying with heat lamps, or vacuum-oven drying. With heat lamps, however, the uncertainty of the drying temperature leads to errors greater than those obtained with closely controlled constant temperature ovens. Overheating should be less with dielectric heating because heat is produced by absorption of the radiation directly by the water in the sample rather than by conduction from the outside. Because water is such an excellent absorber of microwave energy, to some extent heat is produced in proportion to the wetness of the sample. Therefore, as one portion of the sample begins to dry the heat produced within its diminishes, the sample tends to dry uniformly, and the exterior of the samples does not need to be overheated to create the necessary gradient to heat its interior. Stewart (1960) successfully used dielectric heating to dry small soil columns to a uniform water content above oven dryness and Miller *et al.* (1974) proposed it as a rapid method to replace oven drying. The technique Miller *et al.* used, which consisted of placing soil samples in pyrex beakers into a commercial microwave oven, yielded minimum drying times dependent upon the number of samples within the oven, as well as their water content, weight, and soil type. The relationship between minimum drying time and water content for different sized soil samples for two soil types is shown in Fig. 1. Minimum drying times were determined by placing previously weighed soil samples into the oven until they were visibly dry, and then repeating the procedure of reweighing and returning them to the oven for 1-min intervals until the soils reached constant weight. Samples previously dried in the microwave oven did not decrease in weight when placed in the conventional oven for 24 hr, but samples previously dried in the conventional oven for 30 hr lost an additional 0.9% water when placed in the microwave oven. The additional water removed by the microwave oven is probably explained by the fact that when soil samples were removed from the microwave oven, their temperatures were nearly 200°C. If microwave oven drying is to gain general use, further experimentation will be required to verify that the difference between oven drying and microwave oven drying is consistent enough that a standard calibration procedure can be used.

In addition to speed, microwave drying may have an advantage if

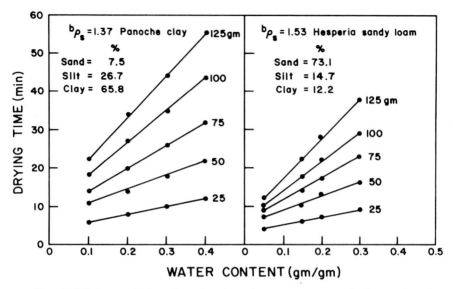

Fig. 1. Minimum drying times by the microwave oven method as a function of initial water content for five sample weights and two soils. Redrawn from Miller *et al.* (1974).

the dry soil samples are to be analyzed chemically. In exploratory work with Pinoche clay loam, Miller *et al.* found the nitrate content after microwave drying was about equal to that in the fresh field soil. Air drying reduced the nitrate content approximately 20% and oven drying reduced it approximately 10%.

Two other thermogravimetric methods are worth mentioning. The first, reviewed by Gardner (1965), involves drying the soil sample by saturating it with alcohol and igniting it. The precision of the method is reported to be about ±0.5 to ±1% water content, which is almost as good as oven drying when elaborate precautions for temperature control and drying time are not taken. The other method, described by Thijssen *et al.* (1954), is based on evaporation of water from a mixture of soil and nonvolatile oil. They claimed that the results were the same as those obtained by oven drying.

2. Miscellaneous Rapid Methods

Numerous methods have been proposed to decrease the time required to determine water content of individual soil samples. These measure the quantity of water in a soil sample by (1) displacing it with a solvent, (2) allowing it to interact with a chemical, or (3) removing the air and weighing the sample while it is immersed in water (Gardner, 1965; Cope and

Trickett, 1965). In all cases, the wet weight of each sample is required. Water content can then be calculated from either the weight of water removed or the weight of dry soil remaining.

Many of these methods approach the accuracy of that possible with the standard oven-drying procedure. One must realize, however, that not all of them measure the same thing. Many salt-affected soils contain considerable quantities of dissolved materials in the soil solution. If a solvent displaces the water from the sample, most of the dissolved materials are removed along with the water. Methods that measure the water content of the leachate (Bouyoucos, 1926; Bowers and Smith, 1972; Hancock and Burdick, 1957) rather than the weight of the dry soil are not subject to this error. Gardner (1965) points out that, depending on the quantities of dissolved substances present, the method of water removal could make a considerable difference in the water content determined.

3. Discussion and Summary

Though rapid methods do decrease the time elapsed between sampling and water content determination, most often, especially where large numbers of samples are involved, the average time spent per sample is of more importance. If waiting overnight for samples to dry causes no serious consequences, the standard oven-drying technique probably takes as little analysis time per sample as any method. This is particularly true where weighing, recording, and calculating is performed automatically. Figure 2 illustrates such automatic equipment consisting of a digital recording balance interconnected with an electronic calculator. Tare weights of the containers are stored in a specific order on a magnetic tape within the calculator. The wet- and oven-dry soil samples are then placed on the pan in turn in the same order, and the calculator records their weights and prints out a table of sample number versus water content. Each sample requires about 1 min of operator time, which includes the calculation of water content and, if desired, statistical analyses of the results.

B. INDIRECT METHODS

Indirect methods measure some property of the soil that varies with water content. One of these properties is matric potential. If the relationship between water content and matric potential is known, any method that measures matric potential can also indirectly measure water content. Unfortunately, soil water content and matric potential are not uniquely related. The matric potential is a function of the history of the water content. This phenomenon, known as hysteresis, is illustrated in Fig. 3. The soil water content for a given energy status will be greater if the soil reaches that point by drying than if it reaches it by wetting. The initial drainage

Fig. 2. Automatic weighing station for determining soil water content. It consists of an automatic balance (Ainsworth Digimetric, Model 30DT) and a digital calculator (Hewlett Packard, Model 9830). From U. S. Salinity Laboratory, Riverside, California.

curve in Fig. 3 gives the relationship between water content and pressure potential that occurs if the soil is drained from saturation, i.e., if the entire pore space is filled with water. The main drainage curve is the more typical relationship that occurs when some air is entrapped in the soil at $\psi_p{}^m = 0$. The main wetting curve is what occurs when the soil is rewet. The primary and secondary wetting and drainage curves are those that occur, for example, if, during drainage, the soil is again rewet before drainage is complete. These scanning curves are all contained within the envelope formed by the main drainage curve and the main wetting curve. Though recent work (Mualem, 1974) indicates that the scanning curves theoretically can be predicted from the envelope curves if certain soil properties are known; in practice, hysteresis is very difficult to take into account. The soil water retentivity curve depends not only on past history, but also to some extent on the time rate of change of water content (Topp and Miller, 1966; Topp, 1969; Smiles *et al.*, 1971; Poulovassilis, 1974).

Some methods based on measurement of matric potential, viz., those based on the measurement of electrical and thermal conductivity and electrical capacitance of a porous block, are often considered as indirect measures of water content. Because the soil water retentivity curve is not

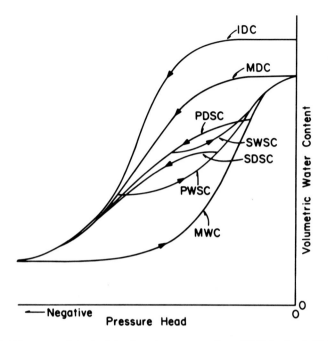

Fig. 3. Diagram of typical hysteresis curves, where IDC is the initial drainage curve, MWC and MDC are main wetting and drainage curves respectively, PWSC and PDSC are primary wetting and drainage scanning curves, and SWSC and SDSC are secondary wetting and drainage scanning curves. From Gillham (1972).

unique, they will be considered in a later section. If these electrical or thermal properties were measured in the soil itself, one would have a more direct method for measuring water content. But attempts to make such measurements have generally failed because of difficulties in providing re-producible boundary conditions between the sensor and the soil.

The indirect methods considered here are based on γ and neutron radiation. The principles of the interaction of both γ and neutron radiation with water are well understood. Basically, the degree to which a beam of monoenergetic-γ rays is attenuated in passing through a soil column depends on the wet bulk density. Thus, for constant dry bulk density, changes in the attenuation of a γ-ray beam represent changes in water content. The attenuation of neutron radiation, on the other hand, is more specific to water. Although all nuclei in the soil scatter neutrons to some degree, hydrogen is by far the most effective.

Two types of nuclear methods for measuring water content—the attenuation method and the scattering method—are used. The first method is based on the degree to which a beam of radiation is attenuated as it passes through the soil. The second is based on the return of scattered radiation to the vicinity of its source as a consequence of backscattering by soil constituents. Because attenuation methods require the source and detector to be separated, two access tubes are needed in the soil—one for the source and one for the detector. Methods based on scattering of neutron or γ radiation require only one access tube.

1. Radiation Attenuation Methods

Following Gardner (1965), the attenuation equation for monochromatic radiation (neglecting air) is

$$N_m/N_o = \exp\left[-S(\mu_s{}^b\rho_s + \mu_w\,\theta)\right] \tag{1}$$

where N_m/N_o is the ratio of the transmitted to incident flux for the moist soil, μ_s and μ_w are the mass attenuation coefficients for soil and water, θ is the volume fraction of liquid (if the density of water is unity, θ is numerically equal to the mass of water per unit bulk volume of soil), ${}^b\rho_s$ is the dry bulk density of the soil, and S is the thickness of soil between the source and the detector. Dividing this equation by the corresponding equation for dry soil (which is identical to Eq. (1) with $\theta = 0$ and N_m replaced by N_d, the transmitted flux for dry soil), assuming the bulk density does not change, yields

$$\theta = -\ln(N_m/N_d)/\mu_w S \tag{2}$$

Although the neutron attenuation method is limited to use in the laboratory because sufficiently high neutron flux is available only at nuclear reactors, it is unexcelled in terms of resolution and precision for nondestructively measuring water content of small soil columns. For example, Stewart (1962) found that for a neutron source yielding 10,000 counts per second through a 1.5-cm-thick dry-soil column, with a collimation slit only 1 mm wide and 10 mm high, the standard deviation in water content measurement for 20 sec of counting was from ±0.0007 to 0.0015 gm cm^{-3}, with the higher value being for higher water content. (For further information on this method, see Gardner and Calissendorff, 1967.)

Bernhard and Chasek (1953) first described the γ-attenuation method for indirectly measuring of wet bulk density of soil. Vomicil (1954) and Bernhard and Chasek (1955) used the same basic technique. The method consisted of placing a small ^{60}Co source into a vertical tube within the soil and a Geiger counter in an adjacent tube. Although Ashton (1956) is often credited with being among the first to use the γ-ray attenuation method to specifically measure changes in water content, his method does not differ significantly from that of Bernhard and Chasek. Ashton (who attributes

his method to Dr. George O. Burr, personal communication, 1953) monitored the water content of pots in which sugar cane was growing. Two
access tubes were installed vertically, 22 cm apart, in the soil pots to re-
ceive the 0.25-mCi ^{137}Cs source and the Geiger tube, each being 6-cm long.
Water content was inferred from counts taken for 5 min and compared
with background and standard counts taken before each set of measurements. Calibration of water content versus count rate, which was strictly
empirical, was based on the assumption that bulk density remained
constant.

a. Field Measurements. Danlin (1955) used a technique similar to that
of Ashton (1956) to measure water content in the field. Van Bavel *et
al.* (1957) replaced the Geiger tube with a scintillation detector, which
made it possible to use electronic discrimination to eliminate scattered
photons, which eliminated all but monochromatic radiation from the count
rate. They demonstrated that if monochromatic radiation only is measured,
bulk density of a dry soil could be determined from Eq. (1), provided
that the mass attenuation coefficient for the specific soil is known. Later,
van Bavel (1959) constructed a double-tube γ-attenuation system designed
for field use, which gave a measurement precision of approximately 0.01
gm cm^{-3}, and a resolution of approximately 1.3 cm, the thickness of the
NaI crystal. Again, although van Bavel was primarily interested in measuring soil density, his technique was the same as that later used for inferring
water content.

Mukhin and Christotinov (1961) improved the method for measurement of water content in the field by collimating the γ beam at both the
source and the detector. Using ^{60}Co with access tubes spaced 50 to 60
cm apart, they found water content measurements made in 10-cm increments to a depth of 60 cm in a stony soil to be more accurate than that
of any other method available. Reginato and van Bavel (1964) designed
a similar apparatus for field measurement of water content *in situ* using
a 5-mCi ^{137}Cs source. With the lower energy ^{137}Cs source, the optimum
spacing of the access tubes was about half that for ^{60}Co. By lowering the
source and scintillation detector into access tubes with holes 3 mm in
diameter spaced at 1-cm increments, they could measure water content
with a precision of about 0.006 gm cm^{-3} and a vertical resolution of about
1 cm when 20,000 counts were taken.

Further improvements in the technique for field measurement of soil-
water content by γ-ray attenuation have centered on eliminating temperature effects in the electronic circuits for the γ-ray detection equipment
(Reginato and Stout, 1970; Reginato and Jackson, 1971), evaluating and
eliminating errors resulting from nonparallelism of the γ-access tubes

(Rijtema, 1969), and more recently, use of dual energy γ-ray sources to detect changes in bulk density of the soil as it shrinks and swells (Reginato, 1974). With these improvements, and if care is taken, commercial equipment can be used to measure water content with a precision of the order of ± 0.015 gm cm^{-3}. Reginato (1974) pointed out that because the γ beam cannot be conveniently collimated in the field, water content measurements by the attenuation technique require the mass attenuation coefficient for each soil to be determined independently. Without collimation, calculated mass attenuation coefficients were lower than those calculated theoretically on the basis of the chemical constituents of the soil.

b. *Laboratory Measurements.* Use of γ-ray attenuation measurements in the laboratory to obtain data for transient analysis of unsaturated flow of water in soils originated with Ferguson and Gardner (1962), who used a stationary 20-mCi ^{137}Cs γ-ray source and a scintillation detector with collimation both at the source and the detector. Water content was followed with time at various positions along a horizontal slab of soil mounted on a dolly, which could be moved in front of the γ-ray beam. The success of this technique for measuring water content changes with distance and time as a means of experimentally verifying various theoretical models for water flow led to its adoption by numerous investigators (Gurr, 1962; Rawlins and Gardner, 1963; Davidson *et al.,* 1963).

Gurr (1962) moved the source and detector rather than the soil column—a technique that has become standard in most laboratories. Almost without exception, ^{137}Cs is used as a γ-ray source, not only because it has a long half-life, but also because it yields maximum resolution for soil columns between 15- and 20-cm thick, a convenient thickness for many laboratory studies. The laboratory, unlike the field, provides sufficient space for collimation at both the source and the detector and for use of high-intensity sources of 100 mCi or more and sufficient temperature control for use of sophisticated counting equipment to permit accurate pulse-height discrimination. Most laboratory systems use scintillation detection systems with thallium-activated NaI cystals and photomultiplier tubes, although ionization chambers and current integraters have also been used (Topp and Miller, 1966; Topp, 1970). Though ionization chambers do not require the usual pulse-height discrimination equipment to restrict counting to primary photons, they are not as highly developed for use in science and industry and therefore require individual modifications before they can be used successfully to measure soil water content.

Gardner and Calissendorff (1967) calculated the theoretical standard deviation for water content resulting from random γ emission for various γ sources as a function of the soil column thickness. For a total count of one million, a bulk density of 1.2 gm cm^{-3}, and for columns between 0 and 30 cm thick, this standard deviation was less than 0.002 gm cm^{-3}. Of course, if the bulk density is not constant, as with soils that shrink or swell, the error will be greater.

From the start of applications of γ-ray attenuation for measurement of water content, it was recognized that bulk density and water content could be measured concurrently by making independent attenuation measurements at two different energies. By solving Eq. (1) twice with the appropriate mass attenuation coefficients for soil and water for each radiation source, the two unknowns of water content and bulk density can be determined independently. It is only a matter of selecting the appropriate energy source and demonstrating the techniques required (Gardner and Fischer 1966; Soane, 1967; Gardner and Calissendorff, 1967). Ideally, one of the sources should be mainly attenuated by soil and the other mainly attenuated by water. However, because of short isotope half-life, the presence of high-energy or closely spaced peaks in the source spectrum, shielding problems, cost of sources, and absorption of low-energy peaks by source materials (Gardner and Calissendorff, 1967), ^{137}Cs has been universally selected as the best available high-energy (0.662 Mev) source, and ^{241}Am as the low-energy (0.060 Mev) source. With low-energy sources, self-absorption is a severe limitation, often making maximum useful source strength low. Gardner and Calissendorff (1967) point out that an ^{241}Am source, with a cross section of 1 cm by 0.1 cm, has a maximum useful source strength of about 230 mCi. By using water-filled glass cells into which thin glass plates could be placed, they simulated a soil-water system with variable bulk density and water content. For counting times sufficiently long to give one million counts through the empty container, they found the expected standard deviation in water content, that resulted from a combination of the error due to the counting and to uncertainty in bulk density as measured using the dual-energy γ technique, to be about 0.004 gm cm^{-3}.

Of course, other sources of error need to be considered and other corrections need to be made to establish the true precision that can be expected for water content determination by the dual-energy γ technique. To prevent Compton-scattered photons from the higher energy source from interfering with the peak from the lower energy source, which occurs when the two sources are combined in tandem in a single beam, Gardner *et al.* (1972) moved the soil column horizontally between two separate γ-ray beams. Bridge and Collis-George (1973) accomplished the same thing by using a single detector, but interchanging sources. Corey *et al.* (1971) could correct for Compton-scattered photons resulting from shooting the Cs γ-beam through the Am source (as shown in Fig. 4) by electronic dis-

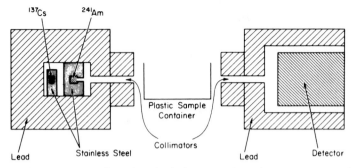

Fig. 4. Schematic drawing of a single-beam dual-energy γ-ray attenuation apparatus for simultaneous measurement of water content and bulk density. From Corey
et al. (1971).

crimination with a multichannel analyzer. Errors can be introduced if the
collimators and source holders are not firmly mounted or if either the
sources or the soil column are not repositioned in the identical location for
each reading. It is also essential to take into account the resolving time
for the detector (Fritton, 1969), any deviation in the mass attenuation
coefficient of water in the soil from that in bulk water (Groenevelt *et al.*,
1969), or any other irregularities, such as nonlinearity in the logarithm
of the count ratio versus the absorbent thickness as found by Gardner *et
al.* (1972). (Stroosnijder and de Swart, 1973, argued that some of the
problems encountered by the latter could be avoided by providing ideal
collimation at both the source and the detector and taking into account
dead time of the detectors for both sources.) By taking all of these precautions and making the appropriate corrections, Mansell *et al.* (1973) found
that with a dual-energy single γ beam from combined 300-mCi ^{137}Cs and
[241]Am sources, errors in measurement of soil-water content could be restricted to standard deviations of approximately 0.01 gm cm[-3]. The capability of measuring water content to this precision, even with soils that shrink
and swell, is a major breakthrough for testing theoretical models for soil-
water interactions in the laboratory (e.g., Collis-George and Bridge, 1973.)
Extension of the method from one-dimensional columns to two-dimensional slabs, such as shown in Fig. 5, should further increase our understanding of the principles of soil-water flow.

2. Radiation Scattering Methods

Measurement of soil-water content by neutron and γ-ray scattering,
initiated by Belcher *et al.* (1950), is based on the interaction of neutrons
or γ photons with soil constituents. This interaction allows some of them

Fig. 5. Two-dimensional soil with γ-ray attenuation apparatus for automatic measurement of water content. Photo courtesy Dr. C. Dirksen, U. S. Salinity Laboratory, Riverside, California.

to return to the vicinity of the source. Although neutrons do not interact strongly with most materials, they do with hydrogen atoms. Having lost most of their energy, neutrons that have been scattered by collision with hydrogen atoms are then free to diffuse back to the vicinity of the source. Because water is the major soil constituent containing hydrogen, the density of the neutron cloud near the source can be used as an indicator of soil-water content.

Belcher *et al.* (1950) used a mixture of Ra and Be as a source of fast neutrons with Rh foil as a detector for slow neutrons. The source, surrounded by the foil, was inserted into an aluminum tube, which was then lowered into a soil access tube to the desired position. After the Rh foil, which selectively captures slow neutrons, had reached equilibrium (20 to 30 min), the β rays emitted from it were measured by slipping it over a Geiger counter mounted inside a lead shield. Belcher *et al.* (1950) were not successful in permanently locating the Geiger counter adjacent to the detector foil, because the intense γ rays emitted from the Ra masked the neutron count. However, in a few tests with a Po-Be source, which had very weak γ-ray emission, the Geiger counter was successfully mounted adjacent to the neutron source in a soil probe. The system was empirically calibrated by plotting the count rate, after it was properly corrected for background, against water content of the soil.

γ-Photons, unlike neutrons, are scattered more or less equally by all components of the soil. Belcher *et al.* (1950) used γ scattering to measure the wet density of the soil with a similar soil probe in which a 4-mCi-radium γ source and a Geiger counter were mounted adjacent to one another, but were separated by a block of lead. With this geometry, only scattered γ photons reached the Geiger tube, which resulted in an increasing count rate with increasing soil density. Though for nonswelling soils the γ-scattering technique could theoretically be used to infer changes in water content, because neutron scattering is so much more sensitive to water content, it has not developed for this purpose. γ-Scattering is used to measure changes in density in conjunction with water content measurements with neutron-scattering equipment.

The nature of the neutron-scattering process imposes an important restriction upon the resolution that can be obtained with it. In wet soils, neutrons travel only a short distance before they collide with hydrogen atoms; in dry soils, they travel a much greater distance. This causes the soil volume over which the water content is averaged to vary with water content.

The soil probe of the neutron apparatus developed independently by Gardner and Kirkham (1952) consisted of a BF_3 slow-neutron counter and a Po-Be fast-neutron source. It was lowered into augered holes in the soil by means of a cable through which counts were fed to a nuclear scaler. With the replacement of Po-Be by Ra-Be, which has a longer half-life, this apparatus was rapidly accepted for indirect measurement of water content. ^{241}Am-Be is now replacing Ra-Be as a fast-neutron source. This change, apparently first suggested by C. Z. Kwast (Perrier *et al.,* 1966), and first implemented by van Bavel and Stirk (1967), reduces the γ radiation and permits higher neutron fluxes. Except for this change, the method has remained essentially the same as originally conceived by Belcher *et al.* (1950) and Gardner and Kirkham (1952). This evolution of the neutron scattering technique from its inception to a standard technique for water-content measurement is documented in well over 100 publications summarized in numerous reviews (Gardner, 1965; Ballard and Gardner, 1965; Cope and Trickett, 1965; Visvalingam and Tandy, 1972). Only brief excerpts of these reviews covering installation and calibration techniques and other factors that influence the neutron count are given here.

Access tubes (closed at the bottom) are usually installed in the soil by driving them into auger holes that are slightly undersized to ensure a snug fit. Aluminum appears to be the general choice of tube material. The probe containing the fast-neutron source and the slow-neutron counter is lowered into the access tube by means of a cable. For measurements near the surface, the probe is either laid horizontally on the soil surface and covered with a hydrogenous shield, or a container of soil is fitted over the access tube to eliminate loss of neutrons from the soil surface.

To calibrate the neutron apparatus, one must locate the center of the sensitivity of the probe. This can be taken to be the source position only when the source is centrally placed on the neutron detector. McCauley and Stone (1972) found that the center of sensitivity of the BF_3 tube was identical to the midpoint of the anode wire within it. Changes in position of the source as small as 0.5 cm from the center can significantly change the calibration curve.

In experimental calibration, the volume of soil should be large enough to be considered effectively infinite relative to the neutron flux and should be homogeneous in structure, density, and water content. These conditions are easily met in the laboratory; but, in the field, care must be taken to select only homogeneous layers of sufficient thickness to include the neutron flux. While there is no universal agreement on the best way to calibrate neutron meters, Visvalingam and Tandy (1972) point out that field operators seem to prefer laboratory calibration, and those with labortaory experience tend to think that field calibration would be more suitable. They suggest that, ideally, it is best to include both field and laboratory tests.

Because all equipment is liable to drift, primary standards should be set up to permit periodic checks to detect changes in the response of the equipment. Although a calibration based on the ratio of counts from a standard absorber and the soil overcomes many of these difficulties, Stone (1972) points out that it will not compensate for detector dead time. Furthermore, for greatest accuracy, the standard absorber should yield counts in the same range as the soil. This eliminates large volumes of paraffin or water as practical standards.

Properties of the soil, other than water content, interact to produce accumulative effects on the count rate. These are soil hydrogen in forms other than free water, neutron-absorbing elements, and soil density. Cohen (1964) found that calibration curves, determined by removing soil samples in 10-cm increments just below the access tube as it was driven into an auger hole, were the same for soils that had identical or similar composition but differed for those that did not. Likewise, Lal (1974) found that for tropical soils it was necessary to calibrate the neutron probe not only for each soil, but also for each horizon. Variations in stoniness, bulk density, and texture limit the usefulness of the neutron meter for measuring absolute quantities of water, but affect its use for monitoring relative changes in water content at a given location far less.

While the count rate is increased by the presence of hydrogen in forms other than free water, such as in organic matter or water that is retained by clay minerals at temperatures above those normally used for oven drying, it is decreased by a concentration of neutron-absorbing elements, such as Mg, K, B, Cl, and Fe. Laboratory- and field-calibration curves have differed as much as 10% as a result of these factors. Like hydrogen in nonwater forms, high bulk density increases the apparent water content because neutron transport in the soil is impeded. The error is greatest at high-water content and can change the slope of the calibration curve by more than 1% for each 0.1-gm cm^{-3} increase in dry-soil density. Differences in bulk density shift the calibration curve, but do not affect relative measurements of water content at a single point, if the bulk density remains constant.

Though soil texture has been proposed as an independent cause for changes in the calibration curves for neutron meters, it is difficult to isolate any such effects from those of bound water, neutron absorbers, or soil density. In early research, the lack of systematic difference between calibration curves for soils with very different textures and parent materials may have been fortuitous. The abnormally high neutron flux that might be expected from soils high in clay minerals, because they contain large amounts of water at oven-drying temperatures, is mediated by the fact that these same minerals also contain an abundance of neutron-absorbing elements. The experience of more recent workers indicates that caution should be taken in using a single calibration curve without an independent test of its validity for the particular soil in question. This is particularly true where access tubes are not aluminum, or if the soil surrounding the tube is compacted during installation.

The neutron method has a major advantage over other methods for measuring soil-water content in its adaptability for automatic recording of water content to depths limited only by the length of the access tube. This is particularly important where measurements of cyclic or seasonal changes are desired. Some, but not all, workers believe the method is more accurate than the gravimetric method. It makes possible measurement of water content in some instances that would be impossible with normal sampling methods. For example, with access tubes inserted into slightly undersized holes to prevent vertical flow down the tube wall, water content can be measured below a perched water table. The neutron method also samples water content over a larger volume, which decreases the standard error. However, the method does have inherent limitations, especially in the measurement of absolute water content as the consequence of inadequate depth resolution and inaccurate calibration curves for the particular soil in question. This introduces serious problems, particularly where studies involve measurements near the surface or in layered soils.

3. Discussion and Summary

Nuclear methods for indirect measurement of water content have made possible many studies that were previously not practical. One area in which particular progress has been made, because such methods have proven so useful for measuring changes in water content, is the study of water balance in the root zone of crops. The balance of mass for water in the root zone can be written

$$dS/dt = (R + U + I) - (E + T + D) \qquad (3)$$

where S is the total amount of water stored in the root zone at any instant
R is the rate of supply from rainfall
U is the rate of upward flow into the root zone
I is the rate of irrigation
E is the rate of evaporation from the soil surface
T is the rate of transpiration
D is the rate of drainage out of the root zone.

Because it is usually not possible to measure E and T directly, other components are measured as accurately as possible and E and T are estimated by difference. I and R can usually be measured, but U and D are difficult to obtain. If U is lacking, D and dS/dt can be estimated in a weighing lysimeter from the volume of drainage and change in weight. But weighing lysimeters (Pruitt and Angus, 1960; van Bavel and Myers, 1962; Hanks and Shawcroft, 1965; Ritchie and Burnett, 1968) are expensive and cumbersome to install. By using neutron or γ-ray techniques to determine dS/dt, the lysimeters need only measure D, and therefore can be simple nonweighing types. Several investigators (for example, McGuinness *et al.*, 1961; Holmes and Colville, 1964) found the precision of seasonal changes in stored water determined with the neutron method to be comparable to that obtained with weighing lysimeters. Giesel *et al.* (1970) used the double-tube γ apparatus to measure dS/dt and estimated D from the soil-water flow equation using tensiometer measurements for the hydraulic gradient. (D cannot yet be directly measured.) Though the absolute error in measurements of water content was about 0.01 gm cm^{-3}, the total error in determinations of the long-term variations in water content was approximately 0.0015 gm cm^{-3}.

Undoubtedly instrumentation will continue to improve, but measurements now being made by nuclear methods are closely approaching the theoretical limits of accuracy. Therefore, future gains in capacity to measure water content indirectly can be expected to be modest compared to those made over the past 15 years. Possibly, however, effective nonnuclear methods will be developed.

Methods based on thermal dissipation (Cornish *et al.*, 1973) or latent heat of vaporization (Chadwick, 1975) show promise for special applications.

III. METHODS OF MEASURING SOIL-WATER POTENTIAL AND ITS COMPONENTS

To use dynamic models of water movement in soils and water uptake by plant roots, one must know the value of the different components of water potential. For example, the main driving force for liquid flow in soils is the gradient of the pressure and gravitational potentials. (This value is sometimes referred to as the hydraulic potential.) Only under special conditions does the osmotic potential gradient influence liquid flow in soils. On the other hand, where semipermeable membranes are involved, such as in the plant roots, all components of the water potential contribute to flow. Vapor transfer is also in response to the total potential gradient.

Considerable progress in both direct and indirect techniques for measuring total water potential and its components has been made recently. Direct methods are those that measure the quantity directly on the basis of its definition. Indirect methods are those that measure some other parameter that can be correlated with total water potential.

Some components of water potential are more easily measured directly than others, simply because they are easily separated. Measurement of gravitational potential, for example, requires only that the elevation of the sample be known relative to an arbitrary reference. Tensiometer pressure potential is also directly measured with the tensiometer or its equivalent, but its components are not always easily separated. Particularly for swelling soils, indirect methods must usually be used to separate the envelope-pressure component from the wetness potential.

Although techniques, such as freezing-point depression, are available to measure the osmotic pressure of solutions, use of these is not practical in soils where the liquid phase is intermixed with the soil matrix. Because it is difficult to separate the soil solution from the soil matrix without altering its composition, measurements such as electrical conductivity are often used as indirect *in situ* measurements of osmotic potential.

Total water potential can be inferred from the sum of each of its components; however, with the development of vapor phase methods, it is often more practical to measure it directly. These methods can also be used to obtain the value of a single component in samples where only one component exists. An example of this is the measurement of osmotic potential of the soil solutions.

A. TOTAL POTENTIAL

The total potential is related to the relative vapor pressure by the well-known equation

$$\psi_t = \frac{RT}{\bar{V}} \ln e/e_o \tag{4}$$

where R is the universal gas constant
T is the Kelvin temperature
$\bar{V}$ is the partial specific volume of water
e is the vapor pressure of water in the system
e_o is the vapor pressure of pure water at the same temperature, reference elevation, and pressure.

Theoretically, it is possible to measure e in the system and e_o over pure water at the reference elevation and pressure to compute the sum of all of the components of water potential (Rawlins, 1971). This has not been practical, however, because of the sensitivity of e to T. At 25°C, the change in e per °C is about 80 times its change per bar of water potential. This makes it essential that e and e_o be measured at the same temperature. This is done easiest by measuring their ratio e/e_o (relative humidity) directly at a single point in the system. Measuring relative humidity at a single point helps to eliminate the temperature problem, but it necessitates an independent evaluation of gravitational and pneumatic potential, because e_o is measured at the system pressure and elevation, rather than at the reference elevation and pressure. However, measurements of elevation and pressure differences can easily be determined.

Early vapor phase techniques for measuring water potential were based on equilibrating samples with a series of test solutions and determining by interpolation the osmotic potential at which the sample neither gained nor lost weight. More recent techniques measure directly the relative humidity in equilibrium with the sample (Barrs, 1968). This approach is preferable not only because it saves equilibration time and decreases the number of subsamples required, but it causes less disturbance to water potential of the sample.

Relative humidity can be measured by absorption hygrometers or miniature wet-bulb psychrometers. (Psychrometers can also be operated to measure dew-point as well as wet-bulb temperature; dew-point temperature is more closely related to vapor pressure than to relative humidity.)

1. Absorption Hygrometers

Absorption hygrometers measure some electrical or thermal property of a hygroscopic medium in vapor equilibrium with the sample. As with psychrometers, absorption hygrometers are calibrated over solutions of known relative humidity.

Absorption hygrometers have not been widely accepted, even though two specially designed types appear to be capable of measuring in the high-humidity range required for soils. Commercial hygrometer elements, on the other hand, are unstable at these high relative humidities (Richards and Decker, 1963).

One absorption hygrometer proposed for use in measuring water potential consists of two screen electrodes embedded in a thin, porous matrix (Bouyoucos and Cook, 1967, 1968). Water content of the element varies with relative humidity and is sensed by the electrical conductivity between the electrodes. Because it requires considerable vapor transfer to wet and dry the element, it is not adaptable for use with small enclosed samples or where rapid response is required, but it shows some promise for use with large samples where water potential changes gradually.

The other absorption hygrometer is based on water-vapor absorption in the natural micropores of the glass coating on a thermistor bead (Richards, 1965a,b). The quantity of water condensed, which depends on the relative humidity, is determined by passing a current through the thermistor and comparing the shape of the resistance versus the time curve with that of a thermistor kept dry, either with silica gel or a hydrophobic coating. Even though the water capacity of the porous medium is extremely small, Towner (1967) found that at relative humidities exceeding 99%, the unit required about 3 days to reach equilibrium when it was returned to the wet soil after calibration. The wetting characteristics also varied considerably among different thermistors, and the calibration of a given thermistor drifted with time.

Notwithstanding the shortcomings of absorption hygrometers, their insensitivity to temperature fluctuations and their relatively simple requirements for electronic instrumentation should not be overlooked in developing instruments for *in situ* measurement of water potential.

2. Psychrometric Dew-Point Hygrometers

Many of the recent developments in psychrometric techniques have been reviewed in detail (Barrs, 1968; Wiebe *et al.,* 1971; Brown and van Haveren, 1972). New techniques are emphasized here.

Psychrometric methods are based on measuring the temperature of, or rate of evaporation from, a water droplet or wet surface. Although theoretically relative humidity can be calculated directly from such measurements (Rawlins, 1966; Peck, 1968, 1969; Scotter, 1972), in practice the instruments are calibrated empirically.

a. Hanging-Drop Psychrometer. Weatherley (1959) and Macklon and Weatherley (1965a,b) developed a technique for measuring relative humidity from evaporation rate from a water droplet to be used with leaf disks, but the method is equally suited for soil samples. A measurement is made by briefly suspending a water droplet from a pipette over the sample in a sealed chamber and, then, withdrawing it to a preset mark and measuring the change in volume. The method is simple and, with precision timing of the droplet exposure and accurate measurement of the water evaporated, water potential can be measured with a precision of about 0.25 bar. Though it requires considerably more sophisticated techniques and apparatus, the laser interferometer technique of Ro *et al.* (1968) for continuously measuring the diameter of a small water droplet might be adapted to improve the precision.

b. Wet-Bulb Psychometers. The most widely used method for determining water potential measures the wet-bulb temperature of an evaporating droplet rather than the actual rate of evaporation from it. The underlying physical principles on which this technique is based have been well understood for many years. As early as the 1930's, biomedical researchers devised techniques for measuring water potential by wet-bulb psychrometry. Hill (1930) used a thermopile constructed by silver-plating constantan wire wound on an insulated brass frame and coating it with varnish. To each of the two sides, he applied a moist filter paper, one soaked in the unknown solution, the other in a solution of known composition. The entire apparatus was kept in a humid chamber at a temperature constant to within $\pm 0.0005°C$. He attributed the method to members of his department, who as early as 1913 realized that the measurements depended on the temperature and the pressure at which they were taken as well as the osmotic potential. These effects of temperature (Klute and Richards, 1962) and pressure (Richards *et al.,* 1964) were rediscovered and reexplained (Rawlins, 1966) almost a half-century later.

Baldes (1934) modified Hill's (1930) technique by using a double-loop thermocouple constructed from 0.1-mm diameter manganin and constantan wires. With this wire, droplets as small as 1 to 2 mm^3 could be measured. By using wires 0.05 mm in diameter, droplets as small as 0.1 mm^3 could be used. If the wires were small enough, the measurements were independent of drop size—a fact consistent with the theory of Rawlins (1966). The precision of these early measurements, approximately 0.1 bar, was limited solely by the sensitivity of the galvanometers available for measuring the output of the thermocouple. Failure of these methods to be adopted universally earlier must be attributed to a lack of hardware, rather than to a lack of understanding of basic principles, as demonstrated by the excellent paper by Baldes (1939).

   i. Thermistor Psychrometers. Brady *et al.* (1951) first used thermistors at the wet bulb, and others have followed (Müller and Stolten, 1953; Kreeb, 1965; Mokady and Low, 1968; Farm and Bruckenstein, 1968; Kay and Low, 1970). The thermistor is more rugged and has a greater response to temperature than other transducers, but it requires current to be passed through it during a measurement. This places a lower limit on the diameter of lead wires and an upper limit on the magnitude of the measuring current that can be used without causing self-heating. Kitchen and Thames (1972) overcame this problem by designing circuitry that applied current to the thermistor in short pulses, but the procedure is complicated. Consequently, psychrometers for use in soils have relied on thermocouples to measure the temperature of the wet bulb.

   ii. Thermocouple Psychrometers. Two types of psychrometers have evolved. With one, water is placed into a loop or ring surrounding the thermocouple to keep the junction wet (Richards and Ogata, 1958). In the other, water is condensed on the wet junction immediately before making a measurement by passing current through it and then cooling it below the dew point by the Peltier effect (Spanner, 1951; Monteith and Owen, 1958). Wet-bulb temperature is measured after the current has ceased flowing and the junction has reached a steady temperature. Because only a small quantity of water usually condenses on the junction, its temperature remains steady only briefly after wetting. The larger drop of water on the wet-loop psychrometer yields a steady output for a longer time.

   The advantage of having a steady output for a longer time for the wet-loop psychrometer is offset for some applications by three inherent problems: (1) the equilibration chamber must be opened to place a droplet on the junction; (2) water is added to the system from the wet junction (Rawlins, 1964); and (3) any parasitic voltages that develop in the measuring circuit after the junction is wetted, or any change in ambient temperature between the reference junctions and the psychrometer chamber result in undetectable errors. These temperature differences between the measuring and reference junctions are accounted for with the Peltier psychrometer by subtracting the emf measured before wetting from that measured after wetting. Because the junction is wetted only briefly, the time for temperature drift is short.

   For *in situ* measurements of water potential, capabilities for wetting the measuring junction without opening the equilibration chamber and for taking into account extraneous voltages caused by nonuniform temperature within the system are mandatory. Psychrometers developed for *in situ* measurement of water potential (Rawlins and Dalton, 1967; Lang, 1968) have the following features: (1) Peltier cooling is used to wet the junction so that temperature drifts can be compensated for and the junction can be wetted remotely. (2) The wet junction is as completely surrounded by the soil sample as possible to provide thermal shielding (Peck, 1968, 1969). (3) The thermocouple wires are small enough that heat flow to the junction through them can be neglected. (4) Soil temperature is measured to correct for temperature in the calibration.

Figure 6 shows the construction of a commercially available* Peltier-effect psychrometer designed for use *in situ*. As with the design of Rawlins

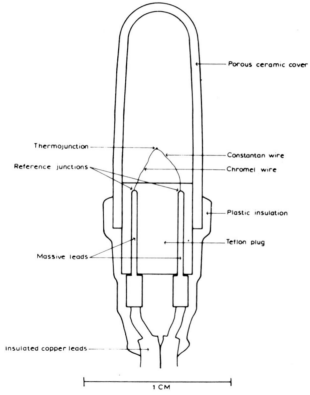

Fig. 6. Diagram of a typical Peltier-effect thermocouple psychrometer for measurement of total water potential *in situ*. From Scotter (1972).

* Wescor, Inc., 459 Main St., Logan Utah. (Company names are included for the benefit of the reader and do not imply any endorsement or preferential treatment of the product listed by the USDA.)

and Dalton (1967), the thermojunction is enclosed within a porous ceramic shield and is mounted in a plug of teflon to minimize vapor absorption. A feature of this design is that the diameter of the thermojunction, approximately 0.019 mm, is about 7.5 times that of the thermocouple wires. The analyses of Dalton and Rawlins (1968) and Peck (1968) call for larger heat sinks at the reference junctions to dissipate the heat produced during the cooling phase than are shown for this psychrometer. But Scotter (1972) has found no error attributable to temperature rise at the reference junctions for this design. He concludes that previous analyses calling for massive reference junctions are misleading because they fail to account fully for heat conduction away from the junctions during cooling.

Results from field experiments with these same psychrometers (Merrill and Rawlins, 1972) showed them to be promising, but not without some faults. Particularly annoying was a diurnal fluctuation in *apparent* water potential as great as ±6 bars measured when the psychrometers were installed vertically in the upper 40 cm of the soil profile. Psychrometers installed horizontally did not show this fluctuation, and those installed upside down gave fluctuations in the opposite direction. From this, it was obvious that the fluctuations were an artifact introduced by the thermal gradient in the soil. Sensors of this design intended for use in the upper 40 cm of the soil profile must be positioned with their axes of symmetry parallel to the soil surface.

Where only mean daily values are needed, Merrill and Rawlins (1972) found that averaging the readings taken at 12-hr intervals helped to eliminate these diurnal variations. Additional data taken by automatic recording and computer processing significantly reduced the standard deviation of measurements even further. Standard deviations for the average of three readings taken 12 min apart were 0.23 bar for the range —2.5 to —5 bars, 0.14 bar for the range —5 to —7.5 bars, 0.08 bar for the range —7.5 to —10 bars, and 0.19 bar for the range —10 to —15 bars. The relatively large error for potentials near zero is probably associated with condensation caused by temperature fluctuations. Calibrations of 33 psychrometers made before and after their use for 8 mo showed both increases and decreases in sensitivity, but the change in about half of the instruments was less than 5%. Only one instrument shifted more than 15%; the median change was 5.3%.

Although averaging readings over time can reduce errors induced by soil temperature gradients, a better approach is to avoid them by making the reference and measuring junctions the same shape and size and placing them adjacent to one another within the psychrometer chamber. An illustration of this arrangement, which is similar to other designs intended for the same purpose (Hsieh and Hungate, 1970; Calissendorff and Gardner,

1972; Meeuwig, 1972; Chow and de Vries, 1973) is shown in Fig. 7. Water is condensed on one junction by passing current through it using the central constantan conductor as one lead. When the cooling current is terminated, the output of the psychrometer is read between the two copper leads. This gives the temperature difference between the two chromel-constantan junctions. Ambient temperature can be read at any time by measuring the emf between the center constantan lead and either copper lead.

iii. Thermocouple Dew-Point Hygrometers. If the temperature of the psychrometer chamber is different from that of the sample whose water potential is to be measured, placing the wet-bulb and dry-bulb junctions at the same location will not eliminate systematic error. For applications where such temperature differences are a problem, for example in the measurement of water potential of plant leaves, Neumann and Thurtell (1972) developed an instrument that detects dew-point depression rather than the wet-bulb depression measured. The dew-point temperature measures e in Eq. (3) rather than the ratio e/e_o. A dew-point measurement is preferable to wet-bulb temperature because, at the dew point, no net water is exchanged at the wet junction. This permits measurements to be made without disturbing the vapor equilibrium in the chamber. Because no vapor

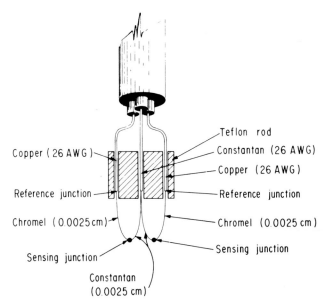

Fig. 7. Detailed illustration of a double-junction temperature-compensated Peltier-effect thermocouple psychometer. After Brown and van Haveren (1972, p. 25).

is exchanged, the measurement is relatively insensitive to factors that influence evaporation with wet-bulb psychrometers, such as the size and shape of the wet surface at the junction. In addition, the relationship of dew-point temperature to water potential is less dependent upon the ambient temperature than is that of wet-bulb temperature.

At equilibrium, e in the chamber is equal to e in the sample, even if the temperature is not the same for both. If Eq. (3) is used to find the water potential of the sample, the parameters T and e_o must be the sample temperature and the saturated vapor pressure. In the dew-point technique, this requirement is satisfied if the dew-point temperature depression is measured relative to the sample temperature.

To measure dew-point depression, Neumann and Thurtell (1972) used four-terminal Peltier-cooled thermocouple psychrometers constructed with crossed 0.025-mm-diameter chromel and constantan wires, spot welded at their junction. This configuration is similar to that used by Millar et al. (1970). Temperature depression of the cooled junction was then measured through one set of leads while cooling current was being passed through the other set. Figure 8 shows typical response curves of dew-point depression versus cooling current for their dew-point hygrometer in dry air and in air in equilibrium with a sample at —9 bars total potential. The curve for dry air is parabolic because Peltier cooling is proportional to the current while the opposing resistive heating of the wires and the junction is proportional to the current squared. If the same junction is operated in a moist environment, when it reaches a temperature below the dew point, water condenses on it. The subsequent latent heat exchange

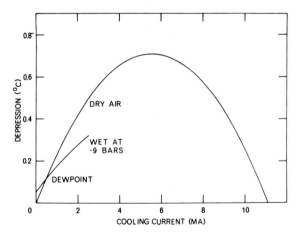

Fig. 8. Typical response curves for the dew-point hygrometers in dry air and in an atmosphere at —9 bars water potential. From Neumann and Thurtell (1972, p. 104).

alters the response as illustrated. In the portion of the wet curve to the right of the dew point, heat released by water condensing on the junction keeps it warmer than it was in dry air at the same current. If the current is lowered so that the wet-junction temperature rises above the dew point, latent heat loss due to evaporation maintains the wet junction cooler than the dry junction, as represented by the portion of the wet curve to the left of the dew-point curve. At the intersection of the curves, the wet and dry junctions are at the same temperature, indicating the latent heat contribution to be zero and the wet junction to be at the dew-point temperature.

The y intercept of the wet curve in Fig. 8 represents the measurement that would be made by a psychrometer and is determined by the slope of the wet curve, since the intersection point of the wet and dry curves is fixed by the dew-point temperature. The slope of the wet curve will be altered by factors affecting the rate at which water can evaporate from the wet bulb. Hence, psychrometric measurements are influenced by the wetting characteristics of the junction and the size and shape of the water droplet formed on the junction, whereas the dew-point technique should be relatively independent of these factors.

A modification of the Neumann and Thurtell (1972) technique made by Campbell *et al.* (1973) makes dew-point measurements possible with standard two-wire thermocouples. Both cooling and dew-point temperatures are measured through the single set of leads by circuitry that time-shares sensing and control functions. By alternately cooling and measuring the thermocouple temperature, this circuitry automatically modulates the time of each cooling period to maintain the thermocouple junction precisely at the dew point. The electronic switching enables the dew-point temperature to appear as a continuous reading on a panel meter or at recorder output terminals.

This dew-point technique reportedly increases the sensitivity for water potential measurement from about 0.47 μV/bar for the standard psychrometer technique to about 0.75 μV/bar, but it is limited to water potentials below -1 bar. Both single-channel and multiple-channel dew-point instruments are now commercially available.

Details of construction, operation, and application of psychrometric techniques are explained in detail in Wiebe *et al.* (1971) and in Brown and van Haveren (1972).

B. PRESSURE POTENTIAL

As previously stated, the basic instrument for measuring pressure potential is the tensiometer. Figure 9 shows its essential components. The

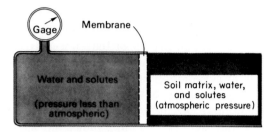

Fig. 9. Schematic diagram showing essential components of a tensiometer system. From "Physical Edaphology: The Physics of Irrigated and Nonirrigated Soils" by Sterling A. Taylor and Gaylen L. Ashcroft. W. H. Freeman and Company. Copyright © 1972.

soil solution within the tensiometer is brought in equilibrium with the liquid phase in the soil through a porous membrane that permits both water and solutes to pass, but excludes the solid phase and air. The gauge pressure indicated by the tensiometer is measured relative to the standard pressure, P_o, so that any deviation of the pressure within the soil atmosphere from this standard (the pneumatic pressure of the soil atmosphere) is measured by the tensiometer. Likewise, if the soil matrix is not rigid so that part of the overburden is borne by the water, its effect is also sensed by the tensiometer. The wetness potential component of the matric potential, arising primarily from adsorptive forces emanating from soil particles, is also sensed as a reduced pressure within the tensiometer.

Tensiometers usually have one or more limitations that restrict their use for measuring pressure potential over the entire range or under all conditions in soils. One of the major limitations is that as the pressure within the tensiometer reservoir approaches −1 bar gauge pressure, air comes out of solution and forms bubbles that release water to flow into the soil or even break hydraulic contact within the tensiometer. Although filling the tensiometer carefully with de-aired water and making certain there are no leaks help to prevent this problem, tensiometer measurements below −0.8 bar are not practical. Another limitation stems from the fact that generally some water must move through the cup to bring the soil solution in it into equilibrium with the liquid phase in the soil. When this water is added to or taken from the soil, its matric potential changes by an unknown amount.

1. Decreasing Tensiometer Response Time

The first tensiometers developed in the laboratory of Willard Gardner at Utah State University (Richards, 1928) consisted of a porous ceramic cup sealed onto a glass tube connected to a mercury manometer. Although, dial vacuum gauges are used instead of the mercury manometer for some applications, until recently, manometers provided the most accurate measurement of tensiometer pressure. Water displaced by the manometer column (or from any other pressure gauge) must move through the porous cup and to or from the soil whenever pressure potential changes. The overall response of the tensiometer to changes in pressure potential will depend on the sensitivity of the gauge, the conductance of the tensiometer cup, and the hydraulic conductivity of the soil in which the cup is embedded. Klute and Gardner (1962), in an analysis of tensiometer response time in terms of these factors, considered one case in which response time was limited by the instrument and another in which it was limited by the soil. Where it was limited by the instrument, a technique for correcting for a nonzero response time using the differential equation of flow of water across a tensiometer cup was developed.

Although this technique could be useful in the laboratory, and possibly in some field studies, it is more desirable to eliminate the problem by increasing the gauge sensitivity. Miller (1951) attempted to do this with a manually operated, null-point tensiometer system. The system included a water manometer and a sensitive null indicator to enable the operator to recognize the correct manometer setting to maintain the tensiometer in equilibrium with the rapidly changing soil pressure potential without permitting flow across the tensiometer cup. Leonard and Low (1962) improved on this design by making it automatically self-adjusting to maintain zero flow. This was an important improvement for some laboratory studies, but the design was still too complicated for routine use.

A major improvement in response time as well as convenience in recording data were obtained by use of strain gauge pressure transducers. Klute and Peters (1962) obtained response times of less than 1 sec. However, these early pressure transducers were temperature sensitive and required about 1°C control. With the develoment of temperature-compensated pressure transducers, tensiometers constructed with integral pressure transducers have gained considerable popularity (Bianchi, 1962; Behnke and Bianchi, 1965; Watson, 1965, 1967a,b; Bianchi and Tovey, 1968, 1970; Strebel *et al.*, 1970).

The major advantage of tensiometers with integral pressure transducers is their capability of measuring pressure potential at any depth, in addition to their ease of recording, rapid response time, and temperature stability as a consequence of having no aboveground components. In contrast, tensionmeters attached by hydraulic lines to pressure gauges at the soil surface always include a gravitational component in the measured pressure. Because the pressure depression is limited to about 0.8 bar, a tensiometer could be used only to about the 8-m depth. But at this depth the gravitational component would cause the maximum pressure depression the tensiometer can operate at, so it could measure no additional matric component. At half this depth, the minimum matric potential the tensiometer could sense would be about —0.4 bar.

But including a pressure transducer with each tensiometer is expensive. Where measurements are to be taken close enough to the soil surface so that the gravitational component does not limit matric potential readings above those required, considerable expense can be avoided by using hydraulic selector switches to attach many tensiometers to a single pressure transducer. Such a system is described by Rice (1969), who points out that, in addition to lower cost, the system has advantages in that (1) the pressure transducer is accessible and can be kept in a watertight box, (2) tensiometers can be refilled when air appears in the system, and (3) calibration and zero reference are easily checked. Disadvantages, in addition to the depth limitation, are (1) there is a response lag when switching from one tensiometer to another, (2) temperature may affect the aboveground components (Watson and Jackson, 1967), and (3) more connections are involved, which increase the chance for leaks.

With the availability of high-quality force-balance-servo pressure sensors such as that shown in Fig. 10, hydraulically scanned tensiometers can measure rapidly changing pressure potentials precisely at a reasonable cost per tensiometer. Force-balance sensors operate on the principle that the force produced by the pressure is counteracted by an electrically caused

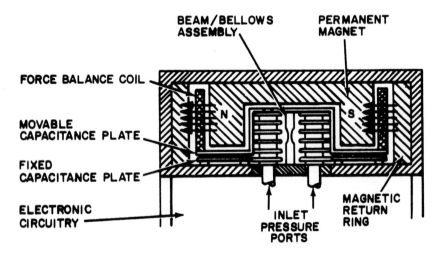

Fig. 10. Diagram illustrating principal components of a force balance pressure transducer. From instruction manual for Model 314D Servo Pressure Sensor (Sundstrand Data Control, Inc., Redmond, Washington).

force to restore the sensing mechanism to its null state. In the sensor shown, a differential pressure applied to the two bellows tilts the beam assembly, changing the electrical capacitance between the fixed and movable capacitance plates. This motion is detected and an amplified current signal is sent through the force-balance coil, which restores the beam to its initial null position. The voltage drop across a precision resistor in the force-balance coil is a measure of the pressure being measured. In practice, amplifier gain and pickoff sensitivity are made very high so that mechanism deflections of only about 0.1 μ occur over the full pressure range. The resulting sensitivity is extremely high. For pressure potential measurement in the soil model shown in Fig. 5, Dr. C. Dirksen is presently switching more than 150 tensiometers into a single pressure transducer of this type. Although the transducer is expensive (about \$800), the cost per tensiometer is reasonable for the precision obtained (reproducibility ±0.2 mbar). The U. S. Salinity Laboratory has installed a similar system in a field experiment near Yuma, Arizona. They inserted the standard 3-mm OD nylon tensiometer tubing coaxially into semirigid polyvinylchloride tubing (approximately 4 mm ID, 6 mm OD). Each tensiometer is purged by forcing water through the inside tube into the tensiometer cup and back through the annular space between the two tubes. As with Dirksen's laboratory model, all data will be automatically recorded on a computer compatible medium for rapid data analysis.

2. Extending Tensiometer Range

The serious limitation of narrow range of measurement suffered by tensiometers can be overcome in special circumstances by effectively increasing the ambient absolute pressure to avoid cavitation of the water in the tensiometer. Two methods have been suggested to accomplish this. The well-known pressure-membrane or pressure-plate method will be discussed only briefly to point out some limitations that are not always well understood. The other method is the osmotic tensiometer.

a. Pressure Methods. The principle of operation of the pressure method for extending the tensiometer range, first introduced by S. J. Richards (1938) and developed into a standard laboratory technique by L. A. Richards and associates (Richards, 1941; Richards and Fireman, 1943), can be seen from Fig. 11. The pressure membrane or pressure plate is equivalent to the tensiometer operating at an increased ambient atmospheric pressure. In this case, however, the pressure within the "tensiometer" remains fixed at ambient atmospheric pressure. The pressure potential of the soil water is varied by increasing its pneumatic potential. Though it is usually assumed that the pressure potential of the soil water is increased in propor-

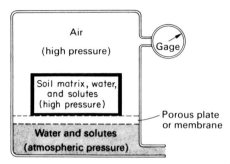

Fig. 11. Schematic diagrams of essential parts of a pressure plate or membrane apparatus. From "Physical Edaphology: The Physics of Irrigated and Nonirrigated Soils" by Sterling A. Taylor and Gaylen L. Ashcroft. W. H. Freeman and Company. Copyright © 1972.

tion to the increased pneumatic pressure within the pressure chamber, this is not the case if the pressure increase changes the geometry of the liquid phase (see definition for pneumatic potential above). Changes in geometry can change the value of the matric potential. Because the tensiometer pressure potential is not changed, the difference must be made up by the pneumatic pressure potential. The basic assumption that pressure potential increases linearly with pneumatic pressure increase is thus invalid.

One way the geometry of the liquid phase can be changed by changing the air pressure, first indicated by Peck (1960) and, apparently, rediscovered by Chahal and Yong (1964), arises from compression of entrapped air bubbles. Though Chahal and Yong eliminated the difference in pressure potential measured by the pressure plate and the tensiometer by removing air from the sample prior to making the measurements, they do not suggest this as a way to prevent the problem. Peck points out that changes in temperature can cause similar errors. Because most soils contain air even at zero matric potential (Fig. 3), the problem is not easily avoided. Fortunately, at matric potentials less than −1 bar, little air is trapped in soils. Thus, if in constructing a water retentivity curve, the tensiometer is used to its full range and the pressure plate is used to extend the curve to lower matric potentials, little difficulty should be encountered. Major discrepancies can be expected, however, if data from both instruments are compared in the pressure potential range above −1 bar.

b. Osmotic Tensiometer. Peck and Rabbidge (1966a,b, 1969) described a technique for overcoming the range limitation of the tensiometer by shifting its pressure in the positive direction by filling it with an osmotic solution that is separated from the soil by a semipermeable membrane. The instrument is shown in Fig. 12. By shifting the zero point, the bubble formation

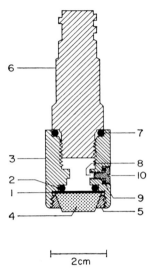

6 ——

3 ——
2 ——
1 ——
4 ——

—— 7

—— 8
—— 10

—— 9
—— 5

|———————|
2 cm

Fig. 12. Cross-sectional view of the osmotic tensiometer proposed by Peck and Rabbidge (1969) showing (1) semipermeable membrane, (2) O-ring seal, (3) brass body, (4) 15-bar bubbling pressure-ceramic, (5) champing ring, (6) strain-gauge pressure transducer, (7) O-ring seal, (8) vent, (9) O-ring seal, (10) filling plug. After Peck and Rabbidge (1969).

normally encountered at pressures below atmospheric is avoided. At zero pressure potential, the tensiometer reads its maximum pressure, which is equal to the osmotic pressure of the offsetting solution. The reading decreases linearly with decreasing pressure potential if the semipermeable membrane restrains only the offsetting solution and not the soil solutes. If the membrane also precludes the passage of soil solutes, the instrument measures the sum of the osmotic and pressure potentials.

Peck and Rabbidge used polyethylene glycol with a molecular weight of 20,000 gm. The pressure in their tensiometer is measured with a permanently mounted transducer. Of course, if desired, a single transducer could scan multiple units. The instrument has a temperature response of about $-1.5\%/°C$ and a zero drift as the result of solute leakage and/or strain gauge creep of more than $1\%/mo$. With recent improvements in pressure transducers, the possibilities of this instrument should be considered.

3. Indirect Methods for Measuring Pressure Potential

With the exception of the centrifuge method, which was introduced by Briggs and McLane (1907) to measure the moisture equivalent (a parameter considered to correspond roughly to the upper limit of available water) and calibrated in terms of pressure potential by Russell and Richards (1938), all indirect methods for measuring pressure potential fall into a single category. All use a porous water absorber. If a block of porous material, such as ceramic or plaster, is placed in contact with soil, then soil solution will exchange between it and the soil until the matric potential in both is equal. If the temperature, pressure, and solute concentration of the block remain constant, then its matric potential will depend solely upon its water content. The pneumatic potential, if present, must be added to measurements with porous blocks to obtain the pressure potential. Although porous blocks exhibit hysteresis, they can be selected to minimize this effect. Thus, if the water content of the porous block can be determined, its matric potential and therefore the matric potential of the soil can be estimated from the water retentivity curve for the block.

The water content of blocks can be estimated by weighing, by measuring the electrical resistance or capacitance between electrodes embedded in them, by measuring dissipation of heat from an embedded source, or by measuring some other physical property of the block such as its permeability to air.

a. Electrical Conductivity. Early interest in electrical conductivity methods (Whitney *et al.,* 1897; Gardner, 1898; Briggs, 1899) was directed toward finding an easy method for measuring soil water content *in situ.* Because of difficulty in obtaining a consistent electrical contact with the soil, however, most of these early attempts were unsuccessful. Balls (1932a,b) found that he could improve the performance of a capacitance sensor by coating it with a stable matrix of plaster of Paris. Bouyoucos and Mick (1940) later modified the electrical conductivity method the same way. They chose plaster of Paris for the stable matrix partly because it was soluble enough to mask moderate variation of salt concentration in the soil solution.

Although Bouyoucos and Mick (1940) and Anderson and Edlefsen (1942a) realized that use of a porous material made the sensors respond to matric potential, not water content, most of the early measurements were reported as water content. This, of course, required a separate calibration for each soil. Haise and Kelley (1946) developed a satisfactory calibration method by installing electrical leads through the wall of a pressure membrane apparatus. With this, the electrical conductivity of blocks brought to a given pressure potential could be read without releasing the pressure in the chamber.

Numerous modifications of electrical conductivity sensors have been proposed. Bouyoucos and Mick (1947) tried different porous materials including plastics, nylon and glass cloth, Hydrocal, and Hydrostone; but, for general use, they found gypsum was best. Though soft and soluble, gypsum blocks could be used for as long as 5 yr in well-drained soils; in water logged or organic soils, they deteriorated beyond use within a year. Slater (1942), concerned that the electrical field was not completely contained within blocks of the original design, developed a block with a caged-screen electrode, but performance was not improved. Bourget *et al.* (1958) tested several types of blocks and also found that below —0.3 bar pressure potential, the standard gypsum units had the best uniformity, the greatest sensitivity, and the least hysteresis. A unit similar to the standard gypsum block, but with a sleeve of woven fiberglass tubing surrounding each electrical lead, was the most satisfactory for use over the entire pressure potential range.

Hysteresis is common to all sensors that use porous absorption blocks, although this effect can be minimized if sensors are used along the same branch of the hysteresis loop they were calibrated on. Tanner and Hanks (1952) found significant errors that were due to hysteresis. For matric potentials in the range from 0 to —8.4 bars, at any given electrical conductivity, the magnitude of the matric potential obtained during desorption of gypsum blocks was nearly double that obtained during sorption.

b. Electrical Capacitance. The capacitance method for measuring the water content of a porous medium is based on the fact that the dielectric constant of water is approximately 40 times greater than that of soil or similar porous materials. If a porous material behaves as an ideal dielectric, capacitance should be a good indicator of the water content.

As with the electrical conductivity method, capacitance measurements were first made with bare electrodes inserted directly into the soil (Balls, 1932a,b; Cashen, 1932; Edlefsen, 1933). Both Balls (1932b) and Fletcher (1939) used a porous matrix to overcome the soil contact problems, but neither seemed to realize that the resulting measurement would correlate with matric potential, not with soil water content. Anderson and Edlefsen (1942b) found electrical capacitance of gypsum blocks ranged from about 0.07 μf at the moisture equivalent to about 0.0003 μf at a water content slightly above the wilting point. They reported reproducibility of measurements was good and independent of salinity. However, in general, partly because mineralized water is far from an ideal dielectric and both dielectric loss and other complicating factors make the measurements difficult (Chernyak, 1964), the method has not yet been adopted for routine use.

c. Thermal Dissipation. Use of thermal dissipation to measure water content of a block is based on the fact that the rate of heat dissipation in a porous body depends on its specific heat, thermal conductivity, and density—all of which increase with water content. Shaw and Baver (1939) used the technique by inserting a heater and temperature sensor directly into the soil. Johnston (1942) enclosed the unit in gypsum to make calibration with matric potential independent of the particular soil for which it was used. A similar thermal sensor was constructed by Haise and Kelley (1946). Although the blocks functioned well, they were insensitive to changes in matric potential below -4 bars, the point at which water loss from the gypsum ceased. Bloodworth and Page (1957) used a thermistor, which served as both heater and the temperature sensor, embedded in blocks of fired ceramic, gypsum, and a dental casting compound. These sensors were independent of soil salinity and were most sensitive at water contents near and below field capacity.

Phene *et al.* (1971a,b) improved the technique by replacing the thermistor by a linear diode temperature sensor. This made corrections of meter readings for changes in ambient temperature unnecessary—a feature lending itself well to automatic recording. Subsequently, Phene *et al.* (1973), using a highly porous ceramic for the absorption medium, found the instrument was capable of measuring and controlling irrigation to maintain matric potential within ± 0.01 bar in the range normally encountered with high-frequency irrigation. Next to the tensiometer, these instruments probably can make the highest precision measurements of matric potential in the field. By using the sensor to control matric potential to a narrow range, the effects of hysteresis should be minimzed.

C. Osmotic Potential

Once soil solutions are separated from the soil, measurement of their osmotic potential is routine. Vapor pressure or freezing-point depression can be used to measure it directly, or it can be inferred from electrical conductivity. Because of the interaction with the pressure potential, only electrical conductivity has proved successful for *in situ* measurements of osmotic potential.

Two useful techniques have recently been developed: One uses a salinity sensor that measures the electrical conductivity in a thin wafer of porous material in equilibrium with the soil liquid phase. The other uses four electrodes inserted directly into the soil.

1. The Salinity Sensor Method

The principle of the salinity sensor technique, first proposed by Kemper (1959) and perfected by Richards (1966), is simple. Electrical conductivity is measured within a thin ceramic disk that has pores fine enough to retain water throughout the matric potential range normally encountered. It is similar in concept to the porous electrical conductivity block used for measuring matric potential; however, the water content is fixed and the salt concentration of the solution within it varies. Figure 13 shows a schematic diagram of the sensor. When buried in the soil, the ceramic wafer equilibrates with the soil solution. The spring forces the ceramic against the side wall of an auger hole to maintain close contact with the soil. The thermistor measures soil temperature, which is required to correct electrical conductivity to standard temperature. If the ceramic remains saturated, the electrical conductance between the platinum electrodes at each of its faces is a unique function of the salinity of the soil solution.

Although the ceramic wafers in these sensors have a bubbling pressure in excess of 15 bars, Ingvalson *et al.* (1970) found that their output decreased with decreasing matric potential. This apparently resulted from dewatering of small cracks that formed near the platinum electrodes during the firing process. Painting electrodes on the surface of the ceramic wafer (see Enfield and Evans, 1969), rather than embedding them within it, might eliminate the problem.

Where matric potential is above −2 bars, this apparent desaturation causes only small errors. Because matric potential is seldom permitted to fall below −2 bars in soils where salinity is a factor, the problem is not as severe as it may seem at first. For matric potentials below −2 bars, the correction procedure outlined by Ingvalson *et al.* can be used.

In a field experiment designed to test the long-time reliability of salinity sensors, Oster and Willardson (1971) found that 78% of the commercial sensors functioned for 2 yr. Accuracy for commercial sensors was estimated to be about ±0.25 bar.

If salinity sensors are to be used in a dynamic system, their response time to changes in soil needs to be known. Wesseling and Oster (1973) found that although only a few hours were required for the sensor to indicate 95% of a step change in salinity in a bulk solution, days were required in a soil. Although response time is normally not a problem in field soils

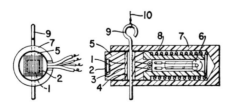

Fig. 13. Front and sectional views of Richards' salinity sensor showing (1) ceramic plate, (2) front-screen electrode, (3) back electrode, (4) nylon tubing, (5) epoxy block, (6) thermistor, (7) outer Lucite case, (8) spring for holding the sensitive element against soil, (9) release pin, and (10) pull wire. After Richards (1966).

where salinity changes slowly, it can be important when sensors are used in laboratory flow models. Using diffusion theory, they developed a procedure for determining a response factor that can be used to correct readings for lag. If salinity changes over periods of more than 5 days, then no correction is needed.

A technique for using automated recording equipment for salinity sensor readings was recently developed by Austin and Oster (1973). A variable frequency oscillator circuit that converts variations in alternating voltage to frequency variations is used. This overcomes the problem of voltage losses with small alternating current signals and long leads. The circuit measures electrical conductivity as a linear function of frequency with a standard deviation of approximately 0.4 mmho/cm.

2. The Four-Electrode Method

While the salinity sensor is useful for many applications in the field, it is limited by its response to only a small, localized volume of soil. Because salinity often varies widely from point to point in the soil profile, numerous sensors would be needed to assess the full range of salinity to which crop roots were subjected. To measure salinity throughout a larger volume, without taking a large number of soil samples or installing a large number of sensors, Rhoades and Ingvalson (1971) adapted the four-electrode earth resistivity technique used for geographical surveys.

Figure 14 shows the arrangement of the four electrodes. The inner

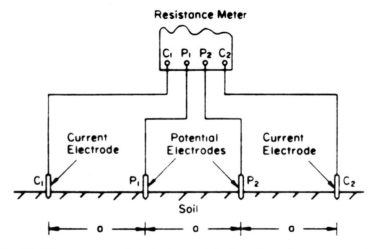

Fig. 14. Wenner array of electrodes used in four-electrode measurement of soil salinity. (*a* represents the inner-electrode spacing.) From Rhoades and Ingvalson (1971).

pair is used to measure the electrical potential induced by a constant current passed between the outer pair. By varying the distance *a*, the distribution of salinity with depth can be estimated. To a first approximation, the depth of current penetration for such an electrode configuration is equal to the spacing *a*. The method is similar in concept to that of Shea and Luthin (1961), who buried electrodes in the same configuration below the surface, but the portable electrodes used by Rhoades and Ingvalson make their method more applicable for salinity appraisal and surveying at many different locations and soil depths with a single set of electrodes. Figure 15 shows apparent electrical conductivity measured with the four-electrode technique as a function of electrical conductivity measured by the saturation extract technique (U.S. Salinity Laboratory Staff, 1954) for a uniformly salinized profile following an irrigation. The agreement is sufficiently close that, at water contents near field capacity, the four-electrode

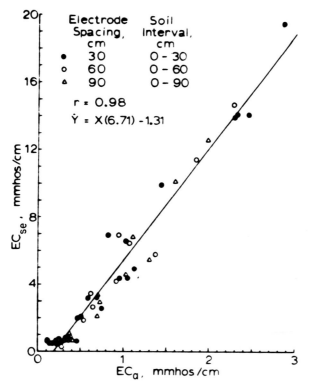

Fig. 15. Relationship between soil conductivity as determined in the field (May 1973) with inner-electrode spacings of 30, 60, and 90 cm and soil salinity expressed as EC_{se} for depth intervals of 0 to 30, 0 to 60, and 0 to 90 cm, respectively. From Halvorson and Rhoades (1974).

measurements give highly useful estimates of the electrical conductivity in the profile.

Halvorson and Rhoades (1974) successfully used this technique to assess the soil salinity and identify potential saline-seep areas near Sidney, Montana. (A saline seep is an area of formerly productive nonirrigated soil that has become too wet and saline for economical crop production because saline subsurface water has seeped to the soil surface. These areas usually develop on hillsides, particularly where slope changes.) Plots of apparent electrical conductivity measured with the four-electrode technique as a function of a gave distinctive types of curves for saline-seep, nonseep, and intermediate sites. These measurements were used to identify and locate encroaching seeps that had not yet appeared at the soil surface.

Although normal spot-to-spot variation of water content in soils near field capacity does not seriously interfere with salinity measurements, the dependence on water content is a significant problem with lower water contents. Results of recent theoretical and experimental research by Rhoades and co-workers (1975) show that the effect of water content and surface conduction can be fully accounted for and included in the calibration of the technique. This may well represent as much of a breakthrough in capability for measuring salinity in the field as was the neutron-scattering technique for measuring soil-water content. The possibility of making detailed salinity appraisals without taking soil samples and making laboratory analyses is encouraging indeed.

D. COMBINATION METHODS

Recent developments in using the thermocouple psychrometer have enabled independent measurement of some of the components of water potential with a single instrument.

Oster *et al.* (1969) used the thermocouple psychrometer in combination with a device equivalent to the pressure plate apparatus to obtain independent measurements of pressure and osmotic potential simultaneously on a single sample in the laboratory. The apparatus is shown in Fig. 16. The psychrometer is similar to that of Rawlins and Dalton (1967), except the ceramic bulb has a bubbling pressure of 15 bars and its interior is vented to the atmosphere. It is mounted through the wall of a pressure chamber, which is packed with the soil whose pressure and osmotic potentials are to be measured. After the ceramic bulb attains equilibrium with the soil, the psychrometer measures the sum of the pressure and osmotic potentials. These components are separated by increasing the pneumatic potential within the soil chamber in increments by air pressure. Each increment of air pressure balances an equal portion of the negative wetness potential. This continues until the magnitude of the pneumatic potential

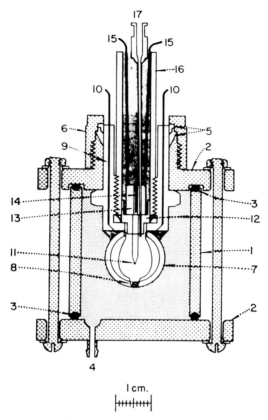

Fig. 16. Cross-sectional view of the pressure chamber and psychrometer assembly showing (1) brass cylinder, (2) brass end plates, (3) O-rings, (4) pressure inlet, (5) Teflon ferrules, (6) compression fitting nut, (7) ceramic bulb, (8) platinum electrodes, (9) acrylic tubing, (10) salinity sensor leads, (11) thermocouple junctions, (12) Teflon holder for heat sinks, (13) copper heat sinks, (14) mica, (15) thermocouple leads, (16) acrylic tubing, and (17) hypodermic needle. The shading represents epoxy resin; the dots represent soil. From Oster *et al.* (1969).

equals the wetness potential. Because at that time the pressure potential is zero, free soil solution exists on the inside surface of the ceramic bulb. At that time, the psychrometer reading is a measure of the osmotic potential of this solution. Further increases in pressure only cause solution to flow into the bulb, but the osmotic potential does not change. Figure 17 shows total potential measured by the psychrometer as a function of applied air pressure for soil samples at three different initial total potentials. At the abrupt change in slope, the osmotic potential is given by the psychrometer reading indicated on the ordinate; the matric potential (numerically equal to but of opposite sign to the pneumatic pressure potential)

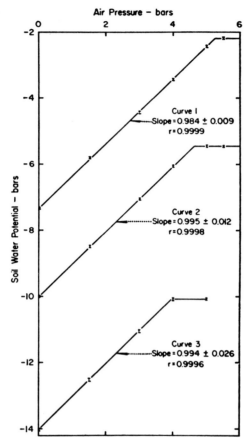

Fig. 17. Soil water potential as a function of the air pressure applied to the soil. The extent of the vertical line centered on each point is equal to two times the standard error of estimate. From Oster *et al.* (1969).

is given by the pressure in the soil chamber. Oster *et al.* (1969) estimated the standard error of measurement of the water potential components with this apparatus to be about ±0.04 bar. The closeness of the slope of the lines of water potential versus air pressure to unity is evidence that the partial specific volume for water did not change appreciably with pressure.

The platinum electrodes shown at the bottom of the ceramic bulb in Fig. 16 function similarly to those in the salinity sensor shown in Fig. 13. Ingvalson *et al.* (1970) designed and used a similar combined thermocouple psychrometer salinity sensor to measure total and osmotic potentials independently, which now makes it possible to infer pressure potential by difference. Estimating pressure potential with this single instrument rather than with separate salinity sensors and soil psychrometers eliminated errors resulting from the large spatial variation of salinity within the root zone.

E. Discussion and Summary

Some of the instruments discussed, notably the psychrometer, the salinity sensor, and the four-electrode salinity apparatus, represent significant recent advances in measurement of total water potential and its components. They have extended the measurement range and made *in situ* measurements possible. Developments that have made continuous measurements possible (for example, measurement of dew point and electrical conductivity) should lead to increased use of these measuring techniques where automation is required to collect large data sets. Technical developments leading to the possibility of automating readings from more traditional instruments such as tensiometers and porous blocks are also important advances. Particularly for studies that involve rapid variations, for example, during the initial stages of infiltration, rapid response of sensors and automatic recording are not only convenient, but essential. Advances in measurement often have resulted primarily from the use of ancillary devices, such as pressure transducers and data acquisition systems that were developed outside the field of soil physics rather than from any specific change in the sensor itself. The potential for future applications, particularly as sensors begin to be used not only for indication but also for control, for example, with automated irrigation, is great.

REFERENCES

Anderson, A. B. C., and Edlefsen, N. E. (1942a). Laboratory study of the response of 2- and 4-electrode plaster of paris blocks as soil-moisture content indicators. *Soil Sci.* **53**, 413–428.

Anderson, A. B. C., and Edlefsen, N. E. (1942b). The electrical capacity of 2-electrode plaster of paris block as an indicator of soil-moisture content. *Soil Sci.* **54**, 35–46.

Ashton, F. M. (1956). Effects of a series of cycles of alternating low and high soil water contents on the rate of apparent photosynthesis in sugar cane. *Plant Physiol.* **31**, 266–274.

Austin, R. S., and Oster, J. D. (1973). An oscillator circuit for automated salinity sensor measurements. *Soil Sci. Soc. Amer., Proc.* **37**, 327–329.

Baldes, E. J. (1934). A micromethod of measuring osmotic pressure. *J. Sci. Instrum.* **11**, 223–225.

Baldes, E. J. (1939). Theory of the thermo-electric measurement of osmotic pressure. *Biodynamica* no. 46, 1–8.

Ballard, L. F., and Gardner, R. P. (1965). "Density and Moisture Content Measurements by Nuclear Methods," Nat. Coop. Highway Res. Progr. Rep. No. 14. Highway Res. Board, Nat. Acad. Sci.—Nat. Res. Counc., Washington, D. C.

Balls, W. L. (1932a). Rapid estimation of water-content in undisturbed soil and in bales of cotton. *Nature (London)* **129**, 505–506.

Balls, W. L. (1932b). Capacitance hygroscopy and some of its applications. *Nature (London)* **130**, 935–938.

Barrs, H. D. (1968). Determination of water deficits in plant tissues. In "Water Deficits and Plant Growth" (T. T. Kozlowski, ed.), Vol. 1, pp. 235–368. Academic Press, New York.

Baver, L. D., Gardner, W. H., and Gardner, W. R. (1972). "Soil Physics," 4th ed. Wiley, New York.

Behnke, J. J., and Bianchi, W. C. (1965). Pressure distributions in layered sand columns during transient and steady-state flows. *Water Resour. Res.* **1**, 557–562.

Belcher, D. J., Cuykendall, T. R., and Sach, H. S. (1950). "The Measurement of Soil Moisture and Density by Neutron- and Gamma-ray Scattering," Tech. Develop. Rep. No. 127. Civil Aeronautics Administration, Technical Development and Evaluation Center, Indianapolis, Indiana.

Bernhard, R. K., and Chasek, M. (1953). Soil density determination by means of radioactive isotopes. *Nondestruct. Test.* **11**, 17–23.

Bernhard, R. K., and Chasek, M. (1955). "Soil Density Determination by Direct Transmission of Gamma Rays," Reprint No. 86. Amer. Soc. Test. Mater., Philadelphia, Pennsylvania.

Bianchi, W. C. (1962). Measuring soil moisture changes. *Agr. Eng.* **43**, 398–399 and 404.

Bianchi, W. C., and Tovey, R. (1968). Continuous monitoring of soil moisture tension profiles. *Trans. Amer. Soc. Agr. Eng.* **11**, 441–443 and 447.

Bianchi, W. C., and Tovey, R. (1970). Soil moisture tension transients associated with evapotranspiration from a water table. *Soil Sci. Soc. Amer., Proc.* **34**, 496–501.

Bloodworth, M. E., and Page, J. B. (1957). Use of thermistors for the measurement of soil moisture and temperature. *Soil Sci. Soc. Amer., Proc.* **21**, 11–15.

Bourget, S. J., Elrick, D. E., and Tanner, C. B. (1958). Electrical resistance units for moisture measurements: Their moisture hysteresis, uniformity, and sensitivity. *Soil Sci.* **86**, 298–304.

Bouyoucos, G. J. (1926). Rapid determination of the moisture content of soils. *Science* **64**, 651–652.

Bouyoucos, G. J., and Cook, R. L. (1967). Measuring the relative humidity of soils at different moisture contents by the gray hydrocal hygrometer. *Soil Sci.* **104**, 297–305.

Bouyoucos, G. J., and Cook, R. L. (1968). Tension of the soil water when permanent wilting occurs, as measured by the gray hydrocal hygrometer method. *Soil Sci.* **106**, 317–322.

Bouyoucos, G. J., and Mick, A. H. (1940). An electrical resistance method for the continuous measurement of soil moisture under field conditions. *Mich., Agr. Exp. Sta., Tech. Bull.* **172**, 1–38.

Bouyoucos, G. J., and Mick, A. H. (1947). Improvements in the plaster of paris absorption block electrical resistance method for measuring soil moisture under field conditions. *Soil Sci.* **63**, 455–465.

Bowers, S. A., and Smith, S. J. (1972). Spectrophotometric determination of soil water content. *Soil Sci. Soc. Amer., Proc.* **36**, 978–980.

Brady, A. P., Huff, H., and McBain, J. W. (1951). Measurement of vapor pressures by means of matched thermistors. *J. Phys. Colloid Chem.* **55**, 304–311.

Bridge, B. J., and Collis-George, N. (1973). A dual source gamma ray traversing mechanism suitable for the non-destructive simultaneous measurement of bulk density and water content in columns of swelling soil. *Aust. J. Soil Res.* **11,** 83–92.

Briggs, L. J. (1899). Electrical instruments for determining the moisture, temperature, and soluble salt content of soils. *U. S., Dep. Agr., Bur. Soils, Bull.* **15.**

Briggs, L. J., and McLane, J. W. (1907). The moisture equivalent of soils. *U. S., Dep. Agr., Bur. Soils, Bull.* **45.**

Brown, R. W., and van Haveren, B. P., eds. (1972). "Psychrometry in Water Relations Research." Utah Agr. Exp. Sta., Logan, Utah.

Calissendorff, C., and Gardner, W. H. (1972). A temperature-compensated leaf psychrometer for *in situ* measurements of water potential. *In* "Psychrometry in Water Relations Research" (R. W. Brown and B. P. van Haveren, eds.), pp. 224–228. Utah Agr. Exp. Sta., Logan.

Campbell, E. C., Campbell, G. S., and Barlow, W. K. (1973). A dewpoint hygrometer for water potential measurement. *Agr. Meteorol* **12,** 113–121.

Cashen, G. H. (1932). Measurements of the electrical capacity and conductivity of soil blocks. *J. Agr. Sci.* **22,** 145–164.

Chadwick, D. G. (1975). Measurement of Soil Moisture by Use of the Latent Heat of Vaporization. Abstract, National Technical Information Service, *Natural Resources,* WGA/48-75/02, p. 16.

Chahal, R. S., and Yong, R. H. (1964). Validity of the energy characteristics of soil water determined with pressurized apparatus. *Nature (London)* 201, 1180–1181.

Chernyak, G. Ya. (1964). "Dielectric Methods for Investigating Moist Soils." Available from U. S. Dept. of Commerce, Clearing House for Federal Scientific and Technical Information, Springfield, Virginia (Russian translation, 1967).

Chow, T. L., and de Vries, J. (1973). Dynamic measurement of soil and leaf water potential with a double loop Peltier-type thermocouple psychrometer. *Soil Sci. Soc. Amer., Proc.* **37,** 181–188.

Cohen, O. P. (1964). A procedure of calibrating neutron moisture probes in the field. *Isr. J. Agr. Res.* **14,** 169–178.

Collis-George, N., and Bridge, B. J. (1973). The effect of height of sample and confinement on the moisture characteristic of an aggregated swelling clay soil. *Aust. J. Soil Res.* **11,** 107–120.

Cope, F., and Trickett, E. S. (1965). Measuring soil moisture. *Soils Fert.* **28,** 201–208.

Corey, J. C., Peterson, S. F., and Wakat, M. A. (1971). Measurement of attenuation of ^{137}Cs and ^{241}Am gamma rays for soil density and water content determinations. *Soil Sci. Soc. Amer., Proc.* **35,** 215–219.

Cornish, P. M., Laryea, K. B., and Bridge, B. J. (1973). A nondestructive method of following moisture content and temperature changes in soils using thermistors. *Soil Sci.* **115,** 309–314.

Dalton, F. N., and Rawlins, S. L. (1968). Design criteria for Peltier-effect thermocouple psychrometers. *Soil Sci.* **105,** 12–17.

Danlin, A. I. (1955). [The measurement of soil moisture by means of gamma rays.] *Pochvovedenie* **7,** 74–83.

Davidson, J. M., Biggar, J. W., and Nielsen, D. R. (1963). Gamma-radiation for measuring bulk density and transient water flow in porous media. *J. Geophys. Res.* **68,** 4777–4783.

Edelfsen, N. E. (1933). A review of results of dielectric methods for measuring moisture present in materials. *Agr. Eng.* **14,** 243–244.

Enfield, C. G., and Evans, D. D. (1969). Conductivity instrumentation for *in situ* measurement of soil salinity. *Soil Sci. Soc. Amer., Proc.* **33**, 787–789.

Farm, R. J., and Bruckenstein, S. (1968). A sensitive and versatile differential vapor pressure apparatus. *Anal. Chem.* **40**, 1651–1657.

Ferguson, H., and Gardner, W. H. (1962). Water content measurement in soil columns by gamma ray absorption. *Soil Sci. Soc. Amer., Proc.* **26**, 11–14.

Fletcher, J. E. (1939). A dielectric method for determining soil moisture. *Soil Sci. Soc. Amer. Proc.* **4**, 84–88.

Fritton, D. D. (1969). Resolving time, mass absorption coefficient and water content with gamma ray attenuation. *Soil Sci. Soc. Amer., Proc.* **33**, 651–655.

Gardner, F. D. (1898). The electrical method of moisture determination. *U. S., Dep. Agr., Bur. Soils, Bull.* **12.**

Gardner, W. H. (1965). Chapter 7. Water Content. *In* "Methods of Soil Analysis," (C. A. Black, Ed.) pp. 82–127. Monogr. No. 9. Amer. Soc. Agron., Madison, Wisconsin.

Gardner, W. H., and Calissendorff, C. (1967). Gamma ray and neutron attenuation in measurement of soil bulk density and water content. *In* "Isotope and Radiation Techniques in Soil Physics and Irrigation Studies." pp. 101–113. IAEA, Vienna.

Gardner, W. H., and Fischer, M. E. (1966). Concurrent measurement of bulk density and water content of soil using two gamma ray energies. *Agron. Abstr. 58th Annual Mtg. Amer. Soc. of Agron.* p. 46.

Gardner, W. H., Campbell, G. S., and Calissendorff, C. (1972). Systematic and random errors in dual gamma energy soil bulk density and water content measurements. *Soil Sci. Soc. Amer. Proc.* **36**, 393–398.

Gardner, W. R., and Kirkham, D. (1952). Determination of soil moisture by neutron scattering. *Soil Sci.* **73**, 391–401.

Giesel, W., Lorch, S., Renger, M., and Strebel, O. (1970). Water-flow calculations by means of gamma-absorption and tensiometer field measurements in unsaturated field soil. *Proc. Panel Isotop. Hydrol.,* pp. 663–672.

Gillham, R. W. (1972). Hysteretic water flow in a porous medium: Experimental study and numerical simulation. Ph.D. Thesis, University of Illinois, Urbana.

Groenevelt, P. H., de Swart, J. G., and Cesler, J. (1969). Water content measurements with 60 keV gamma ray attenuation. *Bull. Int. Ass. Sci. Hydrol.* **14**, 67–78.

Gurr, C. G. (1962). Use of gamma rays in measuring water content and permeability in unsaturated columns of soil. *Soil Sci.* **94**, 224–229.

Haise, H. R., and Kelley, O. J. (1946). Relation of moisture tension to heat transfer and electrical resistance in plaster of paris blocks. *Soil Sci.* **61**, 411–422.

Halvorson, A. D., and Rhoades, J. D. (1974). Assessing soil salinity and identifying potential saline-seep areas with field soil resistance measurements. *Soil Sci. Soc. Amer., Proc.* **38**, 576–581.

Hancock, C. K., and Burdick, R. L. (1957). Rapid determination of water in wet soils. *Soil Sci.* **83**, 197–205.

Hanks, R. J., and Shawcroft, R. W. (1965). An economical lysimeter for evapotranspiration studies. *Agron. J.* **57**, 634–636.

Hill, A. V. (1930). A thermal method of measuring the vapour pressure of an aqueous solution. *Proc. Roy. Soc., Ser. A* **127**, 9–19.

Hillel, D. (1971). "Soil and Water: Physical Principles and Processes." Academic Press, New York.

Holmes, J. W., and Colville, J. S. (1964). The use of the neutron moisture meter and lysimeters for water balance studies. *Trans. Int. Congr. Soil Sci., 8th, 1964* Vol. 2, pp. 445–454.

Hsieh, J. J. C., and Hungate, F. P. (1970). Temperature-compensated Peltier psychrometer for measuring plant and soil water potentials. *Soil Sci.* **110,** 253–257.

Ingvalson, R. D., Oster, J. D., Rawlins, S. L., and Hoffman, G. J. (1970). Measurement of water potential and osmotic potential in soil with a combined thermocouple psychrometer and salinity sensor. *Soil Sci. Soc. Amer., Proc.* **34,** 570–574.

International Society of Soil Science, ISSS. (1974). Report on soil physics terminology. *Trans. 10th, Int. Congr. Soil Sci.* Vol. 12 (in press). [A preliminary version of this report appeared in the *Bull. Int. Soc. Soil Sci.* **44,** 10–17 (1974).]

Johnston, C. N. (1942). Water permeable jacketed thermal radiators as indicators of field capacity and permanent wilting percentage in soils. *Soil Sci.* **54,** 123–126.

Kay, B. D., and Low, P. F. (1970). Measurement of the total suction of soils by a thermistor psychrometer. *Soil Sci. Soc. Amer., Proc.* **34,** 373–376.

Kemper, W. D. (1959). Estimation of osmotic stress in soil water from the electrical resistance of finely porous ceramic units. *Soil Sci.* **87,** 345–349.

Kitchen, J. H., and Thames, J. L. (1972). Pulsed thermistor psychrometer for measuring vapor pressure differential. *In* "Psychrometry in Water Relations Research" (R. W. Brown and B. P. van Haveren, eds.), pp. 113–119. Utah Agr. Exp. Sta., Logan, Utah.

Klute, A., and Gardner, W. R. (1962). Tensiometer response time. *Soil Sci.* **93,** 204–207.

Klute, A., and Peters, D. B. (1962). A recording tensiometer with a short response time. *Soil Sci. Soc. Amer. Proc.* **26,** 87–88.

Klute, A., and Richards, L. A. (1962). Effect of temperature on relative vapor pressure of water in soils: Apparatus and preliminary measurements. *Soil Sci.* **93,** 391–396.

Kreeb, K. (1965). Untersuchungen zu den osmotischen Zustandsgrossen. II. Mitteilung: Eine electronische Methode zur Messung zur Saugspannung (NTC-Methode). *Planta* **66,** 156.

Lal, R. (1974). The effect of soil texture and density on the neutron and density probe calibration for some tropical soils. *Soil Sci.* **117,** 183–190.

Lang, A. R. G. (1968). Psychrometric measurement of soil water potential *in situ* under cotton plants. *Soil Sci.* **106,** 460–464.

Leonard, R. A., and Low, P. F. (1962). A self-adjusting null-point tensiometer. *Soil Sci. Soc. Amer., Proc.* **26,** 123–125.

McCauley, G. N., and Stone, J. F. (1972). Source-detector geometry effect on neutron probe calibration. *Soil Sci. Soc. Amer., Proc.* **36,** 246–250.

McGuinness, J. L., Dreibelbis, F. R., and Harrold, L. L. (1961). Soil moisture measurements with the neutron method supplement weighing lysimeters. *Soil Sci. Soc. Amer., Proc.* **25,** 339–342.

Macklon, A. E. S., and Weatherley, P. E. (1965a). A vapor pressure instrument for the measurement of leaf and soil water potential. *J. Exp. Bot.* **16,** 261–270.

Macklon, A. E. S., and Weatherley, P. E. (1965b). Controlled environment studies of the nature and origins of water deficits in plants. *New Phytol.* **64,** 414–427.

Mansell, R. S., Hammond, L. C., and McCurdy, R. M. (1973). Coincidence and interference corrections for dual-energy gamma ray measurements of soil density and water content. *Soil Sci. Soc. Amer., Proc.* **37,** 500–504.

Meeuwig, R. O. (1972). A low-cost thermcouple psychrometer recording system.

In "Psychrometry in Water Relations Research" (R. W. Brown and B. P. van Haveren, eds.), pp. 131–135. Utah Agr. Exp. Sta., Logan.

Merrill, S. D., and Rawlins, S. L. (1972). Field measurements of soil water potential with thermocouple psychrometers. *Soil. Sci.* **113**, 102–109.

Millar, A. A., Lang, A. R. G., and Gardner, W. R. (1970). Four-terminal Peltier type thermocouple psychrometer for measuring water potential in nonisothermal systems. *Agron. J.* **62**, 705–708.

Miller, R. D. (1951). A technique for measuring soil moisture tension in rapidly changing systems. *Soil Sci.* **72**, 291–301.

Miller, R. J., Smith, R. B., and Biggar, J. W. (1974). Soil water content: Microwave oven method. *Soil Sci. Soc. Amer., Proc.* **38**, 535–537.

Mokady, R. S., and Low, P. F. (1968). Simultaneous transport of water and salt through clays. I. Transport mechanisms. *Soil Sci.* **105**, 112–131.

Monteith, J. L., and Owen, P. C. (1958). A thermocouple method for measuring relative humidity in the range 95–100%. *J. Sci. Instrum.* **35**, 443–446.

Mualem, Y. (1974). A conceptual model of hystersis. *Water Resour. Res.* **10**, 514–520.

Mukhin, L. V., and Christotinov, L. V. (1961). Gamma ray method of determining moisture in coarse fragmentary soils. *Sov. Soil. Sci.* **7**, 807–809.

Müller, R. H., and Stolten, H. J. (1953). Use of thermistors in precise measurement of small temperature differences. Thermometric determination of molecular weights. *Anal. Chem.* **25**, 1103–1106.

Neumann, H. H., and Thurtell, G. W. (1972). A Peltier cooled thermocouple dew-point hygrometer for in situ measurement of water potentials. *In* "Psychrometry in Water Relations Research" (R. W. Brown and B. P. van Haveren, eds.), pp. 103–112. Utah Agr. Exp. Sta., Logan.

Oster, J. D., and Willardson, L. S. (1971). Reliability of salinity sensors for the management of soil salinity. *Agron. J.* **63**, 695–698.

Oster, J. D., Rawlins, S. L., and Ingvalson, R. D. (1969). Independent measurement of matric and osmotic potential of soil water. *Soil Sci. Soc. Amer., Proc.* **33**, 188–192.

Peck, A. J. (1960). Change of moisture tension with temperature and air pressure: Theoretical. *Soil Sci.* **89**, 303–310.

Peck, A. J. (1968). Theory of the Spanner psychrometer. I. The thermocouple. *Agr. Meteorol.* **5**, 433–447.

Peck, A. J. (1969). Theory of the Spanner psychrometer. II. Sample effects and equilibration. *Agr. Meteorol.* **6**, 111–124.

Peck, A. J., and Rabbidge, R. M. (1966a). Note on an instrument for measuring water potentials, particularly in soils. *Pap., C.S.I.R.O. Div. Meteorol. Phys. Conf., Instrum. Plant Environ. Meas.*, September 1966 p. 20–21.

Peck, A. J., and Rabbidge, R. M. (1966b). Soil-water potential: Direct measurement by a new technique. *Science* **151**, 1385–1386.

Peck, A. J., and Rabbidge, R. M. (1969). Design and performance of an osmotic tensiometer for measuring capillary potential. *Soil Sci. Soc. Amer., Proc.* **33**, 196–202.

Perrier, E. R., Stockinger, Karl R., and Swain, R. V. (1966). Scintillation counter for measuring thermal neutrons in a soil-water system. *Soil Sci.* **101**, 125–129.

Phene, C. J., Hoffman, G. J., and Rawlins, S. L. (1971a). Measuring soil matric potential *in situ* by sensing heat dissipation within a porous body. I. Theory and sensor construction. *Soil Sci. Soc. Amer., Proc.* **35**, 27–33.

Phene, C. J., Rawlins, S. L., and Hoffman, G. J. (1971b). Measuring soil matric potential *in situ* by sensing heat dissipation within a porous body. II. Experimental results. *Soil Sci. Soc. Amer., Proc.* **35,** 225–229.

Phene, C. J., Hoffman, G. J., and Austin, R. S. (1973). Controlling automated irrigation with a soil matric potential sensor. *Trans. Amer. Soc. Agr. Eng.* **16,** 773–776.

Poulovassilis, A. (1974). The uniqueness of the moisture characteristics. *Soil Sci.* **25,** 27–33.

Pruitt, W. O., and Angus, D. E. (1960). Large weighing lysimeter for measuring evapotranspiration. *Trans. Amer. Soc. Agr. Eng.* **3,** 13–15 and 18.

Rawlins, S. L. (1964). Systematic error in leaf water potential measurements with a thermocouple psychrometer. *Science* **146,** 644–646.

Rawlins, S. L. (1966). Theory for thermocouple psychrometers used to measure water potential in soil and plant samples. *Agr. Meteorol.* **3,** 293–310.

Rawlins, S. L. (1971). Some new methods for measuring the components of water potential. *Soil Sci.* **112,** 8–16.

Rawlins, S. L., and Dalton, F. N. (1967). Psychrometric measurement of soil water potential without precise temperature control. *Soil Sci. Soc. Amer., Proc.* **31,** 297–301.

Rawlins, S. L., and Gardner, W. H. (1963). A test of the validity of the diffusion equation for unsaturated flow of soil water. *Soil Sci. Soc. Amer., Proc.* **27,** 507–511.

Reginato, R. J. (1974). Gamma radiation measurements of bulk density changes in a soil pedon following irrigation. *Soil Sci. Soc. Amer., Proc.* **38,** 24–29

Reginato, R. J., and Jackson, R. D. (1971). Field measurement of soil water content by gamma-ray transmission compensated for temperature fluctuations. *Soil Sci. Soc. Amer., Proc.* **35,** 529–533.

Reginato, R. J., and Stout, K. (1970). Temperature stabilization of gamma-ray transmission equipment. *Soil Sci. Soc. Amer., Proc.* **34,** 152–153.

Reginato, R. J., and van Bavel, C. H. M. (1964). Soil water measurement with gamma attenuation. *Soil Sci. Soc. Amer., Proc.* **28,** 721–724.

Rhoades, J. D., and Ingvalson, R. D. (1971). Determining salinity in field soils with soil resistance measurements. *Soil Sci. Soc. Amer., Proc.* **35,** 54–60.

Rhoades, J. D., and Prather, R. J., and Raats, P. A. C. (1975). "Effects of Liquid-phase Electrical Conductivity, Water Content, and Surface Conduction on Bulk Soil Electrical Conductivity U. S. Salinity Laboratory, ARS/USDA, Riverside, California (in preparation).

Rice, R. (1969). A fast-response, field tensiometer system. *Trans. Amer. Soc. Agr. Eng.* **12,** 48–50.

Richards, B. G. (1965a). Thermistor hygrometer for determining the free energy of moisture in unsaturated soils. *Nature (London)* **208,** 608.

Richards, B. G. (1965b). A thermistor hygrometer for the direct measurement of the free energy of soil moisture. *CSIRO Soil Mech. Sect., Tech. Rep.* No. 5.

Richards, L. A. (1928). The usefulness of capillary potential to soil moisture and plant investigators. *J. Agr. Res.* **37,** 719–742.

Richards, L. A. (1941). A pressure membrane extraction apparatus for soil solution. *Soil Sci.* **51,** 377–386.

Richards, L. A. (1966). A soil salinity sensor of improved design, *Soil Sci. Soc. Amer., Proc.* **30,** 333–337.

Richards, L. A., and Decker, D. L. (1963). Difficulties with electrical-resistance hygrometers at high humidity. *Soil Sci. Soc. Amer., Proc.* **24,** 481.

Richards, L. A., and Fireman, M. (1943). Pressure plate apparatus for measuring moisture sorption and transmission by soils. *Soil Sci.* **56**, 395–404.

Richards, L. A., and Ogata, G. (1958). Thermocouple for vapor pressure measurement in biological and soil systems at high humidity. *Science* **128**, 1089–1090.

Richards, L. A., and Stumpf, H. T. (1960). Volumetric soil sampler. *Soil Sci.* **89**, 108–110.

Richards, L. A., Low, P. F., and Decker, D. L. (1964). Pressure dependence of the relative vapor pressure of water in soil. *Soil Sci. Soc. Amer., Proc.* **28**, 5–8.

Richards, S. J. (1938) Soil moisture content calculations from capillary tension records. *Soil Sci. Soc. Amer., Proc.* **3**, 57–64.

Rijtema, P. E. (1969). The calculations of non-parallelism of gamma access tubes using soil sampling data. *J. Hydrol.* **9**, 206–212.

Ritchie, J. T., and Burnett, E. (1968). A precision weighing lysimeter for row crop water use studies. *Agron. J.* **60**, 545–549.

Ro, P. S., Fahlen, T. S., and Bryant, H. C. (1968). Precision measurements of water droplet evaporation rates. *Appl. Opt.* **7**, 883–890.

Russell, M. B., and Richards, L. A. (1938). The determination of soil moisture energy relations by centrifugation. *Soil Sci. Soc. Amer., Proc.* **3**, 66–69.

Scotter, D. R. (1972). The theoretical and experimental behaviour of a Spanner psychrometer. *Agr. Meteorol.* **10**, 125–136.

Shaw, B., and Baver, L. D. (1939). An electrothermal method for following moisture changes of the soil *in situ*. *Soil Sci. Soc. Amer., Proc.* **4**, 78–83.

Shea, P. F., and Luthin, J. N. (1961). An investigation of the use of the four-electrode probe for measuring soil salinity *in situ*. *Soil Sci.* **92**, 331–339.

Slater, C. L. (1942). A modified resistance block for soil moisture measurement. *Agron. J.* **34**, 284–285.

Smiles, D. E., Vachaud, G., and Vauclin, M. (1971). A test of the uniqueness of the soil moisture characteristics during transient, nonhysteretic flow of water in a rigid soil. *Soil Sci. Soc. Amer., Proc.* **35**, 534–539.

Soane, B. D. (1967). Dual energy gamma-ray transmission for coincident measurement of water content and dry bulk density in soil. *Nature (London)* **214**, 1273–1274.

Spanner, D. C. (1951). The Peltier effect and its use in the measurement of suction pressure. *J. Exp. Bot.* **2**, 145–168.

Stewart, G. L. (1960). "Uniform Drying of Soil Samples by the Use of Dielectric Heating," Mimeogr. rep. Department of Agronomy, Washington State University, Pullman.

Stewart, G. L. (1962). "Water Content Measurement by Neutron Attenuation and Application to Unsaturated Flow of Water in Soil." Ph.D. Thesis, Washington State University, Pullman (unpublished).

Stone, J. F. (1972). Instrumentation effects on error in nuclear methods for soil water and density determination. *Soil Sci. Soc. Amer., Proc.* **36**, 261–264.

Strebel, O., Giesel, W., Renger, M., and Lorch, S. (1970). Automatische registrierung der bodenwasserspannung im gelande mit dem druckaufnehmer-tensiometer. *Z. Pflanzenernaehr. Bodenk.* **126**(1), 6–15.

Stroosnijder, L., and de Swart, J. G. (1973). Errors in soil bulk density and water content measurements with gamma ray attenuation. *Soil Sci. Soc. Amer., Proc.* **37**, 485–486.

Tanner, C. B., and Hanks, R. J. (1952). Moisture hysteresis in gypsum moisture blocks. *Soil Sci. Soc. Amer., Proc.* **16,** 48–51.

Taylor, S. A., and Ashcroft, G. L. (1972). "Physical Edaphology." Freeman, San Francisco, California.

Thijssen, H. A. C., de Witt, C. T., van Vollenhoven, E., Timmers, H. J., and Admiraal, L. (1954). New instruments for agricultural research. *Neth. J. Agr. Res.* **2,** 209–214.

Topp, G. C. (1969). Soil water hysteresis measured in a sandy loam and compared with the hysteretic domain model. *Soil Sci. Soc. Amer., Proc.* **33,** 645–651.

Topp, G. C. (1970). Soil water content from gamma ray attenuation. A comparison of ionization chamber and scintillation detectors. *Can. J. Soil Sci.* **50,** 439–447.

Topp, G. C., and Miller, E. E. (1966). Hysteretic moisture characteristics and hydraulic conductivities for glass bead media. *Soil Sci. Soc. Amer., Proc.* **30,** 156–162.

Towner, G. D. (1967). A note on the utility of the thermistor hygrometer method for the measurement of the free energy of soil moisture. *Aust. J. Instrum. Contr.* **23,** 108–110.

U. S. Salinity Laboratory Staff. (1954). Diagnosis and improvement of saline and alkali soils. *U. S., Dep. Agr., Handb.* **60.**

van Bavel, C. H. M. (1959). Soil densitometry by gamma transmission. *Soil Sci.* **87,** 50–58.

van Bavel, C. H. M., and Myers, L. E. (1962). An automatic weighing lysimeter. *Agr. Eng.* **43,** 580–583.

van Bavel, C. H. M., and Stirk, G. B. (1967). Soil water measurement with an Am^{241}-Be neutron source and an application to evaporimetry. *J. Hydrol.* **5,** 40–46.

van Bavel, C. H. M., Underwood, N., and Ragar, S. R. (1957). Transmission of gamma radiation by soils and soil densitometry. *Soil Sci. Soc. Amer., Proc.* **21,** 588–591.

Visvalingam, M., and Tandy, J. D. (1972). The neutron method for measuring soil moisture content. A review. *J. Soil Sci.* **23,** 499–511.

Vomicil, J. A. (1954). *In situ* measurement of soil bulk density. *Agr. Eng.* **35,** 651–654.

Watson, K. K. (1965). Some operating characteristics of a rapid response tensiometer system. *Water Resour. Res.* **1,** 577–586.

Watson, K. K. (1967a). A recording field tensiometer with rapid response characteristics. *J. Hydrol.* **5,** 33–39.

Watson, K. K. (1967b). Response behavior of a tensiometer-pressure transducer system under conditions of changing pore air pressure. *Soil Sci.* **104,** 439–443.

Watson, K. K., and Jackson, R. D. (1967). Temperature effects in a tensiometer-pressure transducer system. *Soil Sci. Soc. Amer., Proc.* **31,** 156–160.

Weatherley, P. E. (1959). A new micro-osmometer. *J. Exp. Bot.* **11,** 258–268.

Wesseling, J., and Oster, J. D. (1973). Response of salinity sensors to rapidly changing salinity. *Soil Sci. Soc. Amer., Proc.* **37,** 553–557.

Whitney, M., Gardner, F. D., and Briggs, L. J. (1897). An electrical method determining the moisture content of an arable soil. *U. S., Dep. Agr., Bull.* **6.**

Wiebe, H. H., Campbell, G. S., Gardner, W. H., Rawlins, S. L., Cary, J. W., and Brown, R. W. (1971). Measurement of plant and soil water status. *Utah, Agr. Exp. Sta., Bull.* **484,** 1–71.

CHAPTER 2

STRUCTURE AND FUNCTIONING OF STOMATA

W. G. Allaway

SCHOOL OF BIOLOGICAL SCIENCES, UNIVERSITY OF SYDNEY, N.S.W.

and

F. L. Milthorpe

SCHOOL OF BIOLOGICAL SCIENCES, MACQUARIE UNIVERSITY, NORTH RYDE,
N.S.W. AUSTRALIA.

I. INTRODUCTION

Stomata have continued to intrigue botanists for well over a hundred years, much of this fascination arising from the difficulty of establishing acceptable concepts of stomatal functioning and of the role they play in gas exchange. Appreciable progress in both aspects has been made over the past two decades: in terms of the (asymmetrical) sigmoid curve which describes many biological phenomena, including the degree of our understanding of them, we may now assume that we are moving toward the

point of inflexion in respect of the functioning of stomata and have progressed beyond this in understanding their significance in controlling gas
exchange. It is appropriate, therefore, to attempt a modern synthesis, especially as there have been a number of advances, with changes in some
concepts and strengthening of others, since the publication of the treatise
by Meidner and Mansfield (1968).

In this chapter we will discuss mechanisms—a matter that has been
dogged by the intractability of exploring changes in a few specific cells
surrounded by a large number of other cells differing in structure and physiology—and in the succeeding chapter we will explore the consequences
of these mechanisms—a matter that has been marked as much by the
infelicity of the investigations as by the intractability of the subject. Understanding of mechanisms is, of course, so intimately associated with the responses to various external and internal factors that a degree of overlap
between the material in the two chapters is unavoidable. Generally, we
will cite here the evidence that we believe is immediately concerned with
the elucidation of mechanisms and, in the succeeding chapter that is
directed toward establishing quantitative relationships. This procedure may
well result in omission as well as duplication of significant findings; however, we are not concerned with preparing a compendium of all relevant
contributions, but rather in presenting what appear to us to be the
most appropriate interpretations of available evidence. Moreover, this is
"the year of the stomata review"; an exhaustive cover will be obtained
from the current reviews of Hsaio (1976), Mansfield (1975), Raschke
(1975), and Thomas (1975). *Water Deficits and Plant Growth* is now
well established as a text at advanced level and we will attempt to maintain
that approach. Nevertheless, the singling out of stomata for treatment in
two chapters extends the original terms of reference and tacitly assumes
that stomata play a highly significant role in the control of water deficits.

Functioning is intimately associated with structure and in considering
both over the plant kingdom we may expect to find a common theme with
innumerable variations. This has long been recognized in respect of the
gross morphology of stomata (Copeland, 1902) and in recent years a
corpus of knowledge on fine structure has been accumulating. Study of
the nature and degree of variation in mechanisms has hardly begun—
indeed, knowledge of the system in even a restricted number of species
is still so inadequate that common mechanisms and possible aberrations
cannot be separated and this leads to some confusion in recognizing what
might well prove to be the common theme. Nevertheless, within these constraints, we will aim toward some degree of synthesis.

We use a number of terms for specific parts of the stomatal apparatus;
although none of these are our own, it is necessary to define them. By

"stoma" we mean the stomatal pore together with its two guard cells. In some of the commonly studied species such as *Vicia faba* this constitutes the whole of the stomatal apparatus, since stomata are scattered among epidermal cells apparently undifferentiated from the rest of the epidermis. In many other species, however, each stoma is surrounded by a number of visibly specialized cells quite distinct from the other epidermal cells: these specialized epidermal cells are called "subsidiary cells" and when we write of the "stomatal apparatus" we mean the stoma plus its subsidiary cells. Commonly studied plants with subsidiary cells are several species of *Commelina* and *Zea mays.* Metcalfe and Chalk (1950) distinguished different types of arrangement of the subsidiary cells and listed the families of plants in which the four main types are found, as well as numbers of other types found in only a few genera.

In Fig. 1 are shown a number of structural features of a stoma (see also Fig. 5 for definitions). Lining the stomatal pore are the *ventral walls* of the guard cells. The *lateral walls* are those parallel to the surface of the epidermis, the *outer* lateral wall being adjacent to the outside air and the *inner* lateral wall adjacent to the substomatal cavity. The wall separating the guard cell from the adjacent epidermal cell is the *dorsal* wall. The stomatal pore is not straight-walled, but often has *inner* and *outer* *ridges,* which delimit the *fore* and *rear chambers* of the stomatal pore with the *throat* in between. When we write of width of guard cell or stoma or aperture of the pore (preferably measured at the throat), we mean the measurement made parallel to the epidermis and normal to the pore; length is the measurement along and parallel to the pore and depth that normal to the epidermis.

II. COMPARATIVE MORPHOLOGY AND FINE STRUCTURE

Stomata appeared very early in the history of vascular plants, possibly early in the Lower Devonian, being found, for example, in *Zosterophyllum, Drepanophycus* (Chaloner, 1970) and *Rhynia* (Fig. 2a). They evidently arose more or less simultaneously with the development of vascular systems and cuticles which allowed the emergence of large terrestrial plants. The presence of a cuticle over a photosynthetic organ requires the development of stomata or some similarly variable passage to arrive at a compromise between the conflicting requirements for entry of carbon dioxide and prevention of loss of water vapor. So we should perhaps have expected that stomata, cuticle, and vascular systems would all appear at much the same time in the first large terrestrial plants. Large terrestrial plants are exclusively sporophytic, and stomata are confined to the sporophytic generation.

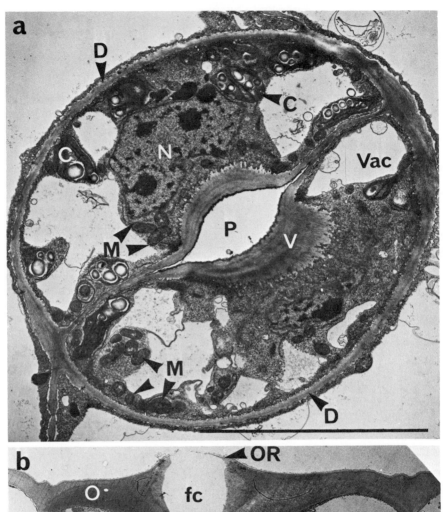

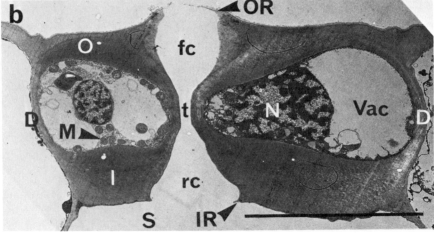

A. BRYOPHYTA

1. Anthocerotae

Stomata have not been observed in liverworts (Hepaticae) but are present in sporophytes of Anthocerotae (von Mohl, 1856). Stomata of *Anthoceros* are large (50 to 70 μm long, Paton and Pearce, 1957) and they consist, like those of most other plants, of two kidney-shaped guard cells. The long axis of the stomatal pore is generally parallel with the long axis of the sporophyte, on which the stomata are irregularly and infrequently scattered; differentiated subsidiary cells are not discernible. There is little information on the structure of *Anthoceros* stomata, but we show an electron micrograph of one in Fig. 2c. The cell walls were not of constant thickness; the dorsal walls of the guard cells were thin, and the ventral walls only slightly thicker at the center; the outer lateral walls were thickened, with the thickest part near the pore, and the inner lateral walls had thickenings at the angle between them and the ventral walls. Outer ridges were present in the pore walls, while the inner ridges were less pronounced. This type of wall thickening is quite common among elliptical stomata. In our specimens the stomata were either difficult to fix properly or had already degenerated so that the cytoplasm was partially disorganized. What remained of the cell contents showed that the guard cells contained a number of mitochondria and at least one chloroplast. [Parihar (1961) stated that *Anthoceros* sporophytes have two chloroplasts per cell.] In our sections the guard cell chloroplast appeared fairly similar to those in other cells of the sporophyte, but was smaller and had fewer grana.

2. Musci

In mosses stomata are usually present on the capsule of the sporophyte, although capsules without them are found in widely divergent groups (Watson, 1964). For some time it was thought that moss stomata show

---

Fig. 1. (a) Stoma of *Brassica rapa* in paradermal section in the electron microscope. V, ventral wall; P, stomatal pore; D, dorsal wall; C, chloroplast with starch grains; N, nucleus; Vac, vacuole; M, mitochondrion. Wall thickenings near the ventral wall are visible in glancing section. Scale: approximately 10 μm. Glutaraldehyde/osmic acid. Electron micrograph by D. Armstrong. (b) Electron micrograph of stoma of *Vicia faba* in transverse section. O, outer lateral wall; I, inner lateral wall; S, substomatal cavity; IR, inner ridge; OR, outer ridge; fc, forechamber; rc, rear chamber; t, throat of stomatal pore. Other symbols as in (a). A common type of wall thickening, very thick inner lateral walls and thickened angles between ventral and outer lateral walls, is present. The outer ridges of the guard cells are seen to be made partly of cell wall and partly of cuticle. Scale: 10 μm. Glutaraldehyde/osmic acid.

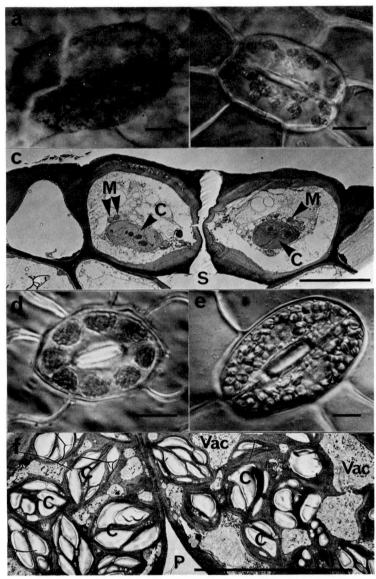

Fig. 2. (a) Stoma of *Rhynia* in a thin section of Rhynie chert. Scale: 10 μm.
(b) Stoma of the capsule (sporophyte) of *Leptobryum pyriforme;* a "long-pored"
moss stoma. Chloroplasts and a faint outline of the guard cells' nuclei are shown.
Scale: 10 μm. Nomarski differential interference contrast. (c) Electron micrograph
of transverse section of stoma of *Anthoceros* sp. sporophyte. Preservation is poor
(there is even a break in one of the guard cell walls). However, one chloroplast
profile (C) and several mitochondrial profiles (M) can be seen in each cell. The

little response to environmental factors other than severe water stress, but recently Garner and Paolillo (1973) reported normal stomatal responses to light, darkness, and abscisic acid in *Funaria hygrometrica* for a few days after capsule expansion. Much of the information given below is from the wide survey of British mosses by Paton and Pearce (1957). Stomata appear generally to be present over areas of green tissue with intercellular spaces—places where CO_2 exchange would be expected to be an important feature of the tissue. Usually there are two kidney-shaped guard cells per stoma (Fig. 2b), but in *Funaria* and some other genera the two guard cells are connected at their ends (Schimper, 1848), giving the appearance of a single oval cell with the elongated pore in its middle; in some species, up to four guard cells per stoma have been observed, but this is abnormal. In *Sphagnum* sporophytes, stomata are only rudimentary and, although the stoma mother cell (if it may be so called) divides, no pore is formed (Parihar, 1961). Stomatal shape is characteristic in many genera: the guard cells may be distinguished as "long," giving an elliptical stomatal pore, or "short," resulting in a more rounded pore. Moss stomata may be up to 70 μm long, although they are generally from 25 to 40 μm in length. Stomata of many moss species have rather uniformly thin cell walls (e.g. *Mnium cuspidatum,* Haberlandt, 1914; Fig. 5a), but in a number of species with "short" guard cells, the walls are very much thickened [e.g., *Buxbaumia aphylla, Fissidens adianthoides, Thamnium alopecurum* (Paton and Pearce, 1957) and *Polytrichum juniperinum* (von Guttenberg, 1971).]

B. Pteridophyta

The stomata of pteridophytes are slightly better known than those of mosses. In *Psilotum, Selaginella* (Fig. 2d), *Equisetum,* and ferns, they are composed of the usual two kidney-shaped guard cells with asymmetrically thickened walls. In a few instances the guard cell walls have lignified bands. There are commonly large amounts of thickening near the pore, in the corners between the lateral walls and the ventral wall; in some spe-

---

guard cell wall looks as though it is not homogeneous, with its inner layers less heavily stained than the outer parts and those of other cells. Note the substomatal cavity (S). Scale: 10 μm. Glutaraldehyde/osmic acid. (d) Stoma of *Selaginella kraussiana* showing 4 large chloroplasts per guard cell. Scale: 10 μm. (e) Stoma of *Nephrolepis exaltata* (fern) showing the very many crowded chloroplasts characteristic of fern guard cells. Nomarski differential interference contrast. Scale: 10 μm. (f) Electron micrograph of paradermal section of *Phyllitis scolopendrium* (fern) stoma. Numerous starchy chloroplast profiles (C). Black lines in some places are folds in section. Other labels as in Figure 1 (a). Scale: 10 μm. Glutaraldehyde/osmic acid.

cies the cell walls are relatively unthickened (von Guttenberg, 1971). We have not found any information on plasmodesmata in pteridophyte guard cells. *Equisetum* is remarkable in that its subsidiary cells grow over the guard cells and have a number of thickened siliceous ridges in the wall next to the guard cell and radiating from the pore (Dayanandan and Kaufman, 1973; Riebner, 1925).

In the fern species we have examined, guard cells were densely packed with numerous small green chloroplasts (Fig. 2e). However, in *Selaginella,* although the chloroplasts packed the guard cells, there were only three to six per guard cell and they were about the same size as those of mesophyll cells (Fig. 2d). The numerous chloroplast profiles, with good grana structure, gave guard cells of the ferns *Polypodium polypodioides* (Stuart, 1968), *Anemia rotundifolia* (Humbert and Guyot, 1972), and *Phyllitis scolopendrium* (Fig. 2f) quite a different appearance in the electron microscope from those of Angiosperms, where chloroplasts are fewer. Large starch grains are present in the chloroplasts. Mitochondria are abundant (Humbert and Guyot, 1972), although mitochondrial profiles are not disproportionately numerous with respect to the number of chloroplast profiles as they are in angiosperms (see below). Endoplasmic reticulum and dictyosomes are present. In *A. rotundifolia,* the nucleus has been reported to change shape from rounded in closed stomata to crenellated in open ones, due to contact with plastids and vacuoles; the many small vacuoles of the closed stomata also appear as larger vacuoles in open stomata (Guyot and Humbert, 1970; Humbert and Guyot, 1972).

C. GYMNOSPERMAE

The stomata of gymnosperms appear to have remained very uniform in structure since their origin during the Mesozoic; this holds even with the Bennetitales whose stomatal development was different from all other gymnosperms (Florin, 1951). Although the guard cells are usually kidney-shaped (see Fig. 3a), their wall thickenings are reminiscent of those of the Gramineae. Like grass stomata, the ends of the guard cells are thin walled, and the middles (next to the pore) are very heavily thickened so that the cell lumen in this part of the cell is very narrow (see Fig. 4a; Florin, 1934). Unlike most other guard cells, the thickenings in the cell walls are usually lignified. The stomata are believed to open by the swelling of the thin-walled ends of the guard cells, forcing the thick-walled central portions apart (von Guttenberg, 1971). In Fig. 3a we have used the stoma of *Ginkgo biloba* to illustrate the structure of the group generally: conifers and cycads have sunken stomata which are difficult to photograph. In conifers, stomata are often at the bottom of a small pit made of the edges of four epidermal cells and this pit is often filled with intermeshed tubes of wax. It has been concluded that about two-thirds of the resistance to diffusion of water vapor and one-third of that to carbon dioxide when the stomata are open is attributable to this wax (Jeffree *et al.*, 1971). Ultrastructural observations on gymnosperm guard cells are few. The guard cells of *Pinus sylvestris* have simple chloroplasts and many mitochondria (Walles *et al.*, 1973). *In vivo* observations indicate that chloroplasts in guard cells of gymnosperms are numerous and green (slightly reminiscent of the stomata of pteridophytes), although they tend to be sparsely distributed in the thickened central part of the guard cells and concentrated at the ends (presumably because of restrictions of space in the narrow central lumen).

D. ANGIOSPERMAE, EXCEPTING THE GRASS TYPE

In this group, the stoma consists of two distinct kidney-shaped guard cells which surround the pore, and are completely separated by cell walls. Guard cells are usually much smaller than subsidiary and other epidermal cells or mesophyll cells in the same leaf (see Table I). The guard cells arise from the division of a protodermal cell rather late in the development of the leaf; however, stomata in any one portion of the epidermis usually are initiated over an extended period so that immature and mature stomata occur side by side. In many species, two to six cells adjacent to each stoma are morphologically distinct from the other epidermal cells; these subsidiary cells may be sister cells of the guard cell mother cell or may arise from the division of adjacent cells, sometimes before the appearance of the mother cell. The development of stomata will not be discussed here; it has been summarized by Esau (1965) and since then contributions have been published by Pickett-Heaps (1969), Landré (1972), and others.

1. Cell Walls

The guard cell of the mature stoma usually has both of its lateral walls greatly thickened; the dorsal wall is thin, and the ventral wall shows some degree of thickening (Fig. 1a,b,5b). The walls are usually not lignified, although they may be in some species (von Guttenberg, 1971). The degree of thickening and the shape of the ventral wall vary greatly between species: outlines of a range of species are given by Copeland (1902). As a result of the variations in shape of the ventral guard-cell wall, the shape of the passage through the pore from inside to outside the leaf also varies greatly. In almost all species there are pronounced outer lips on the pore, and often lips on the inner side; these lips appear to be made of wall and cuticular material.

The cuticle generally covers the outer lateral wall of the guard cells at about the same thickness as over the rest of the epidermis. However, there is an almost infinite number of variations on this pattern, including the formation of a deep cuticular hood with its own fixed aperture as in *Metrosideros excelsa* (Troughton and Donaldson, 1972); some other variations are described by Copeland (1902), Haberlandt (1914), and von Guttenberg (1971). Because of this variation, the structure of the species being investigated should be carefully observed in any detailed investigation of functioning or influence on gas exchange. In most stomata the cuticle extends over the ventral wall and inner lateral walls of the guard cells and often can readily be seen lining the substomatal cavity (see Fig. 3b). It has been stated that all surfaces of mesophyll cell walls where exposed to air are covered by a very thin layer of cuticle (von Mohl, 1856; Scott, 1948; Martin and Juniper, 1970), although this is very much thinner than that lining the substomatal cavity. It seems to be now agreed [cf. chapters by Hallam and Juniper, Holloway, and Baker, in Preece and Dickinson (1971)] that the external cuticle is a matrix of cutin (a polyestolide formed by the cross-esterification of a number of hydroxycarboxylic and carboxylic acids), cellulose, oligosaccharides, and embedded wax with a superficial layer of wax, of which the form, arrangement, and thickness vary greatly between species. It is not known how the cuticle varies through the stomatal pore, although it appears to become thinner and with a higher proportion of oligosaccharides. These issues of structure are important in relation to the permeability of the wall to carbon dioxide and water vapor (cf. Section III,C,2,b).

In the cell wall itself, there have been some observations suggesting that the microfibrils of cellulose are arranged radially from the pore (Ziegenspeck, 1938; Voltz, 1952). These observations were made with polarized light and therefore represent an average orientation of the crystalline lattices of cellulose or micellae; there are likely, of course, in these same walls to be other cellulose microfibrils arranged with less uniform orientation, since these would in no way interfere with detection of the oriented ones, nor would they themselves be detected with the polarized light technique. Radial orientation of microfibrils has been confirmed by electron microscopy (Singh and Srivastava, 1973). Current concepts of cell walls (of parenchymatous, fiber, and tracheid cells) envisage a basic microfibril unit of about 10 nm diameter with a central crystalline core about 4 nm across and less-ordered chains surrounding it, the microfibrils being embedded in a complex matrix of hemicellulose, pectins, and possibly lignin, with intermeshing of some of the outer cellulose chains (Northcote, 1972). The wall may be expected to be less extensible along the axis of the microfibril than across it and in secondary walls successive layers of microfibrils are orientated in different directions, possibly leading to appreciable general reduction in extensibility. However, little is known of the relative degrees of extensibility of the walls of mature cells in the different directions. A good general treatment of this subject is given by Preston (1974). In guard cells, the radial orientation of some microfibrils has been interpreted as indicating that guard cells are hindered from extending across their width but could more readily extend in length (Aylor *et al.*, 1973). We avoid an oversimplified view of the matter at this stage, but return to this issue later (Section III,B).

Since we now know that ion movements are involved in stomatal operation, it has been of interest to look for features known to be prominent in cells specialized for transport. Two of these are "transfer-cell" wall-labyrinths (Pate and Gunning, 1972) and abundance of plasmodesmata (e.g. Ziegler and Lüttge, 1966). No observers have seen transfer-cell wall labyrinths in guard cells, although many have deliberately looked for them. The situation concerning plasmodesmata is more confused. In developing guard cells, plasmodesmata are commonly observed and well developed (Singh and Srivastava, 1973). In mature guard cells plasmodesmata can be demonstrated by staining techniques under the light microscope (Kienitz-Gerloff, 1891; Litz and Kimmins, 1968; Inamdar *et al.*, 1973), but were not easily observed by a number of investigators using the electron microscope (Allaway and Setterfield, 1972; Singh and Srivastava, 1973). However, Pallas and Mollenhauer (1972) clearly demonstrated the existence of plasmodesmata in mature guard cells, and suggested that others failed to find them because they occur only at the ends of the guard cells. The plasmodesmata shown by these authors appear to be convoluted rather than straight as they often are in ordinary cells. It is clear, anyway, that plasmodesmata are not as well developed in guard cells as they are in the classical ion-transporting cells of salt glands, in passage cells of endodermis, or at the boundary of the bundle sheath in C_4 plants.

2. Cell Contents

The fine structure of stomata of a few species is now well known (e.g., Miroslavov, 1966b; Thomson and de Journett, 1970; Allaway and Setterfield, 1972; Pallas and Mollenhauer, 1972; Singh and Srivastava, 1973). Guard cells are usually much smaller than subsidiary and other epidermal cells or mesophyll cells in the same leaf (Table I). Guard cell size has been used as a convenient index of ploidy level in some species; the size of all cells increases with degree of polyploidy (e.g., Francis and Bemis, 1974). Guard cells have small vacuoles in proportion to the size of the cell. [It is not generally possible to fix cells so that the vacuoles remain unchanged in size; changes in vacuolar volume with opening have

TABLE I

Approximate Volumes (μm^3) of Guard and Palisade Cells of
Commelina cyanea and *Vicia faba*[a]

Cell type	C. cyanea	V. faba
Guard cells		
Protoplast volume		
Closed	3,200	1,500
Open to 10 μm	7,200	2,800
Mean volume per chloroplast	17	12
Number of chloroplasts per cell	10[b]	8
Palisade cells		
Protoplast volume	57,000	46,000
Mean volume per chloroplast	64	33
Number of chloroplasts per cell	41	59
Spongy mesophyll cells		
Protoplast volume	48,000	42,000
Number of chloroplasts per cell	28	24

[a] After Pearson and Milthorpe (1974).

[b] Up to 16 observed in guard cells of the same clone grown under other conditions, but other dimensions not measured.

not therefore usually been measurable (but see Guyot and Humbert, 1970), although the vacuolar volume must increase along with the increase in volume of the whole cell as the stoma opens.] The nucleus is much the same size as in other leaf cells and therefore occupies a greater proportion of the cell volume in guard cells.

There are fewer chloroplasts in guard cells than in mesophyll cells (Table I). They are smaller and less well developed, with smaller grana, than mesophyll chloroplasts (Fig. 3c,d). Thylakoids are usually well differentiated and chlorophyll is easily detected. The proportion of the cell volume occupied by chloroplasts appears to be much the same (about 4%) in both mesophyll and guard cells; however, in guard cells much more of the chloroplast volume is often taken up by starch grains, and so there is perhaps proportionately less photosynthetic structure in guard cells per unit volume than in mesophyll cells. In species that form starch, guard cell chloroplasts in electron micrographs usually appear densely packed with starch—although not always as densely as in Fig. 3d. In *Allium,* in which starch is not formed elsewhere, no grains of storage carbohydrate are visible in guard cell chloroplasts. Grana may be quite large in leek (*A. porrum*) (Fig. 3c), although still not as well developed as in mesophyll cells. In the closely related onion (*A. cepa*) grana are not so well developed in guard cell chloroplasts as in leek, and chlorophyll can only be detected with the greatest difficulty; however, Fig. 3e shows that clearly differentiated chloroplasts are present in onion guard cells (cf. Meidner and Mansfield, 1968). Miroslavov (1966b) described "peculiar structures" in chloroplasts of guard cells; these have been shown to be similar to groups of cytoplasmic microtubules (Allaway and Setterfield, 1972), but their functional significance is not known. Another much-mentioned feature of guard cell chloroplasts is the peripheral reticulum round the outside of the chloroplast (Fig. 3c,d); this structure consists of invaginations of the inner of the two bounding membranes of the chloroplast and resembles the peripheral reticulum of chloroplasts of many C_4 (and some C_3) plants. Like the peripheral reticulum in C_4 plants, its function in guard cells is not known, although it was tempting at one stage to suggest that peripheral reticulum is a likely site for the location of phosphoenolpyruvate carboxylase activity (Laetsch, 1971). Pallas and Mollenhauer (1972) suggested a function in sugar transport.

Mitochondria in guard cells are numerous and well developed in all species examined. Mitochondria are too small to count easily in living preparations, so their abundance is judged from the number of mitochondrial profiles found in thin sections in the electron microscope. Use of this count (which does not of course give a real value of the number of organelles in the cell) showed that, while in mesophyll cells there were about equal numbers of mitochondrial and chloroplast profiles, in guard cells mitochondrial profiles were roughly four times as numerous as chloroplast profiles (Allaway and Setterfield, 1972). From such observations, it is reasonable to conclude that mitochondrial activities could play a considerable part in the operation of stomata.

Other organelles in guard cells have attracted little comment. Microbodies appear to be present although infrequent, and this has suggested that photorespiration in guard cells may not be very important in their metabolism. Dictyosomes (Golgi apparatus) are present, but not prominent, after deposition of wall material and maturation of the guard cells have been completed; in this respect, guard cells are like most other plant cells. Spherosomes (Sorokin and Sorokin, 1968) have attracted comment since their content of acid phosphatase seems to be correlated with the phase of stomatal opening. Endoplasmic reticulum and polyribosomes are present, but both are unremarkable in quantity and appearance; cytoplasmic microtubules are also present near the cell walls, but, like dictyosomes, are not numerous except during development (Pickett-Heaps, 1969).

3. Subsidiary and Other Epidermal Cells

In contrast with guard cells, subsidiary and other epidermal cells are usually large, with thin walls usually of even thickness. The layer of cyto-

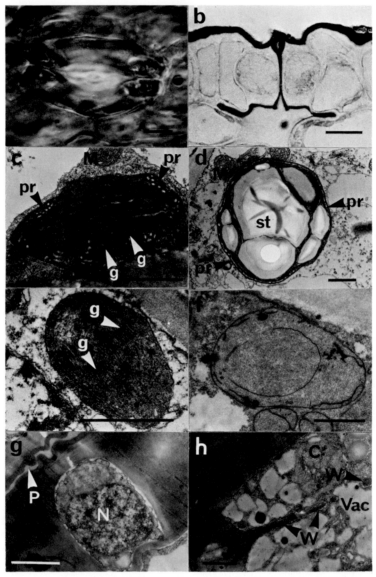

Fig. 3. (a) Stoma of *Ginkgo biloba* showing heavily thickened central guard
cell walls, and cytoplasmic contents visible only through thin walls at ends of guard
cells. General similarity in appearance to a grass stoma (cf. Fig. 4b), but with an
elliptical overall outline. Scale: 10 μm. (b) Transverse section of *Dioscorea cotini-
folia* stoma (from internode) stained with Sudan Black B, so that cuticles are black:
cuticle is shown lining the stomatal pore, covering the inner lateral walls of the
guard cells, and partly lining the substomatal cavity. Scale: 10 μm. Photomicrograph

plasm is very thin and organelles such as mitochondria are few. Chloro-plasts may be absent [e.g., in *Tulipa gesneriana* (Shaw and Maclachlan, 1954) and in most of the epidermis of *Commelina cyanea*] and when present they are usually small, extremely sparse, lacking in starch, and rudimentary [e.g., in the adjacent subsidiary cell of *C. cyanea* (Fig. 3f) and in the epidermal cells of *Vicia faba* (Allaway and Setterfield, 1972)]. In *V. faba* the epidermal cell plastids are green and have slight development of thylakoids and grana; in *C. cyanea* and *A. porrum,* on the other hand, thylakoids and chlorophyll appear to be absent. A quick survey of 18 assorted species showed that all had epidermal cell chloroplasts, although the general rule was for these to be small, few, and only pale green. In contrast with the poor development of chloroplasts in epidermal cells of many terrestrial plants, in the floating leaves of hydrophytes and leaves of many shade plants all epidermal cells contain abundant well-developed chloroplasts (Esau, 1965).

The impression gained from observing epidermis is that usually only the guard cells have a large complement of metabolic machinery. In the *C. cyanea* that we have used, most of the epidermal cells die by the time the leaf reaches full expansion, although the subsidiary cells probably have much the same length of life as the guard cells (Pearson and Milthorpe, 1974). In *V. faba,* in contrast, where the cells surrounding the stomata look the same as the rest of the cells of the epidermis, the epidermal cells remain alive and turgid long into the life of the leaf. From our cursory observations, it seems that living epidermal cells are the rule and *C. cyanea* an exception; however, we visualize guard cells as the only cells in the normal epidermis with an average (or even enhanced) degree of metabolic activity, most of the others being equipped only to maintain a rather basal metabolism; the subsidiary cells if present are somewhat intermediate.

---

by I. von Teichman. (c) Guard cell chloroplast of *Allium porrum* in the electron microscope, showing grana (g) and peripheral reticulum (pr). A mitochondrion (M) of an epidermal cell is shown. Scale: 1 μm. Glutaraldehyde/osmic acid. (d) *Vicia faba* guard cell chloroplast showing large starch grains (st) and peripheral reticulum (pr). Scale: 1 μm. Glutaraldehyde/osmic acid. (e) *Allium cepa* (onion) guard cell chloroplast. Note some well-developed grana. Scale: 1 μm. Glutaraldehyde/osmic acid. (f) *Commelina cyanea* subsidiary cell chloroplast, showing poorly developed lamellar structure. Scale: 1 μm. $KMnO_4$ fixation. (g) Transverse section of middle of guard cell of *Zea mays* (at x-x of Fig. 4b) showing very thick inner and outer lateral walls, nucleus (N), a thin layer of cytoplasm, and the convoluted outline of the pore (P) with cuticular lining. Scale: 1 μm. Glutaraldehyde/osmic acid. (h) Paradermal section of the end of a stoma of *Zea mays* showing incomplete wall (W) between the two guard cells. One small vacuole (Vac) is partly in one guard cell and partly in the other. A chloroplast (C) is labeled. Scale: 1 μm. Glutaraldehyde/osmic acid.

E. The Grass Type of Stoma

A distinctive feature of the Gramineae is the type of stoma consisting of two dumbbell-shaped guard cells with thin-walled ends and thick walls at the middles (or handles of the dumbbells) (Fig. 4b). These stomata are usually "paracytic" - that is, they have one subsidiary cell arranged alongside each guard cell. The shape of the whole stomatal apparatus has been used for taxonomic and diagnostic purposes. Grass-type stomata often occur in longitudinal bands, each with one or more rows of stomata, on the leaf surface. The grass type of stoma seems to be restricted to some of the Monocotyledonae (Metcalfe and Chalk, 1950), although some xeromorphic Dicotyledonae have stomata with very thickened walls which approach it in appearance (von Guttenberg, 1971). It is found, as far as

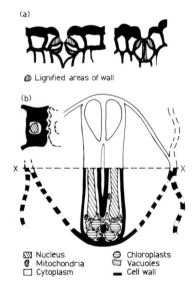

(a)

⬤ Lignified areas of wall

(b)

x --- x

◫ Nucleus   ⊖ Chloroplasts
𝟘 Mitochondria   ▧ Vacuoles
☐ Cytoplasm   ▬ Cell wall

Fig. 4. (a) Transverse sections of stoma of a *Welwitschia mirabilis* leaf. Left, at the middle of the pore; right, through the end parts of the guard cells. Lignified areas of the cell walls are shaded; cuticularization is not shown in the drawing. Note the similarity in central wall thickening between this gymnosperm stoma and the grass stoma (b, left inset) (After Florin, 1934.) (b) Diagrams of a grass stoma (*Zea mays*). Left, inset: transverse section through the middle of a guard cell (at x- - -x), showing positions of heavy wall thickening (principally in inner and outer lateral walls). Main drawing: above the line x- - -x, surface view of stoma as seen on an epidermal strip with the light microscope. Below the line x- - -x, reconstruction of a median paradermal section of the stoma as seen in the electron microscope. Subsidiary cell walls and part of epidermal cell shown in broken lines. Note nuclei located partly in each end of the guard cells; relatively small vacuoles; large pores in the wall between the two guard cells allowing confluence of their cytoplasm.

we know, in the Gramineae, many Cyperaceae, the Lepidocaryoid and a few Arecoid Palmae, Flagellariaceae, Rapateaceae, Marantaceae, Anarthriaceae, and some Restionaceae; and perhaps in Thurniaceae, Heliconiaceae, and Lowiaceae (Metcalfe, 1960, 1971; Tomlinson, 1961, 1969; Cutler, 1969). Detailed information on structure and functioning is, however, restricted to a few species of Gramineae. The development of grass stomata has been examined by Pickett-Heaps and Northcote (1966), Kaufman *et al.* (1970), and others.

The usual concept of the structure of the grass-type guard cell is taken from drawings presented by early workers (notably Schwendener, 1881); with better resolution in light microscopy and new techniques such as electron microscopy, our views of several aspects of this structure have changed in the last few years. To summarize these changes we have drawn a "reconstructed" diagram of a grass stoma using information from a number of workers (Fig. 4b). As mentioned above, the stomatal apparatus consists of the two guard cells with two thin-walled subsidiary cells arranged alongside the guard cells; although the subsidiary cells play a part in the movements of grass stomata, we are not aware of any ultrastructural information on them. The two guard cells together generally have a rectangular appearance rather than the usual elliptical shape of other stomata. The guard cell walls are only sometimes lignified (von Guttenberg, 1971). The walls at the ends of the guard cells are thin, but in the central part they are very much thickened and the lumen connecting the two ends is quite narrow (Fig. 3g). Ziegenspeck (1938) and Setterfield (1957) showed that cellulose micelles in the end parts of these guard cells radiate from the pore and suggested that these walls would therefore be most extensible at right angles to the pore (but recall earlier caution). When the stoma is closed, the thickened walls of the "handles" are pressed tightly together; when the stoma opens, however, they move apart and a more or less parallel-sided pore is formed between them (Meidner and Mansfield, 1968). In transverse sections, convolutions in the surface of the pore walls are often seen; these convolutions presumably aid in tight closure of the pore when conditions demand it (Fig. 3g; Srivastava and Singh, 1972). As in other species, plasmodesmata are not frequently observed in mature guard cell walls, although they are often found in the developmental stages. One of the unique features of grass stomata, however, is that the lumens of the two guard cells are connected together by very large pores in the wall separating them at the ends (Fig. 3h). These pores can be 1 μm or more across, and organelles such as mitochondria and chloroplasts have been observed in them, partly in one cell and partly in the other (Brown and Johnson, 1962; Miroslavov, 1966a; Srivastava and Singh, 1972).

The contents of grass guard cells are also remarkable; firstly, the cell lumens are very small and rather little of this space is occupied by vacuole (Fig. 3h). The cytoplasm contains many mitochondria (Miroslavov, 1966a), as usual for stomata, and some chloroplasts, which are restricted to the ends of the guard cells. The chloroplasts are not always visibly green, (Brown and Johnson, 1962), but in Fig. 3h there is evidence of some development of grana and in living specimens of this species (*Z. mays*) they show some green coloration. Starch grains are generally a prominent feature of grass guard cell plastids, even in species where the main reserve material in the mesophyll is fructosan. The nucleus of each guard cell consists of two bulging ends and a long piece passing through the narrow portion of the guard cell lumen next to the stomatal pore (Figs. 3g, 4b; Flint and Moreland, 1946). This narrow part of the guard cell contains a thin layer of cytoplasm surrounding the strand of nucleus, and occasionally a mitochondrion is seen here.

III. FUNCTIONING

A. Turgor as the Prime Motive Force

Difference in turgor between the guard and subsidiary (or adjacent) cells has been the favored explanation for stomatal movement ever since the experimental investigations of von Mohl (1856). He found that isolated stomata of *Amaryllis* opened when immersed in water and closed in sugar solutions. These observations have been repeated by many investigators. They were supported by direct measurements of the osmotic potential in guard and adjacent cells (e.g., Sayre, 1926) and by the direct experiment of Heath (1938) who showed that puncturing a guard cell led to its immediate closure and puncturing an adjacent cell to wider opening. This approach is being currently extended by H. Meidner (private communication), who is able to open and close stomatal pores by directly increasing or decreasing the hydrostatic pressure within the guard cells. [Some observations have now been published by H. Meidner and M. Edwards (1975). This important paper indicates that the hydrostatic pressures applied to guard and subsidiary cells needed to cause given changes in aperture were appreciably less than those expected from plasmolytically determined osmotic potentials.] The only alternative hypothesis yet advanced, i.e., that stomata are deformed by the differential swelling of various layers of the guard cell wall (Nadel, 1935), has little support. There is overwhelming evidence that an increase in the concentration of osmotically active substances within the guard cells leads to a decrease in the chemical potential of the contained water and, hence, to a net influx of water, thereby increasing the volume of the guard cell and opening of the pore.

1. Water Relations of Guard Cells

In this chapter, we follow the usual convention of representing water potentials (Slatyer and Taylor, 1960). The total water potential ψ consists of the algebraic sum of the following components: that due to the presence of all solutes, solute potential ψ_s; the component due to hydrostatic pressure, the pressure potential ψ_p; that due to the presence of a matrix, the matric potential ψ_τ; and a number of other components which are commonly ignored in plant water relations; i.e.

$$\psi = \psi_p + \psi_s + \psi_\tau + \cdots$$

Here, as in most plant studies, we may use the term osmotic potential, ψ_π, to include both solute and matric potentials and describe the water potential simply as $\psi = \psi_p + \psi_\pi$. In this convention, ψ_π can only be zero or negative, whereas ψ_p is usually positive, but may be zero or negative: it follows from the usual values of these that ψ is negative in almost every conceivable circumstance—that is, that plants and plant cells almost always tend to take up water if pure free water is offered to them. These potentials may be expressed as energy per unit mass (when the appropriate SI unit is J kg^{-1}), energy per mole (J mole^{-1}) or energy per unit volume (J m^{-3}). The last is equivalent to pressure, which is the way these measurements have traditionally been expressed. The appropriate SI unit of pressure is the pascal (1 Pa = 1 Newton m^{-2} = 1 kg m^{-1}s^{-2}), but we prefer to express the potentials as energy per unit mass. The following conversions are equivalent, taking the density of water as 1 g cm^{-3}: 1 J kg^{-1} = 1 kPa = 0.01 bar = 0.00987 atm. In respect of any one cell, we may take $\psi = \psi_\pi$ when $\psi_p = 0$ (zero pressure potential) and $\psi_p = -\psi_\pi$ when $\psi = 0$; and we will assume that $V = V_0 (\psi_p/\epsilon + 1)$ where V and V_0 are the volumes of the cell at turgor potentials of ψ_p and zero, respectively, and ϵ is the modulus of elasticity of the cell wall. We may assume, at least for cells of the type shown in Fig. 1, that the effective force causing deformation of the stoma is a resultant of the net forces acting on the dorsal and ventral walls, i.e.,

$$\text{Effective force} = \psi_p^g (A_i - A_v) - \psi_p^s A_e$$

where the superscripts g and s represent the guard and subsidiary cells, respectively, A_i and A_e the internal and external areas of the dorsal walls and A_v the area of the ventral wall. We will return to a description of stomata of known dimensions later, but here it follows that the opening movement of any given pore is due to changes in ψ_p^g and ψ_p^s and, hence, in ψ_π^g and ψ_π^s. The degree of opening for any given change in these variables will also be influenced by the geometry of the stoma.

2. Changes in Osmotic Potential During Opening

Many observations over the years indicate that the osmotic potential of guard cells decreases during opening by between 200 and 2000 J kg^{-1} (Meidner and Mansfield, 1968). Concomitant changes in the subsidiary or other epidermal cells are usually so small as to be undetectable, reflecting possibly the large volume of these cells relative to that of the guard cells, rather than the absence of any changes in them.

A number of issues must be considered when interpreting these values. Possibly the most important is that all except a few recent measurements have been made by the plasmolytic method, the assumption being that the only fluxes involved are those of water. However, recent evidence indicates that potassium ions efflux readily during such measurements. For example, Fischer (1973) found the estimated values of ψ_π^g were about 500 J kg^{-1} lower when isolated stomata (in contact with broken adjacent cells) were plasmolyzed in sucrose solutions containing 50 mM KCl than when this was absent. He suggests that acceptable measurements can be obtained by measuring the same stomata at equilibrium on two successive external osmotic solutions containing KCl, the systems being exposed to light and CO_2-free air; the change in aperture and volume of guard cells should then be proportional to the difference in osmotic potential of the external solutions. There is evidence that the direction and degree of the net flux of K$^+$ ions during this measurement is related to their external concentration and also possibly to their internal concentration; hence, the effect of these fluxes must be ascertained.

Account must be taken of volume changes, both in the procedure used by Fischer and in the more usual technique of measuring the osmotic potential at incipient plasmolysis. We have a few observations that indicate that the volume at incipient plasmolysis is little different from that when the stoma is closed; however, the volume when open is about twice that when closed (Table I; also Humble and Raschke, 1971).

Further, a close relationship between aperture (or guard-cell volume) and ψ_π^g would be expected; there is little point in determining one without the other. The experimental observations available are consistent with this relationship being linear, although the very approximate nature of all existing measurements must be recognized. Fischer (1973) found the relationship between aperture A μm, and ψ_π^g of isolated stomata of $V.$ $faba$ to be $A = 1 + 50\ \psi_\pi^g$. In similarly prepared material of the same species, but using different methods and without KCl in the plasmolytica, Allaway and Hsiao (1972) obtained data giving $A = 1 + 124\ \psi_\pi^g$. Humble and Raschke (1971) found a change from -1900 to -3500 J kg^{-1} associated with change of aperture from 2 to 12 μm. None of these measurements took into account the changes in volume that occur, and hence are only very approximate. Nevertheless, we may conclude that there are substantial changes in the osmotic potential of guard cells, generating the turgor changes that cause stomatal movements, although the exact degree of these changes in any one species still awaits a more precise study than has yet been made.

B. Modes of Deformation of Guard Cells

1. Elliptical Stomata

In this section we discuss the deformations of the guard-cell walls which accompany the changes in volume. To determine what the deformations are, it is necessary to make very painstaking measurements of, at least, length, width, and depth of the guard cells. Since the walls do not retain their original conformations after fixation (Allaway and Hsiao, 1972), it is not possible to make these measurements on serial sections. This might be possible, after rigorous checks, on freeze-sectioned material, but variation between individuals of the same sample is very large. Measurements *in vivo* are also difficult, and have been avoided by most workers during this century. We rely heavily on the measurements of Schwendener (1881) and Haberlandt (1896) for our information. Of course, measurements of length and width alone (which are easy and therefore commonly made) do not give enough information to determine the conformational changes in the cell walls.

The overall length of the whole stoma does not alter with stomatal opening (Schwendener, 1881; Haberlandt, 1896; Meidner and Mansfield, 1968). However, the individual guard cells must extend in length (except in (*a*), below) and bend for the pore to open. Longitudinal rather than cross-sectional extension may well be favored by the observed radial arrangement of cellulose fibrils in the cell wall, and bending in the right direction would be expected to be furthered by the thickenings of the walls; these ideas are discussed in more detail in the next section.

We find descriptions of changes in stomatal dimensions during opening which can be placed in one of the five following classifications, all with the common property of constant overall stoma length over the full range of deformation.

a. Deformation of the Ventral Wall Only with Constant Width of Stoma. Perhaps the simplest type is that found in *Mnium cuspidatum* (Haberlandt, 1896), in which there is no change in the overall width of the stoma nor length of guard cell. The whole of the opening movement involves the guard cells becoming deeper and narrower, the thin ventral walls changing from a "rounded" state in the closed condition to being more straight-sided in the open condition (Fig. 5a). In this species, the dorsal walls are thicker than the ventral walls. Stomata of many moss species have this type of structure but, as pointed out earlier (Section II,A,2), others have round pores and greatly thickened walls. A type of deformation in some ferns which is similar to that of *Mnium* has also been described by von Guttenberg (1971).

b. Deformation of Dorsal Wall, the Stoma Increasing in Width at the Throat but Not the Mouth. In this type, observed in *Helleborus* by Schwendener (1881), the width of the whole stoma increases in the plane of the throat, but the position at the mouth remains constant (Fig. 5b). The width of each guard cell remains constant, the thin dorsal wall being displaced into the subsidiary cell. There is also a slight increase in the depth of each guard cell, although most of the displacement is in its length and in the change in shape of the dorsal wall.

c. Displacement of the Guard Cell with Constant Cross-Sectional Dimensions, the Stoma Increasing in Width in All Planes. Our own observations on *Commelina* indicate that in this species the cross-sectional dimensions of each guard cell remain constant over the entire range of opening (Fig.

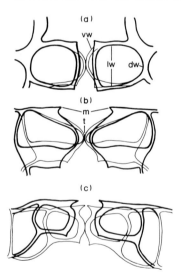

Fig. 5. Diagram showing three types of deformation of stomata. (a) *Mnium* (after Haberlandt, 1914); (b) *Helleborus* (after Haberlandt, 1914); and (c) *Commelina* (our concepts); vw, ventral wall; lw, lateral walls; dw, dorsal wall; m, mouth; t, throat.

5c). The only change is in the length of both guard cells which results in their being pushed into the adjacent subsidiary cells. These become very narrow. It also appears as if the stoma is displaced outward from the plane it occupied in the closed position. The stoma therefore increases in width in all paradermal planes by an amount which is equal to the width of the pore. This is the type of change which is also assumed by Aylor *et al.* (1973) and DeMichele and Sharpe (1973) in their analyses of deformation (see next section).

d. Guard Cell Width Decreases as the Stoma Increases in Width. In *Tradescantia discolor* (Schwendener, 1881), *Allium cepa, Vicia faba,* and *Ranunculus bulbosus* (Meidner and Mansfield, 1968), the guard cells themselves are said to get narrower as the stoma opens. This must represent a change in cross-sectional dimensions from oval to more rounded, the guard cell also becoming deeper as it becomes thinner. There is possibly little change in the circumference of the guard cell and, hence, this type would be consistent with the emphasis placed on the radial orientation of wall microfibrils.

e. Guard Cell Increases in Width and Depth as the Stoma Widens. In *Amaryllis formosissima* (Schwendener, 1881), the width and depth of each guard cell increase as it lengthens during opening, the whole stoma becoming wider — not only by the width of the pore, but by an additional amount due to the changes in each guard cell. In this species, the junctions between the ventral and two lateral walls are thickened, but elsewhere, all the walls are relatively thin; assuming that the wall would stretch readily except where thickened, Schwendener attributed the deformation to the location of these thickenings.

It is easy to see that added thickness of the same material should reduce its extensibility; however, it is not known that the thickenings do have the same properties as the rest of the guard cell walls. A similar reduction in extensibility would be arrived at by arranging the cellulose microfibrils in a longitudinal direction in this thickened part of the cell wall; electron and polarized light microscopy showed no evidence for this in a number of species, but demonstrated a radial arrangement from the pore (Ziegenspeck, 1938, 1954; Singh and Srivastava, 1973). As we have remarked before, the demonstration of the radial arrangement does not preclude that other microfibrils may be arranged so that they run along the cell; it only indicates that the most pronounced orientation is the radial one. The behavior in *Amaryllis* is not consistent with this supposed orientation, but then such orientation has not been demonstrated in this particular species.

The variations described above cover almost the entire range of conceivable forms of deformation. If the observations are valid, and we see no reason to doubt them, then interesting questions concerning the physical bases of these variations arise. Care is necessary in generalizing too widely from observations on any one species and we need to avoid simplified extrapolation from the dominant direction of the crystalline parts of the cellulose microfibrils. There is an immediate need to study a range of species, not only to determine the general qualitative form, but also to provide a clear rigorous analysis based on the established concepts of mechanics. Recently, two groups of investigators have considered these matters.

The first group (Aylor et al., 1973) considers model experiments essentially applicable to isolated stomata, i.e., free of influences from adjacent cells. They emphasize the significance of the radial micellation in guard cell walls, i.e., that the cellulose microfibrils run radially around the guard cells effectively constraining any increase in the cross-sectional area of the guard cell. With increase of turgor pressure, the guard cells increase in length only, and, because of the common wall toward each end of the ventral walls they bend outward. This mechanism would apply to stomata of the type illustrated in Fig. 5c; if we accept the observations respecting the other types, as is reasonable, then issues of the relative moduli of elasticity in all three dimensions arise.

The second group (DeMichele and Sharpe, 1973) accept this basic principle and extend the analysis much further by attempting to take into account the effect of subsidiary cells and the various forces involved. They assume that the cell wall material is uniform and obeys Hooke's Law, that the modulus of elasticity in tension is equal to that in compression, that there are no lateral pressures or shearing stresses between the fibers during movement, that the transverse section of the cell remains in the same plane over the range of apertures, that the neutral axis lies along the outer surface of the ventral wall, that forces transferred between the dorsal and ventral walls are transmitted along the microfibrils, that the subsidiary cells exert an equal pressure in all lateral directions on the dorsal wall of the guard cell, and that the guard cell acts as a rectangular beam forming a straight-sided pore.

Over the more open range of pore sizes, they simplify their more complex relationships without significant loss of accuracy to the following

$$r = 8EI/L_0F \quad \text{and} \quad A = r^2[L_0/r - \sin(L_0/r)]$$

where r is the radius of curvature of the neutral axis, A the area of the pore, L_0 the length of the ventral wall, and E the modulus of elasticity of the cell-wall material (for which no reasonable estimates appear to be available). I, the moment of inertia about the neutral axis, is given by

$$I = \{H_0 T^3 + (H_0 - 2T_\mathrm{L})[T_\mathrm{V}^3 - (T - T_\mathrm{D})^3]\}/3$$

and F, the effective force acting on the ventral wall, by

$$F = \psi_\mathrm{p}^\mathrm{g} H_1 (L_1 - L_2) - H_0 \psi_\mathrm{p}^\mathrm{s} L$$

where H_0 and H_1 are the distances (depths) between the outer and inner lateral walls, respectively; L and L_1 the outer and interior lengths of the dorsal wall; L_2 the interior length of the ventral wall; $\psi_\mathrm{p}^\mathrm{g}$ and $\psi_\mathrm{p}^\mathrm{s}$ the turgor pressures of the guard and subsidiary cells, respectively; T the outer width of the guard cell; and T_D, T_L, and T_V the thicknesses of the dorsal, lateral, and ventral walls, respectively.

The above relationships allow the area of pore and subsidiary components such as the volume of guard-cell protoplast to be calculated from readily measured cell dimensions and estimated turgor pressures. The latter cannot be readily measured directly (however, see Footnote, Section, III,A), but they can be estimated under steady-state conditions from

$$\psi_\mathrm{p}^\mathrm{g} = \psi^\mathrm{g} - \psi_\pi^\mathrm{g} \quad \text{and} \quad \psi_\mathrm{p}^\mathrm{s} = \psi^\mathrm{s} - \psi_\pi^\mathrm{s}$$

We can also take $\psi^\mathrm{g} = \psi^\mathrm{s} = \psi^\mathrm{L}$, the leaf water potential. The issue then resolves to one of either measuring or calculating ψ_π^g and ψ_π^s.

The conductance (or resistance) for diffusion of water vapor through the stomata on a leaf surface can then be calculated from the pore dimensions so generated (cf. Vol I, this treatise, pp. 171–173; Chapter 3, this volume). DeMichele and Sharpe assume the pore to be straight-sided over its entire depth; then the conductance for diffusion, g_s, is given by

$$g_s = nD(H_a/A + \pi/4L_0)^{-1}$$

where n is the stomatal density, D the diffusion coefficient of water vapor, and H_a the effective depth and A the area of the pore, and L_0 the length of the ventral wall. Stomatal pores, however, are not straight-sided (cf. Fig. 1b, 2c, 5) and more accurate estimates of stomatal conductance could be obtained by calculating the area of the pore at its throat and allowing for the varying areas along the channel at any one throat area by using numerical methods such as those employed by Bange or Milthorpe and Penman (cf. Volume II, this treatise, p. 171).

We accept this model and that of Aylor et al. as extremely valuable contributions toward an effective analysis of the mechanics of stomatal deformation. However, we think it may not be completely adequate in that the approximations used to obtain tractable descriptions of the forces involved could possibly introduce significant errors and since there are no available concurrent measurements of essential dimensions, such as areas of the various walls, osmotic potentials, and the modulus of elasticity over a range of apertures. This model does provide the base for a more detailed approach and the values required can be obtained experimentally; hence, this is one very basic component of the stomatal story in which the main features can be measured and which is therefore susceptible to an early solution.

2. *The Grass Type of Stoma*

The opening of the grass stoma has been simply described as resulting from the thin-walled ends of the dumbbell-shaped guard cells swelling under the influence of turgor pressure and moving the ends of the thick-walled central portions apart (Schwendener, 1889). The large pores connecting the end parts of the two guard cells, virtually into one (binucleate) cell, ensure that both guard cells have the same hydrostatic pressure, and if their walls are the same, this should ensure that the pore opens symmetrically. Radial micellation in the walls of the end parts of the cells (Ziegenspeck, 1938; Setterfield, 1957) is consistent with Schwendener's description. It is likely that stomata of conifers operate similarly, because of their particular type of wall thickening, which, as in grasses, would be expected to result in stiff middle parts of the guard cells and extensible ends (cf. Section II,C). Schwendener's explanation of the mechanics of grass guard-cell movement was not challenged between 1889 and 1973. However, Shoemaker and Srivastava (1973) considered, from detailed observation of the wall thickenings in *Z. mays,* that the movements of grass stomata involve bending of the thickened walls, similar to the bending movements in the elliptical stomata of other Angiosperms. These authors examined the properties of grass guard cell walls theoretically [also making a number of assumptions like those of DeMichele and Sharpe (1973)] and found that such a bending system could give rise to reasonable stomatal movements. It is noteworthy here that Raschke (1975) suggests that the movements even of elliptical stomata conform more nearly to Schwendener's suggested scheme for grasses.

C. THE CONTROL OF OSMOTIC POTENTIAL

This brings us to the central issue: what is the mechanism by which changes in the osmotic potential of guard cells are generated? This is, of course, the question that has excited investigators since the time of von

Mohl. We do not propose to enumerate all the hypotheses and speculations that have been advanced over the years, but will proceed immediately to the next step in the chain of events for which we believe there is incontrovertible evidence, i.e., that the changes in osmotic potential are due very largely to changes in concentration of potassium ions.

1. The Role of Potassium and Anions

The accumulation of potassium in guard cells was probably first observed by Macallum (1905) and has been suggested by a number of workers (Imamura, 1943; Yamashita, 1952; Fujino, 1959; Fischer, 1968) as the main agent by which turgor is increased. During stomatal opening, the net influx of potassium ions into guard cells from the external solution or from neighboring cells has been shown for many species in widely different genera: e.g., *C. communis* and *C. cyanea* (Fujino, 1959; Pearson 1975); *V. faba* (Fischer, 1968); *Nicotiana tabacum* (Sawhney and Zelitch, 1969); and *Z. mays* (Raschke and Fellows, 1971). Willmer and Pallas (1973) have shown by a staining method that the potassium content of guard cells of species widely scattered through the Angiospermae increases as stomata open, and Dayanandan and Kaufman (1973) reported similar results with two ferns and two gymnosperms as well as other species (cf. Fig. 6). The dependence of stomatal opening on the presence of K^+ ions in the medium (Fujino, 1959) and calculation that sufficient K^+ was taken up to contribute largely to the change in osmotic potential (Fischer and Hsaio, 1968) led to the development of the concept that K^+ accumulation is the cause of the change in osmotic potential and, hence,

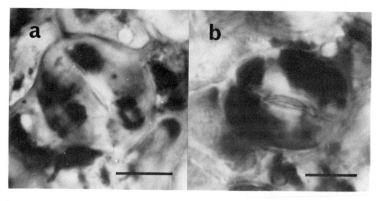

Fig. 6. Stomata of apple leaf stained for potassium by the method of Macallum (1905) and treated with ammonium sulfide so that the location of potassium is black. (a) Originally closed stoma from a dark-treated leaf showing little K in guard cells, some in epidermal cells. (b) Originally open stoma from a light-treated leaf showing heavy K-staining in guard cells. (The staining technique causes the cells to lose turgor so all stained stomata look closed.) Scale: 10 μm.

of stomatal opening. This subject has been extensively reviewed elsewhere (Hsiao, 1975; Thomas, 1975). It has been reported that in one species, *C. communis* (Willmer and Mansfield, 1969), Na^+ is as effective as K^+ in inducing opening of stomata on isolated epidermis, but in this (Willmer and Pallas, 1974; Raschke, 1975), as well as in all other species so far examined, K^+ seems to be the cation accumulated *in vivo*. Na^+, Ca^{2+}, and Mg^{2+} do not appear to move into guard cells to any extent and Na^+, NH_4^+, and Mg^{2+} are much less effective than K^+ or Rb^+ in inducing changes in stomata of epidermal strips. In Table II, we summarize the observations on changes in content of K in guard cells with stomatal opening.

Raschke and Humble (1973) have also shown that stomata of *V. faba*

TABLE II

CONCENTRATIONS OF SOLUTES IN GUARD CELLS OF OPEN AND CLOSED STOMATA

Species	Solute	Concentration (moles m^{-3}) in guard cells when stoma		Reference
		Closed	Open	
Vicia faba	K	77	883	Humble and Raschke (1971)
	K	110[a]	544	Allaway and Hsiao (1972)
	K	144[a]	536	Allaway (1973)
	Na	0	0	Humble and Raschke (1971)
	Malate	0	114	Allaway (1973)
	Cl	No statistically significant difference		Humble and Raschke (1971)
	P, S	No change		Humble and Raschke (1971)
Nicotiana tabacum	K	210[b]	500[b]	Sawhney and Zelitch (1969)
Commelina communis	K	95[c]	448[c]	Penny and Bowling (1974)
Zea mays	K	About 150[d]	About 400	Raschke and Fellows (1971)
	Cl	About 65[d]	About 160	Raschke and Fellows (1971)

[a] No allowance made for changes in volume between open and closed: volume taken to be 5×10^{-15} m³ per guard cell.

[b] No data on volume provided by original authors.

[c] Direct measurements of concentrations.

[d] No allowance made for changes in volume: volume of a guard cell pair taken to be 2×10^{-15} m³.

open quite as widely when only exotic, synthetic, organic and presumably nonabsorbable anions are offered with K^+, as they do when KCl is given. This suggests that uptake of anions is not needed for stomatal opening. Humble and Raschke (1971) showed that a small quantity of Cl^- (about $\frac{1}{20}$ of the amount of K^+) may be taken up when stomata of *V. faba* open; there was no accumulation of P or S, which indicates that anions containing these elements were not accumulating. More recently, Pallaghy and Fischer (1974) showed by double-labeling experiments that isolated stomata of *V. faba* accumulated approximately one Cl^- for every three K^+ ions absorbed. The only grass-type stomata that have been closely examined (*Z. mays*, Raschke and Fellows, 1971) accumulated roughly one Cl^- for four K^+. The evidence available, therefore, clearly suggests that inorganic anion uptake is insufficient to balance the potassium ions taken up; however, the evidence as to the exact quantity of anions, particularly Cl^-, accumulated is still somewhat equivocal, being between $\frac{1}{20}$ and $\frac{1}{3}$ of the amount of K^+; it might even be true that the proportion of Cl^- to K^+ varies according to circumstances.

Raschke and Humble (1973) have shown in *V. faba* that a net efflux of hydrogen ions accompanies the accumulation of potassium ions. Coupled with the above evidence against large-scale net inorganic anion uptake, this suggests that electroneutrality may be maintained in this way. If so, an increase in pH of the guard cells—or a decrease in pH in adjacent cells—might be expected. This is supported by the results of several investigators such as Sayre (1926), Scarth (1932), Pekarek (1934), and Small *et al.* (1942). Using indicators, they found increases in guard cell pH somewhere over the range of about 4 to 7 from the closed to the open state. The pH measured by these methods will generally be that of the vacuole and a large buffering capacity would be expected in the cytoplasm. The results are consistent with extrusion of H^+ ions accompanying stomatal opening, but indicate that the guard cells have (or acquire) considerable buffering capacity even in the vacuole (Raschke and Humble, 1973), since otherwise a very large pH change would be expected (Allaway, 1973). Buffering capacity could be acquired, not only by taking up external inorganic anions (which have been shown to be of limited importance), but by synthesis of organic acids, which thereby generates hydrogen ions that can be extruded in exchange for K^+ ions.

There is a substantial increase in the content of malate in guard cells of *V. faba* during stomatal opening (Allaway, 1973). Increase in the content of organic anions, especially malate, has been shown in epidermis of *C. cyanea* with stomatal opening during the diurnal cycle (Pearson, 1973) and on exposure to light and darkness in CO_2-plus and CO_2-free air (Pearson and Milthorpe, 1974), and a number of organic anions increased in epidermis of *V. faba* with stomatal opening (Pallas and Wright, 1973).

Relations between malate content and stomatal aperture are summarized in Fig. 7. Malate clearly accumulates in sufficient quantity to play an important role in the ionic balance of guard cells; it is likely that other organic acids are also important, but their identity and quantity have not been established.

2. The Metabolism of Guard Cells

Before examining possible mechanisms of potassium uptake and loss, we should explore the degree to which the general metabolism of guard cells is known. This is not great as few direct investigations have been attempted. However, there is a great deal of indirect evidence that has resulted in a wealth of speculative hypotheses. Generally, investigations have depended on the pathways established in other cells and organisms and on exploring the extent to which these exist or are modified within the guard cells.

a. Photosynthesis and Respiration. Action spectra for stomatal opening have been partly consistent with the idea that photosynthesis is important in powering or triggering stomatal opening (Kuiper, 1964; Hsiao et al., 1973); however, the well-established ability of stomata to open in darkness at very low CO_2 concentrations (cf. Meidner and Mansfield, 1968) means that simultaneous photosynthesis is not essential for stomatal opening. As we have seen, chloroplasts are ordinarily a feature of guard cells. Pallas and Dilley (1972) calculated the chlorophyll content to be 0.33 pg per chloroplast and the ratio of chlorophyll a to chlorophyll b to be 4.5 in

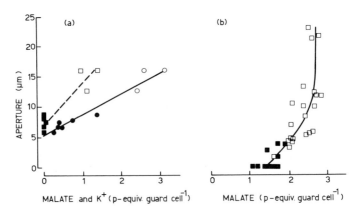

Fig. 7. Relationships between stomatal aperture and concentrations of malate and potassium in the epidermis. (a) In *Vicia faba* (after Allaway, 1973); (b) In *Commelina* (after Pearson and Milthorpe, 1974).

epidermis of *V. faba.* Willmer *et al.* (1973) found this ratio to be between 2 and 3 in epidermis of *C. communis* and *T. gesneriana*, and other determinations range from 1.8 to 5.2. Since values of the chlorophyll a:b ratio seem to be so variable, we cannot use it as an indicator of the main photosystems present. However, Das and Raghavendra (1974) have demonstrated a predominance of Photosystem I in *C. benghalensis* and *Petunia hybrida.*

Guard cells can fix $^{14}CO_2$ in light and dark (Shaw and Maclachlan, 1954). These workers and others (Allaway, 1973; Willmer and Dittrich, 1974) found for several species roughly a twofold stimulation of $^{14}CO_2$ fixation by light, and the rates in light were much lower on a chlorophyll basis than the rate to be expected from photosynthesizing mesophyll cells. Pearson and Milthorpe (1974), in contrast, found a 250 to 800-fold stimulation by light of $^{14}CO_2$ fixation in *C. cyanea* and *V. faba* epidermis. Pearson and Milthorpe used brighter illumination than the others, about one-third the flux density needed for saturation of mesophyll photosynthesis. But this does not account for the difference, since their dark rates were about ten times less than those of the other workers while light rates were twenty or so times higher. These differences in rates between light and dark are consistent with normal photosynthetic behavior, while those of the other workers are not. From Pearson and Milthorpe's data, rates of CO_2 fixation in guard cells in light per chloroplast were roughly twice, and per unit chloroplast volume some five to six times, the rates for mesophyll, which suggests that in spite of their relatively poor structure guard cell chloroplasts are very effective in photosynthesis. Perhaps we may conclude that guard cells can show normal light-stimulation of CO_2 fixation but that in the preparations of Shaw and Maclachlan, Allaway, and Willmer and Dittrich, this had been in some way inactivated; alternatively, there may have been extensive transfer from mesophyll to epidermis in Pearson and Milthorpe's experiments. The contradictions in the evidence are as frequent and mysterious as those concerning other aspects of stomatal physiology.

It is not surprising, in view of the preceding paragraph, that we are not certain about the pathways of CO_2 fixation in guard cells. Material is difficult to prepare, available only in small amounts, and results may be confused as a result of contamination by cells other than guard cells. Dark fixation of $^{14}CO_2$ in guard cells results in accumulation of label almost solely in malate and aspartate (Pearson and Milthorpe, 1974; Willmer and Dittrich, 1974). This would be consistent with the usual dark-fixation in plant cells via phosphoenolpyruvate (PEP) carboxylase. This enzyme is abundant in epidermis preparations, more so (on a chlorophyll basis) than usual for leaf tissue (Willmer *et al.*, 1973) and, in view of the suggested importance of organic anions in stomatal movement (see earlier discussion) this may be a very important enzyme in guard cells. The same authors drew attention to very high activities of malic enzyme and NAD- and NADP-malate dehydrogenases in epidermis and commented on the similarity of this situation to plants showing Crassulacean Acid Metabolism (CAM).

In Willmer and Dittrich's experiments, dark CO_2 fixation went at high rates in epidermis, but there was little stimulation by light. In light, the pattern of $^{14}CO_2$-labeled products was quite similar to that in the dark; it was dominated by malate and aspartate, but with the addition of a small amount of Calvin cycle intermediates. Ribulosediphosphate (RuDP) carboxylase is present in roughly similar activities (on a chlorophyll basis) in epidermis and mesophyll (Willmer et al., 1973); this suggests that stomata have the ability to carry out the Calvin cycle. These observations are consistent with the continuation of dark PEP carboxylase-mediated CO_2 fixation in light with a low rate of RuDP carboxylase-mediated Calvin cycle activity going on at the same time. A pulse-chase experiment (Willmer and Dittrich, 1974) suggested that some label passed from C_4 acids to the Calvin cycle in the light, but the high level of accumulation of label in C_4 acids was not like C_4 photosynthesis (Hatch and Slack, 1970). It was more reminiscent of the little-understood CAM photosynthesis early in the light period (Osmond and Allaway, 1974). In the experiments of Pearson and Milthorpe (1974), light greatly stimulated epidermal CO_2 fixation, and long-term $^{14}CO_2$-labeling patterns in mesophyll and epidermis were similar. These authors suggested that guard cells of C. cyanea and V. faba carry out normal Calvin cycle photosynthesis.

It is difficult to draw consistent conclusions from these data. Guessing at an explanation, with understandable bias, we suggest that guard cells photosynthesize by the C_3 pathway, but that since they are heavily equipped with enzymatic machinery to handle the synthesis and breakdown of organic acids for ionic balance, a larger proportion than usual of CO_2 fixed at any time accumulates in malate and aspartate. The carbon can pass from decarboxylation of malate to the Calvin cycle, but generally the malate is needed for ionic balance (presumably in the vacuole) and, therefore, may be unavailable to act as a photosynthetic intermediate, as it would in C_4 photosynthesis.

Data on guard cell respiration are essentially nonexistent. Many authors have pointed out that mitochondria are usually very numerous in guard cells, and Hsaio et al. (1973) have suggested that the very high effectiveness of blue light in opening stomata may result from a stimulation of respiration. Photorespiration presumably occurs since guard cells possess RuDP carboxylase and therefore, presumably, oxygenase (cf. Lorimer and

Andrews, 1973). Microbodies have been found in guard cells, although they are not very common. There has been controversy over the importance of glycolate metabolism in stomatal opening (Zelitch, 1965; Meidner and Mansfield, 1965); the evidence (mostly from inhibitor studies) suggested that glycolate metabolism was involved in stomatal movement, but a confounding effect of the inhibitor in raising internal CO_2 levels in the leaf (causing stomatal closure) was subsequently found and casts some doubt on this conclusion. Respiration and photorespiration in stomata are areas where research is greatly needed: some indirect evidence (Pearson and Milthorpe, 1974) suggests that the rate of respiration in the light may almost approach that of photosynthesis; both these metabolic pathways appear to function at high rates.

b. Some Aspects of Metabolism including Responses to Some External Factors. Although guard cells can photosynthesize, the generally held view seems to be that direct osmotic products of photosynthesis cannot account for the osmotic changes associated with stomatal opening. Another hypothesis, which was formerly popular, that of starch $\rightleftharpoons$ sugar transformations, also has little support. Both the older and more recent literature show that sometimes starch/sugar changes occur in the required direction, but often even the diurnal variations are not consistently correlated with stomatal movement (Pearson, 1973); and when correlations have been found, they can readily be broken (cf. Meidner and Mansfield, 1968). Changes in sugars have rarely been measured, the evidence usually being based on those in starch. What seems likely is that starch changes in guard cells reflect the resultant between the input from photosynthesis and the rate of use in the respiratory and other pathways.

Possibly the three most significant external factors influencing stomatal movement are light, CO_2 concentration in the substomatal cavity, and water (cf. Chapter 3 for descriptions of response relationships). Much of the light effect has been directly attributed to the accompanying reduction of the internal CO_2 concentration (Raschke, 1975); although this is likely when the conductance is low, as during the initial stages of exposure to light, it is less likely at equilibrium states with a net influx of CO_2 into the leaf. As most of the evidence (cf. Meidner and Mansfield, 1965) indicates that it is the inner lateral walls of the guard cells which are permeable to CO_2 (cf. Section II.D.1 for possible structural reasons), then the CO_2 concentration at this site may well be closer to the ambient rather than the mean internal concentration, if the stomata are open. The results of a number of workers collated in Chapter 3 show usually a small opening effect between CO_2-free air and the CO_2 compensation point (say, 50–80 ppm in C_3 plants). At higher concentrations, either a marked closing response or no response at all has often been found. These results indicate a much less clear-cut response to CO_2 than is often assumed. Nevertheless, there are sufficient observations to indicate that low CO_2 concentrations, in both light and dark, tend to promote stomatal opening (Meidner and Mansfield, 1968; Neales, 1970; Raschke, 1972). Stomata also respond to light in CO_2-free air. There is good evidence that stomatal opening in response to reduced CO_2 concentration in both light and darkness is associated with an influx of K^+ ions (Pallaghy, 1971; Pallaghy and Fischer, 1974). The central metabolic question in this context—what action of low CO_2 it is that brings about changes in cells leading to stomatal opening—is not answerable at all, at present. A simple explanation in terms of reduction of the amount of photosynthetically fixed carbon or of dark-fixed CO_2 as malate does not seem sensible since we should naively expect both of these to lead to stomatal closure in low CO_2 rather than opening. Hypotheses, involving competition for ATP from photophosphorylation between ion uptake and carbon fixation, or the increase at low CO_2 in activity of the glycolate pathway, can be advanced to explain the effect in the light. But there is no evidence which, if either, of these is involved. Similarly, although dark respiration and presumably oxidative phosphorylation are increased in CO_2 free air, the relationships with K^+ uptake have not been explored. It has been suggested that CO_2 might influence permeability of guard cell membranes to water and solutes, but there has been no evidence for this effect in the range of concentrations to which stomata are sensitive.

It is a familiar feature of plant physiology that stomata close when the plant is under water stress and that there is an aftereffect of water stress during which stomata do not open as widely as usual. It is perhaps less familiar that stomata of some species apparently respond directly to changes in atmospheric humidity, by a mechanism that is not understood (cf. Chapter 3). In the context of stomatal response to water stress, a substance that has received much attention in recent years is abscisic acid (ABA). There have been many studies since Mittelheuser and van Steveninck (1969) showed that stomata closed when a 1 μg m^{-3} solution of ABA was fed to cut leaves and Wright and Hiron (1969) found an increase in ABA in wilting leaves which was correlated with stomatal closure. It is clear that ABA has a direct effect on stomata of epidermal strips, the closure being associated with an efflux of K^+ ions (and increase in starch content) (Mansfield and Jones, 1971). CO_2-free air does not reverse this closure (Jones and Mansfield, 1970), nor does the partial pressure of CO_2 influence the concentration of ABA within leaves (Loveys et al., 1973). It would appear, therefore, that both ABA and CO_2 influence stomatal movement through the regulation of K^+ ions, although they are likely to influence different parts of the mechanism. A very thorough review of the current concepts concerning ABA (Milborrow, 1974) has not revealed to us the possible reactions in stomatal metabolism where it may be implicated. Many of the *in vivo* studies are difficult to interpret because they relate stomatal aperture to concentration of ABA in the mesophyll. No estimates have yet been made of the changes in epidermis or guard cells. Where ABA reaches the leaf in the xylem stream it is likely to flow fairly readily to the guard cells (with unknown changes in concentration). But ABA transported in phloem or generated in mesophyll is less likely to reach the guard cells (cf. Section III,D). It is reasonable to assume that most of the ABA and its derivatives measured in the leaf in response to stress from a water deficit or surfeit (e.g., Hiron and Wright, 1973; Loveys and Kriedemann, 1973), in response to photoperiod (Loveys *et al.*, 1974), during chilling (Drake and Raschke, 1974) or with fruit removal (Loveys and Kriedemann, 1974; cf. also Chapter 3) is generated *in situ* in the chloroplasts (Milborrow, 1974) of mesophyll cells. It may be that guard cell chloroplasts also produce ABA under most of these situations, but this could be at quite different rates. It seems likely that much of the stomatal closure associated with lowered water potentials of the leaf may be influenced by ABA production leading to changes in turgor of guard and adjacent cells different from those expected solely as a simultaneous reduction in water content of all cells.

Other growth substances such as cytokinins are believed to encourage stomatal opening (e.g., Livnè and Vaadia, 1965), but no detailed exploration of their effects has been attempted. The fungal toxin fusicoccin (cf. Chapter 6, Volume II, this treatise) which includes wilting of leaves, has been shown to stimulate the entry of K^+ ions and stomatal opening (Turner, 1972; Squire and Mansfield, 1974).

3. The Regulation of Potassium Fluxes

We now return to a consideration of the mechanisms controlling the movement of potassium ions into and out of guard cells. On current evidence either one (or both?) of two types of mechanisms would appear to be possible. In the first, production of organic acids inside the cells, and export of hydrogen ions (together with the import of some Cl^-) could produce a sufficient electrochemical potential inside the cells to cause enough potassium ions to enter passively. K^+ uptake could be controlled, perhaps, by the pool size of organic acids; an active "proton pump" might also be involved. The second possibility involves the uptake of K^+ against an electrochemical potential gradient, mediated by an active K^+ pump.

There are some isolated observations that could be advanced in support of an ATPase-dependent ion "pump." For example, Fujino (1967) and Das and Raghavendra (1974), working with epidermal strips of *Com-*

melina spp, emphasize the role of ATP and ATPase activity. Both groups describe experiments which suggest that ATP is directly involved in potassium uptake and the latter invoke cyclic photophosphorylation as the major means of its regeneration. However, evidence cited earlier on chlorophyll ratios indicates a large degree of disagreement between different workers; we cannot yet accept predominance of Photosystem I in guard cells even of *Commelina*. Fujino suggests that ATPase is responsible for "excretion" of K^+ ions and stomatal movement depends on the balance between ATP and ATPase.

These and other observations do not allow any firm conclusion concerning the mechanism operating *in vivo*. To distinguish between the two possibilities—active or passive accumulation of K^+ in guard cells—it is necessary to measure electrical potential as well as K^+ concentration in the cells. Pallaghy (1968) made electrical potential measurements on *Nicotiana tabacum* guard cells. Although he expressed some reservations about the measurements, his data together with those of other workers indicate that accumulation of K^+ in guard cells during stomatal opening takes place against an electrochemical potential gradient. The only work in which simultaneous measurements of electrical potential and K^+ concentration have been made is that of Penny and Bowling (1974) on *Commelina communis*. Their observations suggest that K^+ moves against an electrochemical potential gradient both during influx to guard cells in stomatal opening and, surprisingly, during efflux in stomatal closure, and also that K^+ movement through the subsidiary cells is against electrochemical potential gradients. These findings imply that at least in guard cells of *C. communis* there is active transport of K^+ in both light and darkness—but in opposite directions—and that subsidiary cells as well as guard cells are involved. These findings are a landmark in that they are the first real evidence for active K^+ movement in guard cells—although it has been suspected for a long time. It will be interesting to see whether active K^+ movement is a feature of guard cells of other species, especially those without specialized subsidiary cells, such as *V. faba*. There is little other firm evidence, although no doubt a number of indirect observations could be marshalled in favor of either of the above speculative hypotheses.

D. RELATIONSHIPS WITH OTHER CELLS AND TISSUES

In the preceding discussion, we have regarded the stomatal complex in isolation. This is an oversimplification because, although positioned at the end of a somewhat remote pathway, it is connected to other tissues, subjected to physical constraints by the other epidermal cells and influenced by the flow of substances within the plant. The guard and adjacent cells are always free from the underlying mesophyll and many of the other epidermal cells have only small areas of contact; the degree of contact, however, varies widely between species and is often less with the abaxial than the adaxial epidermis. Usually it is only in the regions overlying the veins where there is almost complete contact between epidermal cells and underlying tissue. These structural considerations suggest that conductances for flow may be low compared with those elsewhere in the plant.

The most rapid responses would be expected to substances transported within the leaf in the gaseous state. The relations with carbon dioxide were discussed earlier; ethylene should be mentioned also in view of its current popularity as a physiological subject. Available evidence (Pallaghy and Raschke, 1972; El-Beltagy and Hall, 1974) indicates that stomata show no response to ethylene. (The latter authors show that ethylene concentrations in the leaf increase during a water deficit possibly partly because of reduced stomatal conductance and partly because of increased synthesis associated with increased abscisic acid production). We know of no evidence respecting other leaf gases.

Although direct evidence is lacking, the relative diurnal changes in concentrations of sugar and starch in epidermis and mesophyll indicate that exchange of carbohydrates is extremely slow (Pearson, 1973). The same considerations apply to potassium ions, the content of which in the epidermis as a whole stays relatively constant during the marked diurnal readjustments in concentration between guard and other epidermal cells (Pearson, 1975). This conforms with what might be expected from general structural and transport considerations.

There remain water and such substances as may be carried by it. As between other tissues, it would be expected that water flows relatively easily between mesophyll and epidermis in response to differences in water potential. The marked changes in stomatal aperture that follow sudden perturbations in the rate of supply of water to, or loss from, leaves can best be interpreted in terms of turgor changes in guard and other epidermal cells (Chapter 6 of Volume I, this treatise). More recently, many workers have drawn attention to the oscillations in stomatal conductance that are frequently associated with particular environmental constraints (cf. Barrs, 1971). These, which have periods of about 20–100 min, are believed to arise from "feedback" interrelationships in the flow of water through the plant and to the guard cells. Cowan (1972) has reviewed many of the observations and has presented a model that closely simulates the observed behavior. In essence, this describes the interaction between rate of transpiration, water contents of guard and subsidiary cells, and the stomatal conductance: these are considered in relation to perturbations in the potential rate of transpiration (evaporation demand of the atmosphere), the conductance of the roots for water, the water potential of the soil water, and the osmotic potential of the guard cells. Important components (for which approximate estimates are given) are the capacity of each of the tissues for water (i.e., the change in water content per unit water potential) and the conductances of the tissues for water flow. We do not propose here to explore this aspect in any greater depth. It is sufficient for our purpose to note that the water relations of the epidermal cells are influenced by the general flow of water in the plant; these can result in significant changes in stomatal apertures and provide the bases on which the so-called "hydropassive" and "hydroactive" stomatal movements should be approached. The paper by Cowan may be consulted for further details.

One corollary concerns the actual fluxes of water through the guard and subsidiary cells as a result of evaporation into the stomatal pore and substomatal cavity. There are no adequate observations on this matter; we need estimates of the conductances for water vapor of the cell walls of guard, subsidiary, and mesophyll cells, the conductances for liquid flow to the respective evaporating surfaces, and the lengths of the respective pathways. Electron micrographs and staining with Sudan III indicate a thicker osmiophilic layer around the guard and subsidiary cells than over the mesophyll cells bordering on the substomatal cavity; these and similar superficial considerations suggest that the conductance of walls for water vapor is in the order: mesophyll $>$ guard $+$ subsidiary cells $\gg$ external cuticle. It seems likely that the conductance for liquid flow in the epidermis differs from that of the mesophyll by an order of magnitude (Cowan, 1972) and the pathways are much longer. If so, the flux of water through the guard and subsidiary cells is very much less than through mesophyll cells and the transport of substances contained in the water stream from the xylem to the guard cells will be rather slow. Transport of such substances within the epidermis may be more a matter of passive diffusion rather than a consequence of being carried along in a stream of water moving by mass (viscous) flow. [Experiments, such as those of Tanton and Crowdy (1972), in which the degree of accumulation of nondissociated salts purports to indicate the region of evaporation, need to be interpreted with caution. In their experiments, for example, there is likely to be exchange of lead from the lead-ethylenediaminetetraacetic acid chelate used with calcium ions held by the cell walls. Accumulation of lead, therefore, may represent little more than the calcium content of the cell walls]. We are inclined to the view, on present evidence, that transport to all cells of the epidermis is relatively slow and, although significant for general supply, is unlikely to be involved in direct rapid modulation of stomatal movements. We find it difficult, for example, to visualize cytokinins or other growth substances produced in the roots or fruits being transported at a sufficient rate to the guard cells to have a direct effect.

IV. TOWARD THE FUTURE—MODELS OF STOMATAL FUNCTIONING

It may be argued that plant physiology is now in the transitional phase between qualitative and quantitative approaches, i.e., that it is moving from an understanding and documentation of the relevant events toward an assessment of the quantitative relationships concerned and the ways in which these are assembled to make the system. Only when this has been achieved can the interplay between the different component reactions and processes be assessed and the behavior of the system predicted with confidence.

Ideally (and eventually), one would envisage the behavior of the whole plant growing in a variable environment being described in this way, but with the necessary proviso that any one set of statements or model is unable to traverse more than about two levels of organization within the plant. (We use the term "model" in the recent restricted sense of the set of statements describing the particular system in mathematical language with actual values ascribed to the parameters.) Each statement, again ideally, should be a generalization of a detailed quantitative description of events at, say, the next two lower levels of organization, although lack of qualitative, let alone quantitative, understanding often restricts such statements to purely empirical relationships of a "black-box" nature. The limitation to two levels of organization reflects the capacities of both the human mind and the current digital computer. Such models provide an extremely valuable way of summarizing current concepts, of evaluating the relative significance of each of the components, of sharpening rigor, and of emphasizing areas of ignorance.

In concluding this chapter, therefore, it is appropriate to re-enumerate the specific contributions made toward summarizing the features of stomatal functioning through simulation models. The degree of adequacy of these is a measure of our current understanding, and changes therein reflect the rate of progress. In presenting stomatal functioning against a background of structure, which has been the central burden of this chapter, we recognized three major interrelated areas: the mechanics of deformation, the control of osmotic potential within the guard cell, and the influence of water movement through the plant. Simulation of the first is well advanced and might be expected to proceed from the contribution of De-Michele and Sharpe (1973) fairly rapidly and without undue difficulty. Regarding the second, it will be clear that we still lack qualitative understanding of the main metabolic events. Activity here must be directed toward remedying this and much effort will be required before an effective quantitative model can be formulated. The third, reflected by Cowan's (1972) model, is well advanced requiring mainly more effective statements that depend on progress in the first area and on better estimates of various parameters.

There is, however, another group of models—those of the more empirical "black-box" type—that describe stomatal movements (or more usefully, stomatal conductance) in terms of the respective internal and external controlling factors. Among these may be rated the formulations of Lommen *et al.*, (1971), Lemon *et al.* (1971), Penning de Vries (1972), and van Bavel *et al.* (1973). These are important in the wider context of describing stomatal changes in variable environments and assessing the significance of the stomata in controlling gas exchange. This is the subject of the following chapter.

ACKNOWLEDGMENTS

We thank D. Armstrong and I. von Teichman for micrographs 1a and 3b; J. Gregory, G. Kiss and G. Lynravin for help with electron microscopy; B. Lester, A. Stewart, and S. Weyrauch for technical assistance; B. Thorn for drawing, J. Chatfield for typing; and the Australian Research Grants Committee for financial support.

REFERENCES

Allaway, W. G. (1973). Accumulation of malate in guard cells of *Vicia faba* during stomatal opening. *Planta* **110**, 63–70.

Allaway, W. G., and Hsaio, T. C. (1972). Preparation of rolled epidermis of *Vicia faba* L. so that stomata are the only viable cells: analysis of guard cell potassium by flame photometry. *Aust. J. Biol. Sci.* **26**, 309–318.

Allaway, W. G., and Setterfield, G. (1972). Ultrastructural observations on guard cells of *Vicia faba* and *Allium porrum. Can. J. Bot.* **50**, 1405–1413.

Aylor, D. E., Parlange, J.-Y., and Krikorian, A. D. (1973). Stomatal mechanics. *Amer. J. Bot.* **60**, 163–171.

Barrs, H. D. (1971). Cyclic variations in stomatal aperture, transpiration and leaf water potential under constant environmental conditions. *Annu. Rev. Plant Physiol.* **22**, 223–236.

Brown, W. V., and Johnson, C. (1962). The fine-structure of the grass guard-cell. *Amer. J. Bot.* **49**, 110–115.

Chaloner, W. G. (1970). The rise of the first land plants. *Biol. Rev. Cambridge Phil. Soc.* **45**, 353–377.

Copeland, E. B. (1902). Mechanism of stomata. *Ann. Bot. (London)* **16**, 327–364.

Cowan, I. R. (1972). Oscillations in stomatal conductance and plant functioning associated with stomatal conductance: observations and a model. *Planta* **106**, 185–219.

Cutler, D. F. (1969). *In* "Anatomy of the Monocotyledons" (C. R. Metcalfe, ed.), Vol. IV, "Juncales," Oxford Univ. Press, London and New York.

Das, V. S. R., and Raghavendra, A. S. (1974). Role of cyclic photophosphorylation in the control of stomatal opening. *Roy. Soc. N. Z., Bull.* **12,** 455–460.

Dayanandan, P., and Kaufman, P. B. (1973). Stomata in *Equisetum. Can. J. Bot.* **51,** 1555–1564.

DeMichele, D. W., and Sharpe, P. J. H. (1973). An analysis of the mechanics of guard cell motion. *J. Theor. Biol.* **41,** 77–96.

Drake, B., and Raschke, K. (1974). Prechilling of *Xanthium strumarium* L. reduces net photosynthesis and independently, stomatal conductance while sensitizing the stomata to CO_2. *Plant Physiol.* **53,** 808–812.

El-Beltagy, A. S., and Hall, M. A. (1974). Effect of water stress upon endogenous ethylene levels in *Vicia faba. New Phytol.* **73,** 47–60.

Esau, K. (1965). "Plant Anatomy," 2nd ed. Wiley, New York.

Fischer, R. A. (1968). Stomatal opening: role of potassium uptake by guard cells. *Science* **160,** 784–785.

Fischer, R. A. (1973). The relationship of stomatal aperture and guard-cell turgor pressure in *Vicia faba. J. Exp. Bot.* **24,** 387–399.

Fischer, R. A., and Hsiao, T. C. (1968). Stomatal opening in isolated epidermal strips of *Vicia faba.* II. Response to KCl concentration and the role of potassium absorption. *Plant Physiol.* **43,** 1953–1958.

Flint, L. H., and Moreland, C. F. (1946). A study of the stomata in sugar cane. *Amer. J. Bot.* **33,** 80–82.

Florin, R. (1934). Die Spaltöffnungsapparate von *Welwitschia mirabilis* Hook. f. *Sv. Bot. Tidskr.* **28,** 264–289.

Florin, R. (1951). Evolution in Cordaites and Conifers. *Acta Horti Berg.* **15,** 285–388.

Francis, R. R., and Bemis, W. P. (1974). Stomatal size in induced polyploids in *Cucurbita. HortScience* **9,** 138.

Fujino, M. (1959). Stomatal movement and active migration of potassium. *Kagaku* (*Tokyo*) **29,** 660–661.

Fujino, M. (1967). Role of adenosinetriphosphate and adenosinetriphosphatase in stomatal movement. *Sci. Bull. Fac. Educ., Nagasaki Univ.* No. 18.

Garner, D. L. B., and Paolillo, D. J. (1973). On the functioning of the stomates in *Funaria. Bryologist* **76,** 423–427.

Guyot, M., and Humbert, C. (1970). Les modifications du vacuome des cellules stomatiques d'*Anemia rotundifolia* Schrad. *C.R. Acad. Sci., Ser. D* **270,** 2787–2790.

Haberlandt, G. (1896). "Physiologische Pflanzenanatomie," 2nd ed. Engelmann, Leipzig.

Haberlandt, G. (1914). "Physiological Plant Anatomy" (translation of the 4th German edition by M. Drummond). Macmillan, London and New York.

Hatch, M. D., and Slack, C. R. (1970). Photosynthetic CO_2-fixation pathways. *Annu. Rev. Plant Physiol.* **21,** 141–162.

Heath, O. V. S. (1938). An experimental investigation of the mechanism of stomatal movement, with some preliminary observations on the response of the guard cells to "shock." *New Phytol.* **37,** 385–395.

Hiron, R. W. P., and Wright, S. T. C. (1973). The role of endogenous abscisic acid in the response of plants to stress. *J. Exp. Bot.* **24,** 769–781.

Hsiao, T. C. (1976). Stomatal ion transport. *In* "Encyclopaedia of Plant Physiology," New Series (A. Pirson and M. H. Zimmermann, eds.) Vol. 2, Part B, pp. 195–217, Springer-Verlag, Berlin and New York.

Hsiao, T. C., Allaway, W. G., and Evans, L. T. (1973). Action spectra for guard cell Rb⁺ uptake and stomatal opening in *Vicia faba*. *Plant Physiol.* **51**, 82–88.

Humbert, C., and Guyot, M. (1972). Modifications ultrastructurales des cellules stomatiques d'*Anemia rotundifolia* Schrad. *C.R. Acad. Sci., Ser. D* **274**, 380–382.

Humble, G. D., and Raschke, K. (1971). Stomatal opening quantitatively related to potassium transport. Evidence from electron probe analysis. *Plant Physiol.* **48**, 447–453.

Imamura, S. (1943). Untersuchungen über den Mechanismus der Turgorschwankung der Spaltöffnungsschliesszellen. *Jap. J. Bot.* **12**, 251–346.

Inamdar, J. A., Patel, K. S., and Patel, R. C. (1973). Studies on plasmodesmata in the trichomes and leaf epidermis of some Asclepiadaceae. *Ann. Bot. (London)* [N.S.] **37**, 657–660.

Jeffree, C. E., Johnson, R. P. C., and Jarvis, P. G. (1971). Epicuticular wax in the stomatal chamber of Sitka spruce and its effects on the diffusion of water vapor and carbon dioxide. *Planta* **98**, 1–10.

Jones, R. J., and Mansfield, T. A. (1970). Suppression of stomatal opening in leaves treated with abscisic acid. *J. Exp. Bot.* **21**, 714–719.

Kaufman, P. B., Petering, L. B., Yocum, C. S., and Baic, D. (1970). Ultrastructural studies on stomata development in internodes of *Avena sativa*. *Amer. J. Bot.* **57**, 33–49.

Kienitz-Gerloff, F. (1891). Die Protoplasmaverbindungen zwischen benachbarten Gewebselementen in der Pflanze. *Bot. Ztg.* **49**, 17–26 and 49–60.

Kuiper, P. J. C. (1964). Dependence upon wavelength of stomatal movement in epidermal tissue of *Senecio odoris*. *Plant Physiol.* **39**, 952–955.

Laetsch, W. M. (1971). Chloroplast structural relationships in leaves of C₄ plants. *In* "Photosynthesis and Photorespiration" (M. D. Hatch, C. B. Osmond, and R. O. Slatyer, eds.), p. 323–349. Wiley (Interscience), New York.

Landré, P. (1972). Origine et développement des épidermes cotylédonaire et foliaires de la moutarde (*Sinapis alba* L.). Différenciation ultrastructurale des stomates. *Ann. Sci. Natur., Biol. Veg. Bot.* [12] **13**, 247–322.

Lemon, E. R., Stewart, D. W., and Shawcroft, R. W. (1971). The sun's work in a cornfield. *Science* **174**, 371–378.

Litz, R. E., and Kimmins, W. C. (1968). Plasmodesmata between guard cells and accessory cells. *Can. J. Bot.* **46**, 1603–1604.

Livne, A., and Vaadia, Y. (1965). Stimulation of transpiration rate in barley leaves by kinetin and gibberellic acid. *Physiol. Plant.* **18**, 658–664.

Lommen, P. W., Schwintzer, C. R., Yocum, C. S., and Gates, D. M. (1971). A model describing photosynthesis in terms of gas diffusion and enzyme kinetics. *Planta* **98**, 195–220.

Lorimer, G. H., and Andrews, T. J. (1973). Plant photorespiration—an inevitable consequence of the existence of atmospheric oxygen. *Nature (London)* **243**, 359.

Loveys, B. R., and Kriedemann, P. E. (1973). Rapid changes in abscisic acid-like inhibitors following alterations in vine leaf water potential. *Physiol. Plant.* **28**, 476–479.

Loveys, B. R., and Kriedemann, P. E. (1974). Internal control of stomatal physiology and photosynthesis. I. Stomatal regulation and associated changes in endogenous levels of abscisic and phaseic acids. *Aust. J. Plant Physiol.* **1**, 407–415.

Loveys, B. R., Kriedemann, P. E., and Törökfalvy, E. (1973). Is abscisic acid involved in stomatal response to carbon dioxide? *Plant Sci. Lett.* **1**, 335–338.

Loveys, B. R., Kriedemann, P. E., and Leopold, A. C. (1974). Abscisic acid metabolism and stomatal physiology in *Betula lutea* following alteration in photoperiod. *Ann. Bot. (London)* [N.S.] **38**, 85–92.

Macallum, A. B. (1905). On the distribution of potassium in animal and vegetable cells. *J. Physiol. (London)* **32**, 95–118.

Mansfield, T. A. (1975). Mechanisms involved in turgor changes of guard cells. *In* "Perspectives in Experimental Biology, Vol. 2, Botanical" (N. Sunderland, ed) Pergamon, Oxford (in press).

Mansfield, T. A., and Jones, R. J. (1971). Effects of abscisic acid on potassium uptake and starch content of stomatal guard cells. *Planta* **101**, 147–158.

Martin, J. T., and Juniper, B. E. (1970). "The Cuticles of Plants." Arnold, London.

Meidner, H., and Edwards, M. (1975). Direct measurements of turgor pressure potentials of guard cells, I. *J. Exp. Bot.* **26**, 319–330.

Meidner, H., and Mansfield, T. A. (1965). Stomatal responses to illumination. *Biol. Rev. Cambridge Phil. Soc.* **40**, 483–509.

Meidner, H., and Mansfield, T. A. (1968). "Physiology of Stomata." McGraw-Hill, New York.

Metcalfe, C. R., ed. (1960). "Anatomy of the Monocotyledons," Vol. I. Oxford Univ. Press, London and New York.

Metcalfe, C. R., ed. (1971). "Anatomy of the Monocotyledons," Vol. V. "Cyperaceae," Oxford Univ. Press, London and New York.

Metcalfe, C. R., and Chalk, L. (1950). "Anatomy of the Dicotyledons," Vol. I. Oxford Univ. Press, London and New York.

Milborrow, B. V. (1974). The chemistry and physiology of abscisic acid. *Annu. Rev. Plant Physiol.* **25**, 259–307.

Miroslavov, E. A. (1966a). Electron microscopic studies of the stomata of rye leaves *Secale cereale* L. *Bot. Zh. (Leningrad)* **51**, 446–449.

Miroslavov, E. A. (1966b). On the peculiar structures in the plastids of the guard cells of the leaf of *Vicia faba* L. *Bot. Zh. (Leningrad)* **51**, 982–983.

Mittelheuser, C. J., and van Steveninck, R. F. M. (1969). Stomatal closure and inhibition of transpiration by (RS)-abscisic acid. *Nature (London)* **221**, 281–282.

Nadel, M. (1935). On the influence of various liquid fixatives on stomatal behaviour. *Palestine J. Bot. Hort. Sci.* **1**, 22–42.

Neales, T. F. (1970). Effect of ambient carbon dioxide concentration on the rate of transpiration of *Agave americana* in the dark. *Nature (London)* **228**, 880–882.

Northcote, D. H. (1972). Chemistry of the plant cell wall. *Annu. Rev. Plant. Physiol.* **23**, 113–132.

Osmond, C. B. and Allaway, W. G. (1974). Pathways of CO_2 fixation in the CAM plant *Kalanchoë daigremontiana*. I. Patterns of $^{14}CO_2$ fixation in the light. *Aust. J. Plant Physiol.* **1**, 503–511.

Pallaghy, C. K. (1968). Electrophysiological studies in guard cells of tobacco. *Planta* **80**, 147–153.

Pallaghy, C. K. (1971). Stomatal movement and potassium transport in epidermal strips of *Zea mays:* the effect of CO_2. *Planta* **101**, 287–295.

Pallaghy, C. K., and Fischer, R. A. (1974). Metabolic aspects of stomatal opening and ion accumulation by guard cells in *Vicia faba*. *Z. Pflanzenphysiol.* **71**, 332–344.

Pallaghy, C. K., and Raschke, K. (1972). No stomatal response to ethylene. *Plant Physiol.* **49**, 275–276.

Pallas, J. E., and Dilley, R. A. (1972). Photophosphorylation can provide sufficient adenosine 5′-triphosphate to drive K⁺ movements during stomatal opening. *Plant Physiol.* **49**, 649–650.

Pallas, J. E., and Mollenhauer, H. H. (1972). Physiological implications of *Vicia faba* and *Nicotiana tabacum* guard cell ultrastructure. *Amer. J. Bot.* **59**, 504–514.

Pallas, J. E., and Wright, B. G. (1973). Organic acid changes in the epidermis of *Vicia faba* and their implication in stomatal movement. *Plant Physiol.* **51**, 588–590.

Parihar, N. S. (1961). "An Introduction to Embryophyta," Vol. I. Central Book Depot, Allahabad.

Pate, J. S., and Gunning, B. E. S. (1972). Transfer cells. *Annu. Rev. Plant Physiol.* **23**, 173–196.

Paton, J. A., and Pearce, J. V. (1957). The occurrence, structure and functions of the stomata in British bryophytes. *Trans. Brit. Bryol. Soc.* **3**, 228–259.

Pearson, C. J. (1973). Daily changes in stomatal aperture and in carbohydrates and malate within epidermis and mesophyll of leaves of *Commelina cyanea* and *Vicia faba*. *Aust. J. Biol. Sci.* **26**, 1035–1044.

Pearson, C. J. (1975). Fluxes of potassium and changes in malate within epidermis of *Commelina cyanea* and their relationships with stomatal aperture. *Aust. J. Plant Physiol.* **2**, 85–89.

Pearson, C. J., and Milthorpe, F. L. (1974). Structure, carbon dioxide fixation and metabolism of stomata. *Aust. J. Plant Physiol.* **1**, 221–236.

Pekarek, J. (1934). Über die Aziditätsverhältnisse in den Epidermis- und Schliesszellen bei *Rumex acetosa* im Licht und im Dunkeln. *Planta* **21**, 419–446.

Penning de Vries, F. W. T. (1972). A model for simulating transpiration of leaves with special attention to stomatal functioning. *J. Appl. Ecol.* **9**, 57–77.

Penny, M. G., and Bowling, D. J. F. (1974). A study of potassium gradients in the epidermis of intact leaves of *Commelina communis* L. in relation to stomatal opening. *Planta* **119**, 17–25.

Pickett-Heaps, J. D. (1969). Preprophase microtubules and stomatal differentiation; some effects of centrifugation on symmetrical and asymmetrical cell division. *J. Ultrastruct. Res.* **27**, 24–44.

Pickett-Heaps, J. D., and Northcote, D. H. (1966). Cell division in the formation of the stomatal complex of the young leaves of wheat. *J. Cell Sci.* **1**, 121–128.

Preece, T. F., and Dickinson, C. H., eds. (1971). "Ecology of Leaf Surface Micro-Organisms." Academic Press, New York.

Preston, R. D. (1974). "The Physical Biology of Plant Cell Walls." Chapman & Hall, London.

Raschke, K. (1972). Saturation kinetics of the velocity of stomatal closing in response to CO₂. *Plant Physiol.* **49**, 229–234.

Raschke, K. (1975). Stomatal action. *Annu. Rev. Plant Physiol.* **26**, 309–340.

Raschke, K., and Fellows, M. P. (1971). Stomatal movement in *Zea mays*: shuttle of potassium and chloride between guard cells and subsidiary cells. *Planta* **101**, 296–316.

Raschke, K., and Humble, G. D. (1973). No uptake of anions required by opening stomata of *Vicia faba*: guard cells release hydrogen ions. *Planta* **115**, 47–57.

Riebner, F. (1925). Über Bau und Funktion des Spaltöffnungsapparate bei den Equisetinae und Lycopodiinae. *Planta* **1**, 260–300.

Sawhney, B. L., and Zelitch, I. (1969). Direct determination of potassium ion accumulation in guard cells in relation to stomatal opening in light. *Plant Physiol.* **44**, 1350–1354.

Sayre, J. D. (1926). Physiology of stomata of *Rumex patientia. Ohio J. Sci.* **26**, 233–266.

Scarth, G. W. (1932). Mechanism of the action of light and other factors on stomatal movement. *Plant Physiol.* **7**, 481–504.

Schimper, W. P. (1848). "Recherches anatomiques et morphologiques sur les mousses." Strasbourg.

Schwendener, S. (1881). Über Bau und Mechanik der Spaltöffnungen. *Monatsber. Kgl. Akad. Wiss. Berlin* **43**, 833–867.

Schwendener, S. (1889). Die Spaltöffnungen der Gramineen und Cyperaceen. *Sitzungsber. Kgl. Preuss. Akad. Wiss.* **6**, 1–15.

Scott, F. M. (1948). Internal suberization of plant tissues. *Science* **108**, 654–655.

Setterfield, G. (1957). Fine structure of guard-cell walls in *Avena* coleoptile. *Can. J. Bot.* **35**, 791–793.

Shaw, M., and Maclachlan, G. A. (1954). The physiology of stomata. I. Carbon dioxide fixation in guard cells. *Can. J. Bot.* **32**, 784–797.

Shoemaker, E. M., and Srivastava, L. M. (1973). The mechanics of stomatal opening in corn (*Zea mays*) leaves. *J. Theor. Biol.* **42**, 219–225.

Singh, A. P., and Srivastava, L. M. (1973). The fine structure of pea stomata. *Protoplasma* **76**, 61–82.

Slatyer, R. O., and Taylor, S. A. (1960). Terminology in plant and soil-water relations. *Nature (London)* **187**, 922–924.

Small, J., Clarke, M. E., and Crosbie-Baird, J. (1942). pH phenomena in relation to stomatal opening. II–V. *Proc. Roy. Soc. Edinburgh, Sect. B* **61**, 233–266.

Sorokin, H. P., and Sorokin, S. (1968). Fluctuations in the acid phosphatase activity of spherosomes in guard cells of *Campanula persicifolia. J. Histochem. Cytochem.* **16**, 791–802.

Squire, G. R., and Mansfield, T. A. (1974). The action of fusicoccin on stomatal guard cells and subsidiary cells. *New Phytol.* **73**, 433–440.

Srivastava, L. M., and Singh, A. P. (1972). Stomatal structure in corn leaves. *J. Ultrastruct. Res.* **39**, 345–363.

Stuart, T. S. (1968). Revival of respiration and photosynthesis in dried leaves of *Polypodium polypodioides. Planta* **83**, 185–206.

Tanton, T. W., and Crowdy, S. H. (1972). Water pathways in higher plants. III. The transpiration stream within leaves. *J. Exp. Bot.* **23**, 619–625.

Thomas, D. A. (1975). Stomata. *In* "Ion Transport in Plant Cells and Tissues" (D. A. Baker and J. L. Hall, eds.) pp. 377–412. Elsevier, Amsterdam.

Thomson, W. W., and de Journett, R. (1970). Studies on the ultrastructure of the guard cells of *Opuntia. Amer. J. Bot.* **57**, 309–316.

Tomlinson, P. B. (1961). "Anatomy of the Monocotyledons" (C. R. Metcalfe, ed.), Vol. II, "Palmae," Oxford Univ. Press, London and New York.

Tomlinson, P. B. (1969). "Anatomy of the Monocotyledons" (C. R. Metcalfe, ed.), Vol. III, "Commelinales—Zingiberales" Oxford Univ. Press, London and New York.

Troughton, J. H., and Donaldson, L. A. (1972). "Probing Plant Structure." Reed, Wellington, N.Z.

Turner, N. C. (1972). K^+ uptake of guard cells stimulated by fusicoccin. *Nature (London)* **235**, 341–342.

van Bavel, C. H. M., DeMichele, D. W., and Ahmed, J. (1973). A model of gas and energy exchange regulation by stomatal action in plant leaves. *Tex. Agr. Exp. Sta., Misc. Publ.* **1078**.

Voltz, G. (1952). Elektronenmikroskopische Untersuchungen über die Porengrössen pflanzliche Zellwände. *Mikroskopie* **7**, 251–266.

von Guttenberg, H. (1971). Bewegungsgewebe und Perzeptionsorgane. *In* "Handbuch der Pflanzenanatomie" (W. Zimmermann *et al.*, eds.), Vol. V5, pp. 203–219. Borntraeger, Berlin.

von Mohl, H. (1856). Welche Ursachen bewirken die Erweiterung und Verengung der Spaltöffnungen? *Bot. Ztg.* **14**, 697–704 and 713–720.

Walles, B., Nyman, B., and Alden, T. (1973). On the ultrastructure of needles of *Pinus sylvestris*. *Stud. Forest. Suec.* **106**, 1–25.

Watson, E. V. (1964). "The Structure and Life of Bryophytes." Hutchinson, London.

Willmer, C. M., and Dittrich, P. (1974). Carbon dioxide fixation by epidermal and mesophyll tissues of *Tulipa* and *Commelina*. *Planta* **117**, 123–132.

Willmer, C. M., and Mansfield, T. A. (1969). Active cation transport and stomatal opening: a possible physiological role of sodium ions. *Z. Pflanzenphysiol.* **61**, 398–400.

Willmer, C. M., and Pallas, J. E. (1973). A survey of stomatal movements and associated potassium fluxes in the plant kingdom. *Can. J. Bot.* **51**, 37–42.

Willmer, C. M., and Pallas, J. E. (1974). Stomatal movements and ion fluxes within epidermis of *Commelina communis* L. *Nature (London)* **252**, 126–127.

Willmer, C. M., Pallas, J. E., and Black, C. C. (1973). Carbon dioxide metabolism in leaf epidermal tissue. *Plant Physiol.* **52**, 448–452.

Wright, S. T. C., and Hiron, R. W. P. (1969). (+)-abscisic acid, the growth inhibitor induced in detached wheat leaves by a period of wilting. *Nature (London)* **224**, 719–720.

Yamashita, T. (1952). Influence of potassium supply upon properties and movement of the guard cell. *Sieboldia Acta Biol.* **1**, 51–70.

Zelitch, I. (1965). Experimental and biochemical control of stomatal movement in leaves. *Biol. Rev. Cambridge Phil. Soc.* **40**, 463–482.

Ziegenspeck, H. (1938). Die Micellierung der Turgeszenzmechanismen. Tiel I. Die Spaltöffnungen (mit phylogenetischen Ausblicken). *Bot. Arch.* **39**, 268–309 and 332–372.

Ziegenspeck, H. (1954). Das Vorkommen von Fila in radialer Anordnung in den Schliesszellen. *Protoplasma* **44**, 385–388.

Ziegler, H., and Lüttge, U. (1966). Die Salzdrusen von *Limonium vulgare*. I. Mitteilung. Die Feinstruktur. *Planta* **70**, 193–206.

CHAPTER 3

STOMATAL CONDUCTANCE IN THE CONTROL OF GAS EXCHANGE

F. J. Burrows and F. L. Milthorpe

SCHOOL OF BIOLOGICAL SCIENCES, MACQUARIE UNIVERSITY, NORTH RYDE,
N.S.W. 2113, AUSTRALIA

I. INTRODUCTION

Much of the interest shown in stomata as factors modulating the gas exchange of plants arises because they provide a variable transfer component located immediately between the plant and its surrounding atmosphere. They are in the position where they can influence the whole of the internal physical and metabolic relationships, but the degree to which they do so depends also on the sizes of other components of the relevant transport systems. Hence, one of our major tasks is to examine these aspects. Before doing so, it is convenient to explore the responses of the stomata to various external and internal factors, extending outward from the more detailed consideration of structure and mechanism presented in Chapter 2.

The tidy academic mind likes to present a coherent, consecutive, and substantiated picture from the lower to the higher levels of organization— from enzymatic reactions through organelles, cells, tissues, and organs to the organism itself. However, science does not—nor can it—progress in this way; the best possible understanding of all levels is always required and investigations are continually made at all levels. These findings are, of course, closely interrelated and interdependent. Although it was clear from Chapter 2 that many important aspects of structure remain undocumented and of functioning not understood, we still need to predict as accurately as possible the behavior of the stomata over the ranges of circumstances to which they are exposed and the resulting influence on the exchange of water and carbon dioxide. The relationships we have to use in this task are, to a large degree, empirical in the sense that they record experience without a full understanding of the underlying reasons; they are no less valid because of this and, indeed, are essential to the development of an acceptable theory of functioning.

Although we defer detailed consideration of the transport network to a later section, it may be useful here to recapitulate the general concepts of gas transfer. We will confine our attention to water vapor and carbon dioxide, usually within the leaf.

A. General Transfer Functions

The volume rate of transfer of a gas through unit area of a medium, F m^3 m^{-2} s^{-1}, is proportional to the gradient of partial pressure, i.e.,

$$F = -D \, dc/dz \tag{1}$$

where D(m^2 s^{-1}) is a diffusion coefficient and c the concentration. Over a finite length $z(= z_2 - z_1)$ we may write

$$F = g(c_2 - c_1) \tag{2}$$

where $g = 1/r = D/z$ (m s^{-1}), g being a conductance and r a resistance to transfer. It is generally accepted that the conductances for two different gases are related to their free diffusion coefficients as follows:

(a) where the transfer is solely by free molecular diffusion, as within the leaf,

$$g_1/g_2 = D_1/D_2 \tag{3}$$

(b) where the transfer is by fully turbulent diffusion, as between the leaf boundary layer to above the canopy,

$$g_1 = g_2 \tag{4}$$

(c) where the transfer varies from molecular diffusion through a transition region to fully turbulent diffusion, as in the leaf boundary layer (Pohlhausen, 1921; Thom, 1968)

$$g_1/g_2 = (D_1/D_2)^{2/3} \tag{5}$$

Although it is usually assumed that movement within the leaf is solely by molecular diffusion, there are some considerations which suggest that a small difference of absolute pressure is often established between the inside and outside of the leaf (Parkinson and Penman, 1970). These authors suggested that a correction varying from zero to about 80% should be applied depending on the differences in CO_2 concentration across the stomata. Jarman (1974) analyzed the system in more detail, taking into account collisions between the different molecular species. He agrees with the size of the corrections, but on different grounds. He gives a clear description of the system and the corrections necessary to account for concentration differences across the stomata of the different gases involved.

It is convenient in most situations to measure the conductance for diffusion of water vapor and to calculate those for other gases therefrom (care must be taken to ensure that the systems are comparable, of course). In Sections II, III, and IV the values quoted, therefore, will usually be those for water vapor. The free diffusion coefficients are usually related to each other in inverse proportion to the square roots of their molecular weights, i.e.,

$$D_1 = D_{H_2O}(M_{H_2O}/M_1)^{1/2} \tag{6}$$

but exact values of D_{H_2O} (or D_{CO_2}) and the nature of their temperature dependence are uncertain. Information reviewed by Jarvis (1971) suggests that the best estimate of D_{H_2O} is likely to be 0.257 cm^2 s^{-1} at 25°C with a temperature dependence of $(T/273)^{1.8}$, i.e., $D_{H_2O} = 0.219(T/273)^{1.8}$ cm^2 s^{-1} where T is in °K. Equation (6) yields $D_{CO_2} = 0.64\ D_{H_2O}$ whereas two experimental measurements indicate proportionalities of 0.60 and 0.62; therefore, we may take $D_{CO_2} = 0.140 \, (T/273)^{1.8}$ giving a value of $D_{CO_2} = 0.164 \, cm^2 s^{-1}$ at 25°C.

The effective coefficient for diffusion, D_e, varies also with the size of the pore because the molecular slip of diffusing molecules is incomplete at surfaces. From considerations given in Chapter 6 of Volume I of this treatise (cf. Milthorpe and Penman, 1967) we may take

$$1/D_e = 1/D_1 + 1/D^* \tag{7}$$

where D_1 is the free diffusion coefficient and D^* is given by $(8/3)h(\delta/k_1)$ $(2RT/M)^{1/2}$, h being the mean hydraulic radius of the pores, δ and k_1 are shape factors (cf. Carman, 1956), and R the gas constant. This adjustment is necessary when calculating conductances from pore dimensions where the areas are less than about 25 μm^2; with larger pores the correction is small.

B. "Conductance" or "Resistance?"

In a simple homogeneous system, the conductance, which expresses the proportionality between flux and concentration difference [Eq. (2)], would always be used. However, with a complex series/parallel arrangement of components as is interposed in the transport pathways between plants and the atmosphere, it is usually much more expedient to use resistances and almost all measurements in the past two decades have been expressed in this form. This is unfortunate as the functional significance of the relationships tends to be obscured. Biologists interpret direct relationships more readily than they understand inverse ones—perhaps they are more concerned with multiplication than with division! Moreover, approximate linear relationships are transformed to hyperbolas. This point is illustrated in Fig. 1, which describes the courses of conductance and resistance with changing size of pore. Although the true relationship between conductance and aperture may have a greater degree of curvilinearity than emerges from the approximate solutions used, the resistance will always follow the general hyperbolic form shown. Small apparent changes of resistance over the lower range may be extremely important; they also represent an appreciable response by the guard cells. This may seem a minor point to labor, but we believe it wise not to obscure biological reality in the pursuit of computational expediency.

Equally, it is pointless to complicate calculations to preserve an apparent basic simplicity which is soon to be submerged by complex relationships with other components. Hence, the obvious procedure is to use either conductances or resistances according to which is simplest for the purpose

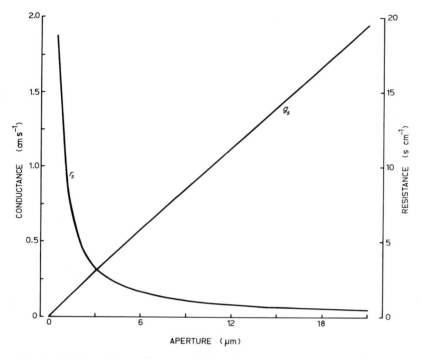

Fig. 1. Relationships with pore width of resistance, r_s, and conductance, g_s, of the stomata for diffusion of water vapor from the abaxial surface of *Commelina cyanea*. The diagram is based on the analysis of DeMichele and Sharpe (1973) and absolute values must be regarded as tentative because of uncertainties in both the model and values of the parameters used. However, the general forms of the relationships are valid.

in hand. Where the emphasis is placed on functional relationships, conductance is the most appropriate; where a number of components are assessed, as in gas exchange systems of the leaf, resistance may be the most expedient. Here, in exploring relationships with various controlling factors, we will use the conductance.

One further procedural point should be mentioned: the variance of stomatal conductance is reasonably constant over the entire range experienced, whereas that of the resistance varies with the mean; hence, for statistical analyses, values of the latter must first be transformed to their logarithms (or reciprocals).

Finally, we again emphasize that the whole concept of conductance (or resistance) only holds when Eq. (2) is valid; this is not so when $dc/dz \to 0$, i.e., when the carbon dioxide concentration approaches the compensation point, for example (Jarman, 1974).

C. AN OUTLINE OF THE TRANSFER SYSTEM

The transport of water vapor from the leaf was described in Chapter 6, Volume I, this treatise, and will be considered more fully in Section V. Here, we remind readers of its main components (Fig. 2). We may rewrite Eq. (2) as

$$F = (g_1 + g_2)\,\delta c \tag{8}$$

where $\delta c = c' - c =$ the difference in water vapor concentration between the evaporating surface in the leaf and that in the ambient air and the subscripts 1 and 2 now refer to the adaxial and abaxial surfaces, respectively. Then,

$$g_1 = 1/(r_{a_1} + r_{l_1}), \quad g_2 = 1/(r_{a_2} + r_{l_2}) \tag{9}$$

For any one leaf in a given environment $r_{a_1} = r_{a_2}$ and

$$1/r_{l_1} = g_{l_1} = g_{c_1} + g_{x_1}, \quad 1/r_{l_2} = g_{l_2} = g_{c_2} + g_{x_2} \tag{10}$$

where $g_{x_1} = 1/(r_{s_1} + r_{i_1} + r_{w_1})$ and similarly for g_{x_2}.

Most of the published observations belong to one of the following categories:

i. Measurements of r_{l_1} and r_{l_2} are made separately (by diffusion porometers) but rarely published separately; more usually they are given as R_l where $1/R_l = 1/r_{l_1} + 1/r_{l_2}$. This composite leaf resistance (or conductance) cannot be combined with the appropriate value of r_a and translated accurately to other situations unless $r_{l_1} = r_{l_2}$. The better procedure is to publish estimates

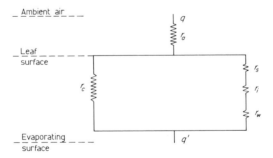

Fig. 2. Arrangement of resistances in the path of water vapor diffusing from one surface of the leaf. q' and q are the vapor concentrations at the evaporating surface and in the ambient air, respectively, and r_a, r_c, r_s, r_i, and r_w the resistances of the boundary layer, cuticle, stomata, substomatal cavity, and cell walls, respectively. (Redrawn from Fig. 8, Chapter 6, Volume I, this treatise.)

of r_{l_2} (or preferably g_{l_2}) and the ratio of the conductivities $\alpha = g_{l_1}/g_{l_2} = r_{l_2}/r_{l_1}$. Fortunately, with many species, α is close to unity, but there are also others where this is not so. (With hypostomatous species, α is close to zero, say 0.02–0.04; with amphistomatous species with widely differing numbers of stomata on the two surfaces its value lies between 0.1 and 1.)

ii. Estimates of R are obtained as $\delta c/F$ from entire leaves enclosed within leaf chambers. From knowledge of r_a, a composite value of R_l is derived. These estimates will always contain errors unless $\alpha = 1$.

iii. Estimates of a composite leaf resistance to viscous flow or from pore dimensions reduced to estimates of g_s (cf. Section II,B).

In the following pages, we compare the accumulated experience of a number of investigators. To provide a common basis for comparison, we have expressed the results as either g_l or g_s, i.e., per unit area of surface rather than as the composite parallel conductance G_l expressed per unit projected leaf area. Unless information to the contrary was available, we have assumed $\alpha = 1$, i.e., $g_l = G_l/2$ and $r_l = 2R_l$. We chose area of surface rather than area of leaf as the unit of comparison, as it is the more useful base from which further observations may extend. However, comparisons at this juncture can be no more than extremely crude and subject to many sources of variation. For interpretative purposes, we have usually assumed that $g_s \doteq g_l$ whereas it is usually 0.01–0.02 cm s^{-1} less than g_l; and we have neglected r_i and r_w, each of which may be about 0.2 s cm^{-1}, i.e., conductances which are some 5–10 times greater than that of the stomata.

II. MEASUREMENT OF STOMATAL AND LEAF CONDUCTANCE

The methods devised for measuring the stomatal contribution in gas exchange may be grouped into three broad categories:

A. Direct measurements of the rate of diffusion of water vapor from the leaf to give estimates of the leaf conductance;

B. Methods in which knowledge of the critical dimensions of the pores is required before the initial measurements can be reduced—nevertheless, these methods provide estimates of stomatal rather than leaf conductance;

C. Methods in which the leaf conductance is estimated from flux and other measurements in a crop.

In recent years, some or all of these methods have been discussed in reviews by Barrs (Chapter 8 of Volume I, this treatise), Slatyer and Shmueli (1967), Meidner and Mansfield (1968), Jarvis (1971), and Stigter (1972); of these, that by Jarvis is the most comprehensive and critical and we are much indebted to it.

A. Direct Measurement of Leaf Conductance

1. Diffusion Porometers

These are no doubt the most popular instruments currently being used to measure stomatal changes, there being extensive developments in their technology over the past decade. Measurements of the relative rates of diffusion of water vapor from a leaf have long been made by the classical cobalt chloride paper method (Milthorpe, 1955). Over the past decade, starting with the early models of Wallihan (1964) and van Bavel (1964), there have been extensive developments of more sophisticated instruments giving direct quantitative values of the diffusive conductance of the leaf. All depend on measuring the rate of change of water vapor in a small chamber attached transiently to the leaf. The air in the chamber is dried and the time taken for the humidity to change over a set range, the transit time Δt, is recorded. Humidity is measured by an electric hygrometer; these usually employ either lithium chloride (e.g., Kanemasu *et al.,* 1969; Stigter, 1972), sulfonated polystyrene (Stiles, 1970; Beardsell *et al.,* 1972), or aluminum oxide (Parkinson and Legg, 1972) as the hygroscopic medium. Thermocouples are mounted in the cup and preferably also on the leaf. In some models, a small fan circulates air within the chamber (Jarvis, 1971; Stigter *et al.,* 1973) to reduce the boundary layer and cup resistance; this refinement appears to be unnecessary, provided the hygroscopic sensor has a plane surface close to and parallel with the leaf and the relative humidity is maintained between 50 and 70%. Possibly the most satisfactory version currently in use is that developed by Monteith and colleagues (1974); it is commercially manufactured. It allows for repeated cycles of aspiration with dry air followed by the measurement of the time required for a predetermined increase in the relative humidity of a small chamber with a plane sensor. All switching is automatic and the transit time is displayed digitally.

Both stirred and nonstirred porometers may be calibrated by placing a series of known resistances between a free evaporating surface and the sensing element (Kanemasu *et al.,* 1969; Monteith and Bull, 1970; Stigter *et al.,* 1973; Stigter and Lammers, 1974). These must not absorb water and are preferably made of metal, such as the nickel membranes used by the last-named authors, and the calibrations must be made with great care at constant temperatures. Then, in theory,

$$\Delta t = K(r_p + r_l) \tag{11}$$

where r_p is the porometer and boundary layer resistance and r_l the calibration (leaf) resistance, with

$$K = (V/S)\ln[(q_l - q_p)_{t_i}/(q_l - q_p)_{t_f}]$$ (12)

where V and S are the volume and surface area of the cup and q_l and q_p are the absolute humidities of the evaporating surface and sensor, respectively, at the beginning, t_i, and end, t_f, of the transit time. Equation (12) may be simplified (Turner and Parlange, 1970; Stigter et al., 1973), in practice, to

$$K = (V'/S)\Delta q_p/(q_l - q_{p_i} + \Delta q_p/2)$$ (13)

where V' is an effective volume which takes into account the absorption of water vapor by the sensor and wall of the cup, hysteresis effects, slow response times and other difficulties—most of which are temperature dependent—that have dogged the use of these machines (cf. Meidner, 1970b; Morrow and Slayter, 1971a,b; Stigter et al., 1973, for ways of minimizing these). The use of a series of tubes differing in length (Monteith and Bull, 1970) or of injecting known quantities of saturated air into the chamber (Byrne et al., 1970) have not proved to be satisfactory calibration techniques.

Approaches intermediate between the "still-air" porometers mentioned before and the "flowing-air" leaf chambers described next have been used by Beardsell et al. (1972), who developed a porometer in which steady-state conditions are obtained by feeding dry air into the chamber at a measured rate to obtain a balance between the flux of water from the leaf and the flow of moist air out of the chamber, and by Parkinson and Legg (1972). The latter used an Al_2O_3-hygrometer mounted in a temperature-controlled chamber separated from the porometer cup to detect changes in the water vapor content of a stream of dry nitrous oxide passing over the leaf.

This general group of diffusion porometers provides a range of valuable instruments that allow the diffusive conductances of each leaf surface to be measured separately. They cannot be used when the leaf surface is wet, as by dew or rain, and this imposes severe limitations on their use in some environments. Moreover, they are all subject to errors arising from water absorption by sensors and the cup, by hysteresis effects, and by other difficulties. If used without caution, they can give very misleading results; if used and calibrated by the same person with care, they give results with acceptable precision.

2. Leaf Chambers

The total conductance of a leaf or portion of a leaf contained within a chamber may be derived using Eq. (2). Chambers may be designed for use in either field or laboratory and to operate under closed, semiclosed or open conditions (Eckardt, 1968). Usually fluxes and concentrations of both water vapor and carbon dioxide are measured, the former by thermocouple psychrometry, dew-point hygrometer, or an appropriate infrared gas analyzer and the latter by infrared gas analysis or a conductivity method. Many different combinations of ancillary apparatus have been evolved and are widely used; these are described in the papers cited before and will not be dealt with here.

These methods are used, of course, to explore detailed aspects of water vapor and carbon dioxide fluxes in which the stomatal (and leaf) conductances are but a large and variable part of the whole. They would not be used to measure only the leaf conductance. Conversely, because the leaf conductance cannot readily be estimated from measurements of carbon dioxide only, water vapor measurements are now almost universally made in experiments primarily concerned with photosynthesis. Estimation of the components, including leaf conductance, requires a detailed analysis of the whole system; this is considered in Section V.

B. Calculation from Stomatal Dimensions

The relationship between conductance and stomatal dimensions was dealt with at some length in Chapter 6 of Volume I (pp. 171–173), and there is little point in recapitulating that treatment. We know of no further contributions since then and believe that it is adequate for most purposes, although several aspects require to be further explored. This relationship is becoming increasingly important as the horizon of simulation modeling continually brightens, thereby allowing a more rigorous summary of current concepts and a reduction of sheer empirical effort. It is central to understanding the whole subject of stomatal modulation of gas exchange.

1. Direct Measurements

Nevertheless, even with perfect understanding of this relationship, direct measurement of stomatal dimensions, such as pore width, would rarely be employed as a routine method. Direct observation *in situ* or on epidermal strips (cf. Chapter 8, Volume I), even when adequately fixed as by Heath's (1947) reagent, are slow and laborious, subject to large sampling errors because of the variation between individual pores, and limited by the resolution of the light microscope—conductances less than about 0.2 cm s^{-1} cannot be measured. Similar considerations apply to impressions made of the leaf surface with collodion or silicone rubber; these have the further disadvantage of uncertain penetration and measurement of throat diameters.

2. Infiltration Techniques

There is, however, a further group of methods within this category, i.e., those involving the ease of infiltration by a series of liquids differing in surface properties (cf. Chapter 8, Volume 1; Schönherr and Bukovac, 1972; Parlange and Waggoner, 1973; Hack, 1974). The observations are essentially subjective and many ways have been used to arrive at an index of infiltration; nevertheless, successful calibrations between the index and pore size or viscous flow resistance have been made (Burrows, 1969; Hack, 1974). Schönherr and Bukovac (1972) attempted to describe the physical system involved, but were unable to find any strong relationship between infiltration and any one of several components explored such as surface tension, viscosity, capillary rise, contact angle, and diameter of pore. For example, movement of liquids into a capillary results from the pressure deficit, ΔP, developed at the concave meniscus. With a capillary of circular section, $P = 2\gamma \cos \theta / r$, where γ is the surface tension, θ the contact angle and r the radius of curvature at the meniscus. In a converging circular capillary, $P = 2\gamma \sin (\phi_1 + \theta)/r$ and, in a diverging capillary (i.e., as in a circular pore after passing the throat), $P = 2\gamma \sin (\phi_2 - \theta)/r$ where ϕ_1 and ϕ_2 are the angles of the wall in the plane of the advancing drop. Pores will fill readily provided $180° > \phi_1 + \theta > 0°$; at $\phi_1 + \theta = 90°$, $P = 2\gamma/r$ and the walls of the pore are wetted completely. When $\phi_2 < \theta$, P becomes negative and a positive expulsion pressure must be exerted before the drop can penetrate. However, stomatal pores often are elliptical or rectangular in cross section, the shape varying with degree of opening and the pore usually varies along its length in a more complicated manner than envisaged above (cf. Chapter 2, this volume). The required analysis must therefore be very much more complex than outlined and may not be warranted by the issues at stake; probably, sheer empirical calibrations against an accepted method of measuring stomatal (or leaf) conductance will serve in those field situations where this method is used (say, in the absence of a diffusion or viscous flow porometer or with wet leaves).

3. Viscous Flow Porometers

Instruments in which the flow of air through the leaf, or some colligative property thereof, in response to a difference in total pressure, is measured have played a valuable role in developing understanding of the functioning and significance of stomata. The principle was probably first employed by Dutrochet as long ago as 1832, but the modern family of porometers stems from Darwin and Pertz (1911) progressing, with many modifications, through the resistance porometer of Gregory and Pearse (1934) and its younger brother, the Wheatstone Bridge porometer (Heath and Russell, 1951), to a simple field version (Alvim, 1965; Bierhuizen *et al.*, 1965). As these porometers no longer play a significant part in laboratory investigations, we shall confine remarks to the last-mentioned version. It is still a most significant instrument having fewer technological incapacities and being usable over a wider range of conditions than the diffusion porometer; moreover, it is the simplest method of following stomatal rather than leaf conductance in the field and allows a large number of readings to be made very rapidly, thereby reducing sampling errors.

When using the Alvim porometer, a cup is transiently attached to the leaf and connected to a reservoir of volume V containing air at a pressure P_1 above atmospheric pressure p. The time t for the air to flow through the leaf and for the excess pressure to fall to a new value P_2 is measured. Then, the volume flowing per unit time

$$dq/dt = kP/p = (V/p)dP/dt$$

and hence

$$k = V \ln(P_1/P_2)/t \qquad cm^3 \ s^{-1} \qquad (14)$$

where k is the specific conductivity for viscous flow of the area of leaf contained within the cup. Values of specific conductivity may be reduced to values of stomatal conductance by employing the theory developed by Penman (1942). This theory uses analytic solutions of a complex flow system and, hence, a number of simplifying assumptions that could introduce appreciable errors. The development in approximate methods such as the finite element method over recent years provides a means of obtaining more exact solutions, but has not yet been applied. In the Penman analysis, the specific conductance for viscous flow per unit area of lower epidermis may be extracted from the measured conductivity of the leaf, if one assumes known values of resistance to flow along the leaf, that the resistance across the leaf is zero, and ratios of conductivities of the two surfaces. The first may be measured (Heath, 1941); in wheat, sugar beet, and potatoes, it is about 3.2 s cm^{-3} per cm^2 leaf (Milthorpe and Penman, 1967; Burrows, 1969). The ratio of the conductivities of the two surfaces is usually taken in proportion to the stomatal densities, but the advent of the diffusion porometer now allows the accumulation of empirical experience on this issue; recent experiments suggest that the ratio varies substantially during opening and in response to light and leaf water potential (Kanemasu and Tanner, 1969a,b; Turner, 1970, 1973) and it is only crudely related to stomatal densities (cf. Section III). The degree of error introduced by these considerations is uncertain; in the very few investigations where flux measurements have also been made (e.g., Krizek and Milthorpe, 1973), estimated stomatal conductances agree well with those expected.

The conductance for viscous flow may then be reduced to conductances for diffusive flow from considerations of general flow relationships and stomatal geometry and dimensions (Waggoner, 1965; Jarvis *et al.*, 1967; Milthorpe and Penman, 1967; Parlange and Waggoner, 1970). Because errors in dimensions or geometry influence both components in much the same way, they are much less important in this derived relationship than in estimating stomatal conductance directly from pore measurements.

C. Micrometeorological Methods

Estimates of conductances for gaseous exchange may be extracted from measurements of properties of the atmosphere above or within a crop. Ideally, knowledge of the energy balance and of changes with height of temperature, vapor concentration, and wind speed are required (cf. Chapter 4, Volume 1; Denmead, 1970; Denmead and McIlroy, 1971; for details of theory and instrumentation).

1. Above-Canopy Measurements

Approaches used by Monteith (1963, 1973) and Hunt *et al.* (1968) combine estimates of the canopy surface temperature and fluxes with those of the wind profile and energy balance, respectively. These allow an indirect estimate of the mean stomatal conductance of the canopy. In the elegant, but controversial, method used by Monteith, the position of a hypothetical plane crop surface is located by extrapolating the profile of wind speed, u, above the crop to the height where $u = 0$; i.e., at $z_0 + d$ where z_0 and d are the roughness length and zero-plane displacement, respectively. The temperature, T, and saturation vapor concentration of this surface, q', are similarly determined; then the conductance of the canopy, g_C, is given by

$$1/g_C = (q' - q)/F - 1/g_A \qquad (15)$$

where F is the vapor flux above the canopy, q the vapor concentration at the reference point above the canopy, and g_A the aerodynamic conductance of the canopy. Implicit in this approach are the assumptions that the transfer coefficients for heat, vapor, and momentum are identical and that these can be described by the logarithmic profile. It is assumed (Monteith, 1973) that the sources of latent and of sensible heat are similar, which is likely in a dense well-watered canopy, but not in a sparse canopy, nor one in which there is a high capacity for sensible heat such as a forest. The sources of heat and water vapor are usually found at a lower position within the canopy than the sink for momentum; hence, g_C is not strictly independent of g_A. The conductance of an additional boundary layer, assumed to be the source for heat as for water vapor, can be determined from the logarithmic profile of temperature; this has a value around 5 cm s^{-1}. Despite the above objections, canopy conductances have been estimated which are in good agreement with those expected from readings by a diffusion porometer (Monteith, 1973).

More simply the mean leaf conductance of the canopy can be estimated as

$$1/g_C = (\epsilon + 1)(E_p/E - 1)/g_A \qquad (16)$$

where E is the evaporation flux obtained from a combination or similar relationship (e.g., the Penman function or one of its variants), E_p the potential evaporation (i.e., estimated from the same weather parameters but with $g_C = \infty$), and ϵ the increase in latent heat content with increase in sensible heat content of saturated air at the temperature of the ambient air.

These approaches do not take into account, of course, the variations in time and space within the canopy of the sources and sinks for heat and water vapor. These may be large (Denmead, 1970) and are influenced by changes of stomatal conductance within the canopy (cf. Sections III and IV). Moreover, they are unlikely to be applicable during the early stages of growth and development nor during the later stages of rapidly decreasing leaf area of annual crops. Despite these objections and if used with due caution, the approach can yield values that have physiological meaning and assist in understanding the response of the crop as a whole.

2. Within-Canopy Measurements

Many of the objections raised above can be overcome if measurements are made of the changes with height in the canopy of net radiation, air and plant temperatures, humidity and/or carbon-dioxide concentration. The canopy is then treated as a series of layers, and the source and sink strengths within each layer determined. (Measurements of conductances using measurements of carbon dioxide are extremely difficult because of the additional physiological components involved.) An outline of this approach was given in Chapter 6, Volume I (p. 182); further details are given by Cowan (1968), Denmead and McIlroy (1971), and Jarvis (1971).

III. RELATIONSHIPS WITH CHANGES IN THE EXTERNAL ENVIRONMENT

Changes in the stomatal (or leaf) conductance in response to the range of environmental factors may be described by the generalized expression

$$g = f(Q, \psi_L, q, T, u, N, [CO_2], \ldots) \tag{17}$$

where $Q, \psi_L \ldots$ represent specific functions $[f(Q), f(\psi_L) \ldots]$ relating g to light flux density, leaf water potential, humidity, temperature, wind speed, mineral nutrition, carbon dioxide concentration, etc. Precise formulation of Eq. (17) is not yet possible although there is some information on the nature of the individual functions. Even with these, there are, as yet, too few observations to establish with acceptable precision values of the parameters in each function for any one species (cf. with what follows). Nevertheless, we should simultaneously be working toward ways of combining these to provide adequate prediction. Of the three ways in which the individual functions may be combined—additive, multiplicative, and reciprocal—it is likely that we can learn a useful lesson from the studies relating growth to the supplies of mineral nutrients where reciprocal functions appear most useful (cf. Milthorpe and Moorby, 1974). That is, we write

$$1/g = f'(Q) + f''(\psi_L) + \ldots \tag{18}$$

If so, then it may emerge that functions of either resistances or conductances turn out to be additive(!); this is a matter yet to be explored.

A. LIGHT FLUX DENSITY

Stomata on intact leaves open in response to increasing flux density of visible radiation over the entire sensible range (Fig. 3). There are usually two phases in the response curve: a rapid increase over low flux densities followed by a gradual, but still substantial, increase over higher ranges. (These changes are usually disguised when resistances are plotted.) These responses have been observed using diffusion porometers in sorghum (*Sorghum*), maize (*Zea*), and tobacco (*Nicotiana*) by Turner (1970, 1973), in sorghum by McCree (1974) and in beans (*Phaseolus vulgaris*) by Kanemasu and Tanner (1969b) and in leaf chamber experiments with *Acacia harpophylla* (van den Driessche *et al.,* 1971), bean, *Hyocyamus,* and tomato (Kuiper, 1961), *Capsicum* (Hall, 1975), citrus (Kriedemann 1971), soybean (*Glycine max*) (Sambo, 1974), tobacco (Clough, 1971), a range of C_3 legumes and C_4 grasses (Ludlow and Wilson, 1971a), and the C_4 grass *Pennisetum typhoides* (McPherson and Slatyer, 1973). In reducing the observations shown in Fig. 3 to a common basis, we will have introduced some errors by inexactitude in measuring from small published diagrams, in converting units of irradiance that were not always explicitly stated, and possibly sometimes using a value of $D_{CO_2}:D_{H_2O}$ slightly different from that used by the authors. These are likely to be small compared

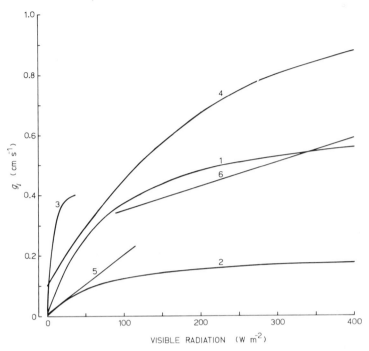

Fig. 3. Relationships between leaf conductance for water vapor per unit area of surface, g_l, and the flux density of visible radiation. Curve 1, abaxial surface of tobacco (Turner and Begg, 1973); Curve 2, average surface of maize (ibid.); Curve 3, abaxial surface of *Phaseolus vulgaris* (Kanemasu and Tanner, 1969b); Curve 4, average surface of *Pennisetum purpureum* (Ludlow and Wilson, 1971a); Curve 5, average surface of *Glycine max* (Sambo, 1974); Curve 6, average surface of *Pennisetum typhoides* (McPherson and Slatyer, 1973).

to effects arising from variation between leaves, prehistory of the leaves used, hysteresis effects where the same leaf was exposed to a sequence of irradiances, and the proportion of the sensible range of light flux density actually explored. Curve 6 is consistent with Curve 1, for example, and Curve 4 may well have shown lower values at low flux densities if the leaf had experienced a different sequence of exposure.

  Two issues arise from this figure: (i) the form of the response curve and (ii) the significance of the variation between the different curves. Concerning the first, the response curve of resistance always follows a hyperbolic form and is often assumed to be a rectangular hyperbola expressed as

$$r_l = a + b/I \qquad\qquad (19)$$

where I is the flux density of visible radiation, a the minimum resistance when $I = \infty$ and the parameter b may be equated to aI_x where I_x is the flux density when $r_l = 2a$. This is a very conservative function and observations can usually be constrained to fit this function; however, it does not describe Curves 4 or 5 and some other data. Moreover, it can only be approximate as r_l is never zero in darkness. Even if r_s replaces r_l in Eq. (19), it is still approximate because it is doubtful if stomata ever close completely in the dark. However, accepting it as a useful approximation, we can write

$$g_l \doteq I/(aI + b) \tag{20}$$

Possibly, a more useful function which gives an equally acceptable (or often better) fit and which describes the system more exactly is

$$g_l \text{ (or } g_s) = A\{1 - \exp[-(c + dI)]\} \tag{21}$$

where A is the maximum conductance ($\equiv 1/A$) and c is a constant numerically equal to the cuticular plus stomatal conductance in darkness. However, this is not an issue that warrants quibbling at this level of detail; the accuracy of available observations and the precision required do not merit such finesse. We had hoped to use some such empirical expression and order the results in a meaningful way, but have been sadly disillusioned in this endeavor, as we have not been able to find any consistent order.

We will therefore discuss the variation evident in Fig. 3 in more general terms. First, we should note the less tidy nature of the relationships chosen here than those presented by Turner (1974a) which conform to a "textbook" pattern. (Turner's conductances are expressed per unit leaf area and, hence, are twice those given in Fig. 3; we believe his irradiance scale is that of incoming short-wave radiation and, hence, is again about twice that given in Fig. 3.) Turner suggests that conductances of amphistomatous C_3 species are consistently higher than those of amphistomatous C_4 species: this contention is supported by the response of the C_4 species, *Atriplex spongiosa*, and the C_3 species, *A. hastata*, when grown under the same conditions, maximum values of g_l of 0.5 and 1 cm s^{-1} being recorded (Slatyer, 1969); it does not hold more generally as shown by the high conductances of the two C_4 species of *Pennisetum* (Curves 4 and 6). We believe that available evidence does not allow us to generalize further than to state:

i. That those amphistomatous species, whether C_3 or C_4, in which $\alpha \doteq 1$ usually have higher conductances than species in which $\alpha \ll 1$. Within this group, there are large differences between species: maize and sorghum (cf. references cited by Turner, 1974a; Gifford and Musgrave, 1970; and others) have consistently low conductances, whereas Hodgkinson's (1974) and our unpublished observations on alfalfa (*Medicago sativa*) suggest that it has extremely high conductances (1-2 cm s^{-1}) and is very responsive to low light flux densities, following an even more exaggerated form of Curve 3. Some other clover and medic species may also be very responsive, but this is not true of legumes generally (cf. Curve 5). It may be noted that α, the ratio of the conductivities of the two surfaces, is only very crudely related to the ratio of numbers of stomata, n_1/n_2; it tends to unity with $0.5 < n_1/n_2 < 1.3$. This partly arises from the stomata on the surface with the lower density being the larger, but it seems that differential response mechanisms within the guard cells are also involved.

ii. Conductances of hypostamatous leaves as a whole are often less than those of amphistomatous leaves, i.e., the high conductance of the abaxial surface is unable to compensate fully for the low cuticular conductance of the adaxial surface. However, species differences even override this generalization.

iii. Conifers are usually regarded as showing the lowest conductances (cf. Turner, 1974a), which is attributed to the sunken stomata and the intermeshed wax fibrils in the stomatal pit (cf. Chapter 2). However, it should be noted that Ludlow and Jarvis (1971) obtained values of 0.6 and 0.37 cm s^{-1} with glasshouse- and forest-grown plants of Sitka spruce (*Picea sitchensis*), respectively. Allowing for the hypostamatous leaves, this still gives effective conductances from the leaf as a whole higher than that of maize and possibly similar to that of soybean (Curve 5) in Fig. 3.

Finally, we should note that true differences between species can only be separated by a much more systematic set of observations than we have considered here. Very large differences within a species occur with age (Section IV,B), conditions under which the plants have grown, immediate history, and other factors, and these often far exceed responses which can be attributed to inherent differences between species.

B. Leaf Water Status

Figure 4 illustrates the general type of response that is found when leaves experience a water deficit. Much the same type of relationship is found whether the deficit is recorded in terms of water potential, ψ_L, or relative water content, S; this is not surprising as the relationship between ψ_L and S is usually close to linear. As usual, any one set of observations rarely covers the full range of experience and this influences the function that may be fitted to any given set.

We believe that most observations are consistent with the view that there is a range of high water potentials over which there is no influence. In this range, the variation is largely due to the operation of the many other factors that influence stomatal movement. Further reduction below this "critical" level leads to rapid closure over a relatively narrow range. At a crude level, there is effectively an "on-off" situation—the stomata

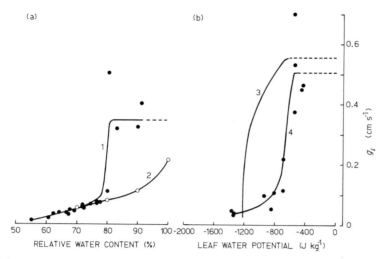

Fig. 4. Relationship between leaf conductance, g_l, and (a) relative water content or (b) leaf water potential. Curve 1, average surface of cotton (Troughton, 1969); Curve 2, average surface of *Lolium perenne* (Wilson, 1975); Curve 3, abaxial surface of *Phaseolus vulgaris* (Kanemasu and Tanner, 1969a); Curve 4, average surface of *Beta vulgaris* (Biscoe, 1972).

are either open or relatively closed. At a somewhat finer level of comparison, we see evidence of differences in the form of the curve; sometimes, this decreases exponentially and sometimes it is closer to a linear decrease (cf. for example, Raschke, 1970). There is also rarely an exact "off" position; the sudden fall in conductance over a small range is followed by a continued slow decrease over a much wider range of lower potentials. In the semiarid shrub, *Acacia harpophylla,* van den Driessche *et al.* (1971) found an exponential decrease in conductance over the entire range of ψ_L from —500 J kg^{-1} in well-watered plants to about —7000 J kg^{-1} in severely droughted plants. The change over this last part of the response curve almost certainly includes some changes in the stomatal component and it possibly also reflects a decreasing cuticular component, although we do not know of any investigations on this point.

The position of the response curve also varies widely between different experiments and possibly species (cf. Fig. 4, references cited above, Mc-Cree, 1974, and review by Turner, 1974a). Turner draws attention to a number of aspects. First, it is fairly clear that field-grown plants are less responsive than glasshouse- or chamber-grown plants. For example, Jordan and Ritchie (1971) found the critical ψ_L for beginning of closure of leaves on glasshouse-grown plants was about —1400 J kg^{-1}, but field plants did not respond at ψ_L as low as —2800 J kg^{-1}. McCree's (1974) observations also show that hardened plants were less responsive than well-watered plants. Second, stomatal responses to water deficits will be exerted through the turgor pressure of the guard cells, which, in turn, is a resultant of at least two components—water supply as reflected by ψ_L and the guard-cell osmotic potential which is very likely to be strongly influenced by abscisic acid (ABA) varying the concentration of potassium ions (cf. Chapter 2). Turner collated the results from a number of workers to derive relative relationships between g_l and leaf turgor; these were much more consistent than those with leaf water potential, but there were still substantial differences in both the form and position of the curves. These might well be explained by differences between guard and mesophyll cells in respect of modulus of elasticity influencing the relationship between turgor pressure and water potential, in differences in ABA production, and other aspects. As pointed out in Chapter 2, in terms of water supply only, stomatal closure may not occur until the turgor pressure of guard cells is close to zero and much of the response to water deficits which is found may be exerted through ABA production and modification of the osmotic potential. There is need for a much fuller investigation of this whole subject, with attention paid to the guard and subsidiary cells, rather than relying on measurements made on bulk leaf tissue.

Finally, we should draw attention to the marked response in the relation between g_s and ψ_L which occurs both with age of a leaf on vegetative plants and during development (cf. also Section IV,B). The latter is particularly shown by cereals; stomata on the flag leaf, for example, become less responsive to water deficits as the plant proceeds from ear emergence through anthesis to grain filling, with the leaf having a lower water potential before closure commences and a very much lower potential before the conductance reaches an approximate equilibrium value (Table I). Similar

TABLE I

VALUES OF LEAF WATER POTENTIAL, ψ_L, AND LEAF CONDUCTANCE, g_l, AT THE BEGINNING AND END OF THE CLOSING PHASE OF THE FLAG LEAF OF WHEAT ON PLANTS AT DIFFERENT GROWTH STAGES DURING DRYING OF THE SOIL[a]

Stage of growth	At beginning of closure		At end of closure	
	ψ_L(J kg^{-1})	g_l(cm s^{-1})	ψ_L(J kg^{-1})	g_l(cm s^{-1})
Preanthesis	−1510	0.49	−1860	0.02
Anthesis	−1740	0.28	−2680	0.02
Early grain filling	−1940	0.33	−3180	0.04

[a] Unpublished data of J. M. Morgan.

responses have been noted by Turner (1974a) and are supported by considerable indirect evidence from measurements of rates of transpiration. These are certainly stomatal rather than cuticular responses, but there is no evidence on which components of the stomatal apparatus are involved.

C. Humidity

There is good evidence that stomatal conductance is influenced by the concentration of water vapor in the ambient air. As well as older information (Meidner and Mansfield, 1968), Slavík (1973) observed a much lower conductance of maize leaves grown at 45% than those grown at 85% relative humidity, with no difference in stomatal densities or sufficient variation in other structural components to explain the difference. Whiteman and Koller (1967b) found a curvilinear relationship between transpiration of sunflower and water vapor gradient between leaf and air, which suggests that stomatal conductance was varying. It is, of course, essential in indirect experiments of this nature to ensure that leaf temperatures are measured; provided ψ_L is greater than about -2000 J kg^{-1}, it is acceptable to take the vapor concentration in the leaf as equal to the saturation vapor pressure of pure water at the same temperature (Jarvis, 1971). However, leaf temperatures may vary widely from ambient temperatures and some effects attributed directly to humidity may reflect responses to temperature. This may well apply to the results of Camacho-B et al. (1974) where humidity changes were confounded with temperature variations between 20° and 35°C, but not to those of Aston (1973) where the temperature ranged from 20° to 25°C. Both sets of results, as well as those of Schulze et al. (1972), can be approximately represented by logarithmic relationships over ranges of vapor concentration differences greater than about 10 μg cm^{-3} (Fig. 5). However, an element of curvilinearity is apparent in these relationships and the relationship may well depart from this form over the range of low vapor concentration differences. For example, observations by Barrs (1973) on maize, cotton, and tobacco suggest that the conductance may increase with increasing vapor concentration differences up to 6–8 μg cm^{-3}.

We have used the vapor concentration difference between leaf and air rather than ambient humidity or other concomitant variables because it provides the most consistent relationship. Possibly the vapor pressure concentration around the inner walls of the guard cells would be more appropriate. It is difficult to visualize effects arising from humidity other than by influencing the water relations of the guard cells; interpretation would be greatly assisted if these could be measured in experiments on this factor.

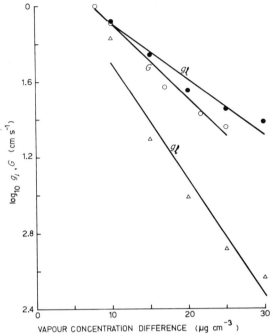

Fig. 5. Relationships with vapor concentration difference between leaf and air of the leaf conductance of *Phaseolus vulgaris* (triangles) and *Helianthus annuus* (black circles) from the observations of Aston (1973) and of total conductance (i.e., including boundary layer conductance) of *Prunus armeniaca* (white circles) from the observations of Schulze *et al.* (1972).

D. TEMPERATURE

As pointed out earlier, the close coupling between temperature and humidity makes it very difficult to ascribe effects to one factor or the other. Most experiments that purport to explore the effects of temperature have usually also been confounded with changes in humidity components.

Recognizing these limitations, the responses recorded in the literature are of three forms:

i. Optimal relationships as shown in Fig. 6 for onion (*Allium*) (based on diurnal measurements in the field and hence also confounded with light flux density—but differences between shaded and sunlit leaves were small) and the semiarid shrub *Acacia harpophylla* (phyllodes contained in controlled chambers). Similar relationships have been found with *Pinus halepensis* (Whiteman and Koller, 1964).

ii. Increase in conductance over the entire range, but with a transient closure over a small range in the higher region (Fig. 6, Curve 3). This

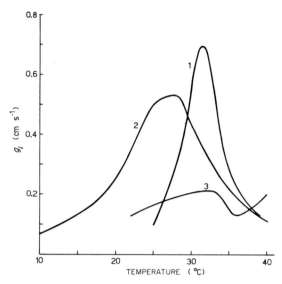

Fig. 6. Relationships of g_l with temperature for onions in the field (Curve 1, after Goltz and Tanner, 1972), phyllodes of *Acacia harpophylla* in growth chambers (Curve 2, after van den Driessche *et al.,* 1971), and maize in growth chambers (Curve 3, after Gifford and Musgrave, 1972).

type of response appears to be reasonably consistent with maize: increased conductance was found by Moss (1963) using temperatures of 14°, 20°, 30°, and 40°C only in a chamber in which the vapor concentration difference between leaf and air was maintained constant. He measured stomatal aperture only and obtained greater responses with CO_2-free air than with air of 200 or 400 ppm CO_2. Increased conductance with temperature has also been found in *Xanthium* by Drake *et al.* (1970) and several desert species by Schulze *et al.* (1973).

iii. Decreasing conductances with temperature over the range 20°–40°C were found by Wuenscher and Kozlowski (1971) using *Quercus*. They maintained a constant vapor concentration in the chamber over the range of temperatures explored; hence, the vapor concentration difference between leaf and air would have varied from about 3.5 to 38 μg cm^{-3}.

These findings are not necessarily in conflict as different species may well have widely differing temperature optima. However, it is certain that these responses include direct effects of temperature on metabolism of the guard cells and also indirect effects exerted through the water relations of the stomata and possibly solubility of CO_2 in water. Perhaps, the most direct experiments are those done in a water-saturated atmosphere. For example, Stålfelt (1962) placed halves of leaves of *Vicia faba* in small chambers with adequate water and measured pore width after 3 hr. He found increased opening, in almost a linear function, between 5° and 35°C in CO_2-plus air and light and up to 40°C in CO_2-free air and darkness. At 45°C, there was slight closure in both treatments. Raschke (1970) used leaf sections of *Zea mays* in a complex chamber with a small but constant vapor concentration difference between leaf and chamber air. His measured function of stomatal conductance indicated no change or even closure over the range from 7° to about 15°C, followed by increased opening up to about 40°C (the maximum explored). There was evidence of time-dependent temperature inactivation at the high temperatures and this indeed would be expected even with a high-temperature species such as maize. Brunner and Eller (1974) found that stomata of *Piper betle* open as widely in darkness as in light when temperatures are raised above 36°–38°C to within the range 40°–45°C. Conductances were higher at high than low humidity, and there was no evidence of temperature inactivation.

A further complication in attempting to unravel the responses to temperature is the effect often called preconditioning. For example, Ludlow and Wilson (1971b) found a marked increase in the stomatal-light response curve at 30°C between leaves grown at 20°C and preconditioned at 30°C, during the 14-hour dark period preceeding the experiment, and leaves transferred directly from 20°C. Other workers have also found effects of the environment of the previous day on stomatal responses. It is unlikely that these can be explained or even predicted satisfactorily until the stomatal mechanism is much more clearly understood. As membrane functioning is undoubtedly highly important in guard-cell responses, we might expect responses associated with change of phase of the phospholipids; possible implications of these have not yet been explored.

E. Wind

Very few direct studies of the effect of wind on stomatal conductance have been made. In a recent study, Davies *et al.* (1974) showed that stomata of 2-year-old seedlings of *Fraxinus americana* and *Acer saccharum* had lower conductances in wind of speeds between 0.6 and 2.7 m s^{-1} than in "still" air, those of *Acer* being the more sensitive. A more extensive range of indirect observations on the effect of wind on transpiration rates also suggests that responses are usually in this direction. For example, transpiration rates are often lower at high than low wind speeds: such response will be the resultant of increased boundary layer conductance tending to increase transpiration—usually lower leaf temperatures tending toward a decrease of transpiration—and stomatal changes; so unless all components are measured, it is not possible to extract the stomatal response. Nevertheless, we can accept provisionally that stomatal conductance decreases with increased wind speeds, even if we cannot formulate the relationship concerned. This response may arise from a higher carbon-dioxide concentration around the guard cells, effects through the water relations, or even possibly vibrational effects on stomata arising from leaf flutter.

F. MINERAL NUTRIENT SUPPLY

There is a marked response of stomata to supply of potassium in the plant; indeed, documentation of the now well-attested central role of potassium ions in stomatal movement arose from initial observations that stomata were much more closed in K-deficient than in normal plants (cf. Chapter 2). We can accept that the maintenance of an adequate supply of K^+ ions in the epidermis is essential for stomatal functioning, but we still have a long way to go in understanding the details of the system. For example, there is as yet only imperfect understanding of the recycling of potassium between leaves, the rate of exchange between epidermis and mesophyll, or indeed of the quantitative relationship between stomatal functioning and concentration of potassium within the epidermis. Much of the age response in stomatal functioning (Section IV) could well be associated with changes in potassium concentration in the epidermis, and we might expect this to be accentuated in plants with a restricted supply to the roots; however, the system is complex and so inadequately documented that it is not yet possible to predict responses except in a very general way.

Similar considerations apply to nitrogen and phosphorus. For example, Shimshi (1970a,b) found higher transpiration rates from high- than from low-nitrogen plants of *Phaseolus vulgaris* with high soil moisture, but the opposite response at low soil moisture. These results suggested that stomata on leaves of high-nitrogen plants were more responsive; there were no differences in stomatal density, but apertures were different. Further, Ryle and Hesketh (1969) found that the stomatal conductance of maize leaves increased with time when plants were grown with ample nitrogen, but decreased when the supply was low.

G. CARBON DIOXIDE

There is a great deal of evidence that stomata are very responsive to the concentration of carbon dioxide in the air surrounding the inner walls of the guard cells (cf. Chapter 2, Section III,C,2). The concentration in this region, i.e., on the lateral walls of the guard cells bordering on the substomatal cavity, is determined by the size of all other components of the system including the conductance of the stomata. There is, therefore, a high degree of "feedback" within the system. Moreover, we need to distinguish equilibrium situations from those where the system is responding to a sudden perturbation, such as sudden exposure to light following a period of darkness. Two estimates of CO_2 concentration are readily available: that in the ambient air and a mean internal concentration throughout the whole leaf. Where the stomatal conductance is very small, the concentration at the lateral walls of the guard cells is likely to approximate that of the mean internal concentration; when it is large and, generally, when there is a net inward flux, it will approximate the ambient concentration. In the present context, we are mainly concerned with the last situation and we will express relationships in response to the ambient concentration.

The range of experience is illustrated in Figs. 7,8. No clear conclusion emerges, which makes it very tempting to assume that under field conditions, where the ambient concentration is likely to vary between, say, 250 and 350 ppm, the response is too small to be taken into account. This is a reasonable conclusion to adopt when predicting values of stomatal conductance to be used in wider systems such as calculation of rates of evaporation or photosynthesis. Considering the results in finer detail and

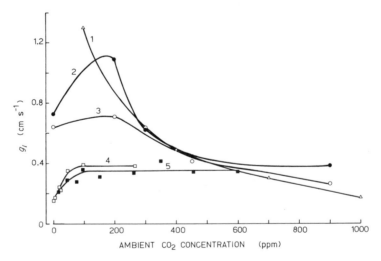

Fig. 7. Relationship between stomatal conductance per unit leaf surface, g_l, and ambient CO_2 concentration. Curve 1—one line of *Zea mays*—another studied in the same experiment gave values less than half those shown at 100 and 400 ppm (after Gifford and Musgrave, 1973); Curve 2, *Sorghum almum* and Curve 3, *Vigna luteola* (after Ludlow and Wilson, 1971a); Curve 4, kale (after Parkinson, 1968); and, Curve 5, *Picea sitchensis* (after Ludlow and Jarvis, 1971).

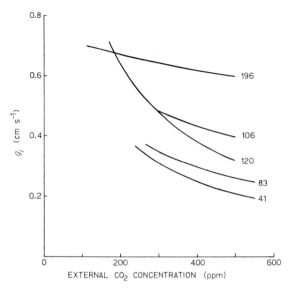

Fig. 8. Relationships between g_l and ambient CO_2 concentration in leaves of *Pennisetum typhoides* under flux densities of visible radiation (nE cm^{-2} s^{-1}) shown. 1 nE cm^{-2} s^{-1} $\doteq$ 2.17 W m^{-2}. (After McPherson and Slatyer, 1973.)

taking into account the ranges explored by the different investigators (cf. also Whiteman and Koller, 1967a), it might be argued that the findings are consistent with the contention that there is an opening effect between CO_2-free air and the internal CO_2 compensation point, i.e., the internal concentration at which influx equals efflux under high light intensity conditions. This is somewhere in the regions of 0–5 and 50–80 ppm with C_4 (maize and *Sorghum almum*) and C_3 (other species shown), respectively. At higher concentrations, there is generally increasing closure (i.e., lower conductances) with increasing concentration; however, Bierhuizen and Slatyer (1964) found no variation in cotton between 200 and 2000 ppm. The suggestion that maximum opening occurs at the CO_2 compensation point is supported by the usual observation that lower conductances are usually found when a mass of leaves is illuminated in a small volume of air than when exposed directly to CO_2-free air.

Some experience of the interaction between CO_2 concentration and light flux density is shown in Fig. 8. Despite some internal inconsistency, which may reflect differences in age or other components of the leaves used, there was generally a decreasing response to CO_2 concentration with increasing irradiance. This family of curves could be represented by a relation of the form $g_l = (a + bC)^{-1}$ where C is the CO_2 concentration, i.e., there was a linear relation between leaf resistance and CO_2 concentration.

The observations shown in Curves 1–3 of Fig. 7 at concentrations above 100 ppm could also be approximated by this relationship.

The different responses to CO_2 concentration illustrated in Fig. 7 may possibly be explained by reports that sensitivity to CO_2 is induced by other factors. It has been claimed that plants grown in a "well-watered" environment in which the leaf never experiences a water deficit are not sensitive, but can be made to respond by subjecting to a moderate deficit by chilling, or by provision of abscisic acid in the transpiration stream (Raschke, 1975). It is also clear that stomatal behavior varies with many internal factors that are not yet understood (see next section). All of the plants for which observations are presented would be expected not to have experienced water deficits, but there may well be other complexes of factors involved. As with many other aspects of stomatal behavior, we yet know too little about the whole system to invoke rigid polarized explanations.

H. OTHER GASES

1. Oxygen

It is generally agreed that oxygen is necessary for stomatal opening to occur and that lack of oxygen accelerates the response to closing treatments such as darkness or CO_2 (Heath and Orchard, 1956; Zelitch, 1965). The effect of deprivation of oxygen on open stomata in an otherwise unaltered environment has given variable responses. For example, Ludlow (1970) found no response on changing the oxygen concentration from 21 to 0.2% around leaves of tropical species of grasses and legumes with open stomata exposed to light; Zelitch (1965) records a similar experience with tobacco. However, Heath and Orchard (1956) did find some closure with wheat and Hall (1974) recorded rapid closure in *Capsicum,* but not in bean and sunflower, on exposure to O_2-free air. This is now a widely used method of estimating rates of photorespiration; obviously, the stomatal response of each particular species must be carefully explored before the method can be used to estimate photorespiration.

2. Sulfur Dioxide

The effect of increasing the SO_2 concentration of the atmosphere is to increase the stomatal conductance (Biscoe et al., 1973b). The minimum concentration to which the stomata of *Vicia faba* responded was 72 $\mu g\ m^{-3}$, suggesting that the responsive sites were rapidly saturated. The response was greater and more rapid in old than in young leaves. The effects were such that the transpiration rates of field crops (in English climates) could be increased by over 20%.

3. *Air Pollution Oxidants*

Duggar and Ting (1970) reviewed a number of observations showing generally that stomata close in response to oxidants such as ozone and peroxyacetyl nitrate in the atmosphere. Closure on exposure to ozone has also been used as the basis of selection of onion varieties resistant to damage, although in most species, there seems to be little relationship between degree of injury and stomatal opening. As far as we can ascertain, there have been no attempts to establish quantitative relationships between stomatal conductance and concentration of the different oxidants in the atmosphere.

IV. RELATIONSHIPS WITH INTERNAL FACTORS

A. GROWTH SUBSTANCES

Although the closing effect of abscisic acid is now well documented and there have been several reports of opening responses to cytokinins and gibberellic acid (cf. Chapter 2), quantitative relationships between stomatal conductance and guard-cell (or epidermal) concentrations of growth substances have not been established. The concentration of abscisic acid (and associated derivatives) in leaves of well-watered mesophytes seems generally to be about 10–40 μg (kg fresh weight)$^{-1}$ (Milborrow, 1974; Hiron and Wright, 1973), but may be as much as 100–200 $\mu g\, kg^{-1}$ in *Vitis vinifera* (Kriedemann and Loveys, 1974; Loveys and Kriedemann, 1974). It may increase some 10- to 40-fold during a water deficit, although possibly even a doubling of the concentration is associated with appreciable stomatal closure. Changes in the guard cells themselves have not been measured. Solutions of concentration 0.1–10 $\mu g\, ml^{-1}$ applied to the leaf or absorbed by the transpiration stream will cause closure.

Effective concentrations of kinetin solutions resulting in opening within about 2 hr—compared with response times of a few minutes to abscisic acid—appear to be about 10^{-6} to 10^{-5} M (Meidner, 1967).

Many of the responses to be described in the following section are of the type in which mediation by growth substances is usually invoked. However, little evidence of their role therein has yet been sought. Concentrations in guard cells or even the epidermis have not been explored and rarely have measurements of the three groups of substances likely to be involved—abscisic acid, cytokinins, and gibberellins—been made in the same experiment. We still have little understanding of the flow of growth substances through the plant and, as the guard cells are relatively isolated (cf. Chapter 2), transport to them even from the associated mesophyll is likely to be slow.

B. CHANGES DURING GROWTH AND DEVELOPMENT

1. *Species and Age of Leaf on Plants in the Vegetative Phase*

Attention was drawn in Chapter 2 to the large variation in structure, dimensions, and density of stomata between species. Equally, variation in functioning, and hence differing values under comparable conditions, might be expected. Minimum values of stomatal resistance, i.e., the most open state reached during the diurnal cycle by leaves of well-watered plants under high illuminance, of some species were given in Volume I of this treatise and in Altman and Dittmer (1971); further records have been summarized in the preceding pages. From this experience, only the broadest of generalizations can be made; indeed, the conclusions enumerated under response to visible radiation (Section III,A) represent the limit to generalization that can be made.

Variation within species certainly seems to be greater than variation between species, which reflects the large responses to a range of environmental factors. There is also another source of variation—that associated with the development of the plant.

Each leaf as it expands is often accompanied by changing stomatal behavior. For example, leaves of debudded decapitated plants of *Xanthium* showed increasing maximum conductances from the time they were large enough to measure; they reached their highest values when the leaf was about three-quarters of its final size (Fig. 9). These were maintained until the leaf had reached full size; thereafter, the stomata opened less and less widely as the leaf aged. Although these plants were abnormal in that all apical meristems were removed, the pattern shown is fairly typical of most leaves. The increase in maximum conductance as the leaf expands is almost invariably found, but the duration of the phase before values commence to decrease varies widely. Sometimes this is short as shown in Fig. 9 and by Kriedemann (1971) in *Vitis vinifera,* Ludlow and Wilson (1971c) in *Sorghum almum,* McPherson and Slatyer (1973), Woledge (1972), and others. Sometimes it is maintained for a long time, as shown by Ludlow and Wilson (1971c), Osman and Milthorpe (1971), Holmgren *et al.* (1965), Brown and Rosenberg (1970), Ludlow and Jarvis (1971), Turner (1974a), and others; there is then a rapid decrease in the maximum stomatal conductance in the later stages of leaf senescence.

The maximum value of stomatal conductance reached during each diurnal cycle is also accompanied by changing values in the minimum conductance attained during the night. Far fewer observations are available

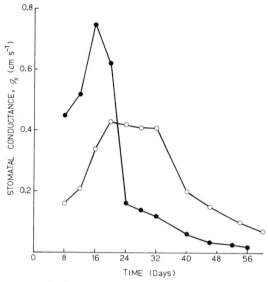

Fig. 9. Changes with time in maximum stomatal conductance, g_s, of Leaf 9 of *Xanthium* plants debudded and decapitated above Leaf 9 on Day 0. White circles indicate plants kept in short nights (uninductive for flowering) and black circles plants in long nights (flower-inducing). Both leaves were about 0.25 cm² on Day 0 and fully expanded by about Day 30. (After Krizek and Milthorpe, 1973.)

on this aspect, so there is still much uncertainty. We agree with Turner and colleagues (cf. Turner, 1974a) that the maximum daytime conductances found in near to fully expanded leaves are also accompanied by the minimum nighttime conductances; i.e., at this stage, the stomata open more widely and close more tightly than at other times. We also agree that during the senescent phases the decreasing maximum conductances are accompanied by increasing minimum conductances, i.e., the stomata neither open as widely nor close as tightly. We disagree concerning behavior during leaf expansion; we find the stomata then close as tightly as during the mature phase, whereas Turner finds that they close less tightly. Ludlow and Wilson (1971c) found decreasing minimum conductances over the whole range of leaf age. This disagreement may well be due to our measurements being of *stomatal* conductance using the viscous-flow porometer, whereas those of Turner were of leaf conductances with the diffusion porometer and those of Ludlow and Wilson from leaf chambers.

This experience is supported by observations on *Capsicum* (Fig. 10a), where the deflorated plant may be regarded as a vegetative plant, with new leaves continuing to be produced, while they are transiently halted in fruiting plants. The leaf conductance in this species over a whole range of experiments was extremely low, being similar to that of maize (Fig. 3); we do not know whether this is an intrinsic feature of this species or peculiar to these experiments.

The differences associated with age appear to be evident over the whole range of responses to external factors. This has been documented in respect of irradiance by Ludlow and Wilson (1971c), McPherson and Slatyer (1973), and others. Other age responses have been noted in the respective sections.

There are, as yet, no suggested explanations for this behavior and it is unlikely that any will be forthcoming until we have a much clearer concept of stomatal functioning in general and the way in which the guard-cell components change during development. In this context, we may accept the general concept established for cells of other tissues that there is a continually changing pattern of enzyme activity during the life of a cell and that guard cells are among the last to be formed and the last to die in the leaf, whereas other epidermal cells often die relatively early. There could also be a decrease in the supply of potassium ions in the epidermis during senescence, if these follow a similar pattern to the concentration in the mesophyll, and no doubt many other possible explanations.

2. Effect of Flowering and Fruiting

A very marked increase in stomatal conductance was also found in *Xanthium* as a response to the long dark periods which induce flowering (Fig. 9). This increase in responsiveness was maintained for only a short period; they then became much less responsive as the leaf senesced. It is likely that the surgical treatments used also caused wider opening than is found in intact plants (Meidner, 1970a), but there is little doubt that the reactions in the leaf that lead to the invocation of flowers also influence stomatal behavior. There is a reasonable amount of indirect evidence, such as increased transpiration rates around the time of flower initiation and early development, which supports this contention.

These results were obtained with debudded plants on which flowers and fruits were not allowed to develop. There is also evidence that fruiting leads to increased stomatal opening, the high conductances generally being maintained for a longer period than in leaves from vegetative plants (Fig. 10). The response shown by *Capsicum* is possibly more exaggerated than in many other species because of their generally low conductance and the complete transient inhibition of leaf growth during the period of fruit growth. If flowers are removed, the stomata do not open as widely as on a leaf at the same position on fruiting plants. Moreover, removal of a fruit from a plant at the time of maximum fruit growth leads to an appreciable decrease in the rate of photosynthesis (e.g., Burt, 1964; King *et al.,* 1967;

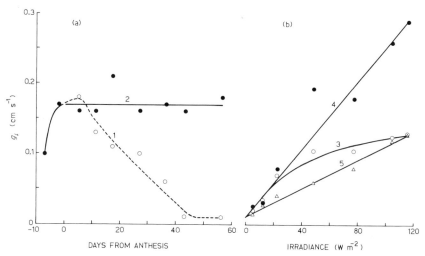

Fig. 10. Relationship of leaf conductance of the first leaf above the flower of *Capsicum* with (a) time and (b) flux density of visible radiation. Curve 1, deflorated plants; Curve 2, fruiting plants; Curve 3, plants at anthesis; Curve 4, fruiting and, Curve 5, deflorated plants, 42 days after anthesis. (After Hall, 1975).

Neales and Incoll, 1968). Hall (1975) has explored this effect rather fully in *Capsicum* and has shown that most of the decrease in rate of photosynthesis is accounted for by stomatal closure; changes in the intracellular component, reflecting intensity of the biochemical reactions, are very much less. Stomatal responses (and photosynthetic rates) could be manipulated by changing the light regimes to which defruited plants were exposed, which suggests that most of the controls of the stomatal reactions were at the level of each individual leaf, with this being related in some unknown way with the concentration of assimilate within the leaf. The actual concentrations did not change much: conditions leading to accumulation (high photosynthesis and low translocation) were reflected by greater stomatal closure and opposite conditions leading to greater opening.

We visualize both the above responses, i.e., (i) a "flowering" response leading to enhanced opening until about the stage of full flower differentiation and followed by increasing closure and (ii) a "sink-strength" effect related to the rates of photosynthesis and translocation—occurring in normal fruiting plants. Possibly they will often work in contrary directions and both may reflect changes in guard-cell metabolism itself or some control mechanism related to the activity of the mesophyll. Attention was drawn in Section III,B to the marked change in the relationship between stomatal conductance and leaf water potential associated with flowering and fruiting, especially of cereals. There are similar, but less marked, changes in response to radiation (Fig. 10b), with defruited plants being much less responsive than fruiting plants at all irradiances and less responsive than plants at anthesis at low irradiances.

The responses described here indicate that fairly extensive changes in stomatal behavior occur during the lives of the successive leaves on a plant. The amount of evidence that has been accumulated to date, however, is insufficient to allow prediction of changes except in a very cursory way; equally, there is no firm evidence of the metabolic components involved in these changes.

C. Organs Other Than Leaves

Stomata occur in most epidermal tissue, but there is very little information on conductances of organs other than leaves. Biscoe *et al.* (1973a) showed that the awns of barley had less responsive stomata to radiation and to water supply than did leaves; nevertheless, the conductance varied from the (high) values of 0.15–0.2 in darkness to about 0.2–0.25 with irradiances (visible radiation) of about 400 W m^{-2}. Conductances of this order are a significant component of the photosynthesis taking place in the ear, which, in turn, can make an important contribution to dry weight increase during grain filling. No attempts have been made to measure stomatal or cuticular conductance of glumes, stems, or leaf sheaths of cereals, although several measurements have been made of photosynthesis rate (*cf.* Evans, 1975).

Although several excellent studies have been made of photosynthesis by pods [notably those of Flinn and Pate (1970) and Kumura and Naniwa (1965)], the only attempt we know to partition its components is that of Sambo (1974). Sambo found that stomata on pods of soybean responded to light in much the same way as those on leaves, the conductance varying from about 0.2 in the dark up to 0.5 cm s^{-1} at 200 W m^{-2}. They also responded to ambient CO_2 concentrations, the maximum conductance under high irradiance being at about 150 ppm.

This seems to be the extent of published observations, and we can do little more than deplore the lack of attention paid to stems, fruits, and other organs, especially as these appear to be significant contributors to both photosynthesis and transpiration under certain circumstances. Detailed studies are urgently needed.

V. SIGNIFICANCE OF STOMATA IN CONTROLLING GAS EXCHANGE

A significant body of experience has been summarized in the preceding pages and this allows some assessment of the role of stomata in the two major processes of gas exchange—transpiration and photosynthesis. Reference to Fig. 2 reminds us of the reasonably complex network involved and it follows that the significance of any one component cannot be assessed without knowledge of the values of all others. To do so would simply perpetuate the nonsense found in most textbooks of plant physiology, ever since Brown and Escombe ignored the external resistance in their classical studies. Unfortunately, classical errors appear to be transmitted from generation to generation by unthinking compilers of textbooks, whereas attention drawn to the error is ignored. Thus, we find the sequence: error by Brown and Escombe in 1900, correction by Renner in 1910, rediscovery of the error by Sayre in 1926, its correction by Maskell in 1928, a whole range of texts in 1930–1950 with distorted descriptions, correction with commendable clarity and precision by Penman and Schofield (1951); yet, even in the recent and generally excellent text of Salisbury and Ross (1969), we still find the same misplaced emphasis and erroneous interpretation! Before proceeding to more detailed matters, we crave indulgence to belabor this aspect solely for the benefit of elementary students and especially writers of elementary texts. Let us take the example given by Salisbury and Ross (1969, p. 84): they found the evaporation rate over 15 hr at 25°C from an open pan of 4 cm^2 filled with water to be 2.4580 gm. Over the same time in the same environment, evaporation from a dish of similar area, but covered with a multiperforate septum (49 pores with a total pore area of 4.55 mm^2), was 0.2735 gm. They then calculated the water lost per "unit area for evaporation" [as $(2.4580 \times 4.55)/(400 \times 0.2735)$] and found the first "rate" to be about one-tenth the second. They conclude ". . . evaporation through the pores was *far more efficient*." This is, of course, a completely spurious comparison. The valid comparison is between the rate of water movement through the septum with the rate through a length above the water surface equal to the thickness of the septum. Above both systems, there is an external conductance representing diffusion of water vapor away from the surface to the ambient air; i.e., we write Eq. (2) for the open surface as $E_O = \rho g_a \delta q$, where ρ is the density of air, g_a the external conductance, and δq the difference in specific humidity. Assuming the relative humidity (not stated) to be 50%, we find g_a to be 0.98 cm s^{-1}. Describing the septum surface by $E_s = \rho G \, \delta q$, we find $G = 0.109 = g_a g_l / (g_a + g_l)$, giving a value of g_l for the septum of 0.12 cm s^{-1} or 10.8 cm^3 (cm^2 pore area)$^{-1}$ s^{-1}. A comparative value of g for the equivalent length of the septum in the open water system should be given by D/z, where the diffusion coefficient D is 0.257 cm^2 s^{-1} and the equivalent length z is 0.0762 cm. This gives an estimate of 3.37 cm^3 cm^{-2} s^{-1}. Taking account of the external conductance reduces the apparent "efficiency" from 10 times to 3 times, but it does not explain why the difference is still as large as it appears to be. If the air in the room were drier than half-saturated and if the systems were not completely isothermal, the difference would be reduced still further. Lack of information on these aspects allows us to proceed no further; we do not find it sufficiently convincing to deter us from relationships established on firm physical principles and confirmed by fully documented experiments. These show that the conductance of a multiperforate septum is given by $Dna/(l + 2x)$ cm s^{-1} where there are n pores of effective area a and effective length l per cm^2 and x is an end correction. As x is always positive, diffusion through the septum is always less than through an equivalent length of free air, i.e., evaporation through pores is "less efficient."

The estimation of effective areas and lengths, end corrections, and diffusion coefficients of stomata was discussed in Chapter 6 of Volume I (cf. also Parlange and Waggoner, 1970). We do not propose to reiterate that presentation and we have little to add concerning many uncertainties in estimating conductances from stomatal dimensions. However, we do emphasize the value of analyzing the system in terms of the various components because these vary independently of each other.

A. INDIVIDUAL LEAVES

1. Transpiration

As in Chapter 6 of Volume I (cf. Monteith, 1973) we find it most convenient to describe the transpiration rate per unit *projected area of leaf, E* kg m^{-2} s^{-1}, by a combination formula, which may be expressed as

$$E = \{\epsilon r_h H/\lambda + 2\rho(q' - q)\}/(\epsilon r_h + r_a + r_l) \tag{22}$$

where ϵ is the increase of latent heat content with increase of sensible heat content of saturated air at the temperature of the ambient air, $H(\text{J m}^{-2}\,\text{s}^{-1})$ the net rate of absorption of radiant heat by the leaf, $\lambda(\text{J kg}^{-1})$ the latent heat of vaporization of water, $\rho(\text{kg m}^{-3})$ the density of air, q' and q, respectively, the saturation and actual specific humidities of the ambient air, and r_h, r_a, and r_l (sec m^{-1}), resistances to diffusion of heat and of water vapor away from and within the leaf, respectively, expressed per unit area of *surface*. From the Pohlhausen analysis of heat and mass transfers from a plate in laminar parallel flow, we may take $r_h = 1.12r_a$ (Kays, 1966; Cowan and Troughton, 1971).

Equation (22) holds only when the leaf resistances of the two surfaces are more or less equal. Where they are widely different, the term

$(r_a + r_l)$ must be replaced by an appropriate term taking into account the differences. Before examining this, it is convenient to explore the effects of the various components of r_l.

a. Components of Leaf Conductance other than Stomatal. An attempt was made in Chapter 6 of Volume I to give estimates of the various components of g_l ($= 1/r_l$) [cf. Fig. 2, Eq. (10)] and these were reviewed recently by Stigter (1972). We are not aware of any further studies in this area and we must still assume the general values presented there: i.e., $r_w \doteqdot 0.2$, $r_i \doteqdot 0.2$ sec cm^{-1}, and r_c varying between 20 and 400 sec cm^{-1} depending on species, say, 50 and 200 sec cm^{-1} for mesophytes and xerophytes. There are no difficulties in estimating r_i and r_c and such should be attempted in any detailed study. The cell-wall resistance is much more difficult to determine and much uncertainty still exists of its value in turgid leaves over a range of species and of the extent to which it changes with leaf water potential. In the most recent reexamination of this issue, Jarvis and Slatyer (1970) concluded that "incipient drying" and capillary and osmotic effects on the surface water potential were extremely small, but they did detect wall influences which they attributed to hydraulic resistance to liquid flow; these, however, would be small under normal outdoor conditions.

In the most general sense, then, we should be able to write

$$\text{Mesophytes:} \quad g_l = 0.02 + 1/(r_s + 0.4)$$
$$\text{Xerophytes:} \quad g_l = 0.005 + 1/(r_s + 0.4) \tag{23}$$

where $\infty > r_s > 0.5 - 5$ sec cm^{-1}. These estimates are consistent with the observations presented previously: minimum values of g_l appear to be about 0.01–0.02 cm s^{-1} (i.e., r_l is 50-100 sec cm^{-1}) and maximum values vary from about 0.2 (r_l of 5) with some varieties of maize and *Capsicum* to greater than 1 ($r_l < 1$ sec cm^{-1}) in *Medicago sativa*.

b. The External or Boundary Layer Conductance. In Chapter 6 of Volume I, we pointed out that the boundary layer resistance is related to the diameter of broad leaves (or width of narrow leaves), b cm, and wind speed, u cm s^{-1}, by the approximate function $r_a = a(b/u)^{0.5}$ where a appears to be about 3.4 in laminar flow (possibly typical of quiet indoor conditions) and about 1.3 in outdoor and most growth-chamber conditions. In view of the imprecision in estimating r_a, there is little point in distinguishing between r_h and r_a (Eq. 22); i.e., r_h can be taken as equal to r_a in most situations. Most experience would be covered by values of b between 1 and 9 cm and of u between 0.4 and 4 m s^{-1}; hence, the range of g_a is usually within that shown in Table II.

TABLE II

EXPECTED VALUES OF EXTERNAL CONDUCTANCE
(cm s^{-1}) UNDER MOST CONDITIONS

Wind speed (cm s^{-1})	Width of leaf (cm)		
	0.4	4	9
40	7.7	2.4	0.6
100	12.2	3.8	2.6
400	24.3	7.7	5.1

c. *Significance of the Ratio of Conductances of the Two Surfaces.** Equation (22) may be rewritten as

$$E = \frac{\epsilon H/\lambda + 2\rho g_a(q' - q)}{\epsilon + 2g_a/G} \tag{24}$$

where $G = 1/(r_a + r_{l_1}) + 1/(r_a + r_{l_2})$, the subscripts 1 and 2 referring to the adaxial and abaxial surfaces, respectively (Section I,C). When $r_{l_1} = r_{l_2}$, Eq. (24) reduces to Eq. (22). However, to explore the effects when $r_{l_1} \neq r_{l_2}$, we may put $\alpha = g_{l_1}/g_{l_2}$ and $\eta = g_a/g_{l_2}$; then

$$2g_a/G = \frac{2}{\alpha/(\alpha + \eta) + 1/(1 + \eta)} \tag{25}$$

The ratio of stomatal densities between the two surfaces, n_1/n_2, of most land plants varies from zero (hypostomatous leaves) to about 20 (e.g., in *Lolium perenne*). With leaves in which $n_1/n_2 > 0$, α seems to be only crudely related to n_1/n_2. In equilibrium "open" conditions, α is usually close to unity in leaves in which $0.3 < n_1/n_2 < 3$, although it must be appreciated that these are extremely crude limits and that they have not been explored with any precision. The ratio n_1/n_2 also varies appreciably over the surface of any one leaf, between cultivars of a species, and with the environmental conditions under which the leaf develops. We usually assume that $\alpha = 1$ when stomata are in the completely closed condition, but this is another area that has not been explored to any extent. During stomatal opening in most species that have been examined, α decreases rapidly in the very early stages and then gradually increases to

---

* We are most grateful to Dr. M. Thorpe of Long Ashton Experiment Station for valuable discussions on this topic.

near unity in the equilibrium condition (Clough, 1971; Kanemasu and Tanner, 1969a,b; F. L. Milthorpe, unpublished observations). These authors also noted differing values of α in response to irradiation and leaf water potential. In one cultivar of tobacco grown in a growth chamber ($n_1/n_2 = 0.75$), Clough found $\alpha = 1$ ($g_l = 0.014$ cm s^{-1}) in darkness; during opening, it was 0.45 after 9 min, 0.79 after 42 min and reached 1 after 110 min (g_l then 0.63 cm s^{-1}). In a field-grown cultivar, Turner and Begg (1973) found α to range from 0.16 through 0.38 to 0.75, as the flux density of incoming visible radiation varied from 10 through 100 to 400 W m^{-2}. Calculations from Eqs. (24) and (25) show that the transpiration rate is linearly related to α when the conductance of the lower surface is reasonably low (up to about 0.2 cm s^{-1}), but has an increasing curvilinear component with higher conductances.

It is a common practice by many investigators to calculate the leaf resistance to diffusion from measurements of the overall resistance to transpiration (i.e., without separating the fluxes from the two surfaces) and then subtracting the bulk boundary layer resistance. This can lead to considerable errors where α is less than about 0.3 (Gale and Poljakoff-Mayber, 1968; Moreshet et al., 1968). It is much more exact to determine α for the particular leaf and environment and generally less confusing to consider the two surfaces separately. The errors are small if the ratio of the two surfaces is greater than 0.3, as with many crop plants, and therefore this may not be a very significant issue in general terms; indeed, we hope not as many of the results presented here have been interpreted without knowledge of α.

d. Significance of Stomatal Conductance in Controlling Transpiration. The type of experience that may be expected in growth chambers and glasshouses is illustrated in Fig. 11. We have calculated the expected transpiration rates, using Eq. (24) with rather "low" and "normal" net rates of absorption of radiation in humid and dry atmospheres, from hypostomatous and amphistomatous ($\alpha = 1$) leaves over a range of conductances of the stomata on the abaxial surface and at two values of the external conductance.

We see that there is a significant influence of stomata on transpiration over this range of usual experience. The exact relation, however, varies with the values of the other parameters and the type of curve found in a particular experiment will vary with the proportion of the possible range covered. Generally g_{s_2} will be related to H and we have also assumed that the leaf water potential is maintained at a high value. We might also assume that many cultivars of maize have values of g_{s_2} not exceeding 0.3, that tobacco extends to 0.6 and alfalfa to 1, all with $\alpha = 1$. With

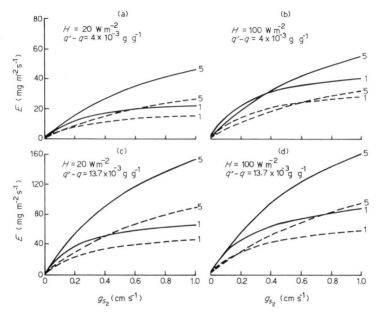

Fig. 11. Expected relationships between transpiration rate, E, and stomatal conductance of lower surface, g_{s_2}, of hypostomatous (broken lines) and amphistomatous leaves with $\alpha = 1$ (continuous lines). Values of net radiation, H, and specific humidity deficit, $q' - q$, are shown and the numerals 1 and 5 against each curve refer to the value of the external conductance (cm s⁻¹). The cuticular conductance is taken as 0.01 cm s⁻¹ and the ambient temperature as 25°C. Note that the ordinate scales in (a) and (b) are twice those in (c) and (d).

hypostomatous leaves, α decreases from 1 when $g_{s_2} = 0$ to about 0.014 with $g_{s_2} = 1$; in some deciduous trees, $g_{s_2} < 0.4$ and the relationship would then be close to linear especially with high irradiance. Most features of these curves will be readily interpreted, but we may draw attention to one: that is, that transpiration rates are sometimes less at a high, than at a low, external conductance when the stomata are in the more closed state [cf. (b), for example]. This arises, of course, from the partitioning of the available energy between latent and sensible heat; if evaporation is retarded, then a higher proportion of the absorbed radiation is lost as sensible heat.

2. Assimilation

The net rate of photosynthesis per unit area of lamina, P_N, may be written

$$P_N = [1/(r_a' + r_{l_1}') + 1/(r_a' + r_{l_2}')](C_a - C_s) = (C_s - \Gamma)/2r_i' \qquad (26)$$

where C_a and C_s are the concentrations of CO_2 in the ambient air and at the cell surface respectively, Γ is the CO_2 compensation point, and the primes indicate resistances to diffusion of CO_2 of comparable parts of the pathway as with water vapor. The new resistance, r_i', is an intracellular resistance which includes the diffusion of CO_2 into the cell and activities of the biochemical reactions in the leaf; it indicates the intrinsic photosynthetic efficiency of the leaf with stomatal and external influences on diffusion eliminated.

It is customary to assume that $r_{l_1}' = r_{l_2}'$, i.e., $\alpha = 1$; Eq. (26) then becomes

$$P_N = 2(C_a - C_s)/(r_a' + r_i') = (C_s - \Gamma)/2r_i' \qquad (27)$$

where r_a' is obtained from r_a using Eq. (5). Similarly, r_l' can be calculated from r_l using Eq. (6), although the pathways are not exactly the same: that for CO_2 via the cuticle is very much longer and hence the resistance is much higher, but the errors that arise from including r_c in the calculations are usually extremely small (Jones and Slatyer, 1972). The errors arising when $\alpha \neq 1$ have been considered by Koller (1970) and Jones and Slatyer (1972). These papers should be consulted for exact analytical procedures. The latter show that although there is a substantial vapor phase resistance to CO_2 uptake, via a surface with few stomata, not accounted for by the water vapor diffusion analog, errors are almost undetectable if $\alpha > 0.3$ and rarely exceed 10% in most circumstances. It should be remembered that Eqs. (26) and (27) only apply when CO_2 is limiting photosynthesis; this condition may not be satisfied with concentrations above about 1000 ppm or at low irradiance. Results based entirely on a leaf area basis should also be interpreted with caution; Charles-Edwards and Ludwig (1975) have drawn attention to this and have shown that leaf thickness—hence, size of the biochemical system—can influence the interpretation of the significance of the component conductances.

Despite these cautions, we will here use Eq. (27) to examine some issues at a rather general level. The relative importance of the stomata in influencing the rate of photosynthesis depends, of course, on the relative sizes of the three conductances. A range of empirical evidence is summarized in Fig. 12 taken from Ludlow and Wilson (1971c). In the tropical C_4 grasses and tropical C_3 legumes with which they worked, there was a very close relationship between photosynthesis rate and both g_l' and g_i' over a range in which both parameters were less than 0.2 cm s^{-1}. Most of the grasses had values of g_i' greater than this; hence, there was a close relation between P_N and g_l' over the range found (0.1–0.4 cm s^{-1}). On the other hand, the legumes showed a much more variable relationship with g_l', both because of the wider range covered (0.1–0.8) and the lower values of g_i'. Most C_4 species appear to have values of

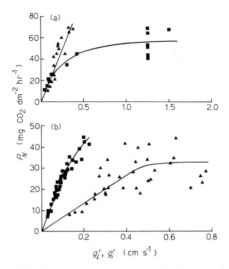

Fig. 12. Relationships between net photosynthesis rate, P_N' and stomatal (g_l', triangles) and intracellular (g_i', squares) conductances for CO_2 in leaves of (a) tropical grasses and (b) tropical legumes. Note differences in scales. (Redrawn from Ludlow and Wilson, 1971c.)

g_i' around 0.4–0.8 cm s^{-1} (cf. also Gifford and Musgrave, 1970) and values of g_l' between 0.12 and 0.4 cm s^{-1} (Section II,A); hence, photosynthesis rates will always be closely related to stomatal conductance. The same holds for *Acacia harpophylla* (van den Driessche *et al.*, 1971) and *Capsicum* in which $0.01 < g_l' < 0.12$ and g_i' was about 0.06 cm s^{-1} (Hall, 1974). In the tropical legumes explored by Ludlow and Wilson (1971c), g_i' was always less than about 0.2 cm s^{-1} and, hence, it played a dominant role. In alfalfa, in which g_l' often approaches 1 cm s^{-1}, g_i' varies from 0.03 to 0.2 cm s^{-1} and, hence, the relationship between P_N and stomatal conductance is very poor (Hodgkinson, 1974).

3. Relative Role of Stomata in Transpiration and Assimilation

Cowan and Troughton (1971) compared the ways in which transpiration and assimilation depend on stomatal (i.e., leaf) conductance. They calculated $e = E/E_0$ (Eq. 22) and $a = P_N/P_{NO}$ (Eq. 27) where E_0 and P_{NO} were the rates of transpiration and assimilation which would be expected if g_l and g_l' were infinite. They showed that

$$e = a/[\beta + (1 - \beta)a] \qquad (28)$$

where $\beta = (g_l'/g_i' + 0.5)/(0.66\epsilon + 0.58)$. That is, the relationship between the relative rates of transpiration and assimilation depends on the ratio of the leaf conductance to the intracellular conductance and on the temperature—which determines ϵ (cf. Eg. 22). The implications of this relationship are too complex to discuss in detail; however, it may be pointed out that with leaves in which $g_i' > 0.25$ cm s^{-1}, the rate of transpiration will increase relatively more than that of photosynthesis as the stomata open, whereas at values of g_i' less than this the converse is more likely. Cowan and Troughton also point out that antitranspirants may not reduce transpiration relative to assimilation as much as is sometimes supposed; the extent depends, of course, on the values of the relative components.

B. CANOPIES

The application of the information summarized in the preceding pages to the real world of crops and stands of natural vegetation requires detailed analyses of each particular system. This area of activity is now reasonably fully understood (cf. Evans, 1975; Milthorpe and Moorby, 1974; Monteith, 1973; Šetlík, 1970; Vols. I, II, and III of this treatise). One shortcoming has been lack of understanding of the quantitative changes in stomatal conductance and other internal parameters such as the intracellular conductance. This chapter has been an attempt to bring together the information on stomatal conductance; clearly, however, the understanding is not as extensive nor the behavior as predictable as is required. We are still somewhat in the same state which H. L. Penman expressed some years ago as needing "facts to fit formulae." We also need some formulae in order to predict the quantitative changes in stomatal conductance with acceptable precision. Nevertheless, this is an area receiving attention; measurements are being made and there is a widespread general appreciation of the significance of stomata in exchange processes.

Coupled with this search for "stubborn irreducible facts," we have some development in our understanding of stomatal functioning (Chapter 2) and laudable attempts are being made to integrate the information in simulation models. A number of attempts to formulate satisfactory models were listed in Chapter 2 and we do not propose to repeat that listing. Those and a number given in Šetlík (1970) refer specifically to exchange processes and attempt to account for stomatal behavior. We hope the information given here will be of assistance in improving the precision of these attempts to predict.

We also strongly support the attempts to simulate crop growth in the wider context, while taking into account all relevant environmental and growth parameters (cf. Evans, 1975). In these, prediction of stomatal behavior necessarily plays a very significant role. With a strong background in elucidation, more attention can now be paid to integration; the much closer interlocking of both activities should lead to much better understanding and advice. With this optimistic note, we should add a general corollary concerning the general attenuation of potential advantages in plants—and indeed the influence of any separated component—as we move up the various levels of organization.

This has been beautifully illustrated by Gifford (1974) in comparing the various parameters of C_3 and C_4 species when moving from the level of primary carboxylation enzymes (advantage of $C_4:C_3$ about $\times$ 50) to grain yield (advantage effectively zero). Gifford also brings together valuable information on the more particular topic that has concerned us and cautions against uncritical extrapolation from the laboratory to the field situation. We have kept this in mind, but regret that we have not illustrated the multitudinous field situations as fully as desirable.

REFERENCES

Altman, P. L., and Dittmer, D. S., eds. (1971). "Respiration and Circulation." Fed. Amer. Soc. Exp. Biol., Bethesda, Maryland.

Alvim, P. de T. (1965). A new type of porometer for measuring stomatal opening and its use in irrigation studies. *Arid Zone Res.* **25**, 325–329.

Aston, M. J. (1973). Changes in internal water status and the gas exchange of leaves in response to ambient evaporative demand. *In* "Plant Responses to Climatic Factors" (R. O. Slatyer, ed.), pp. 243–247. UNESCO, Paris.

Barrs, H. D. (1973). Controlled environment studies of the effects of variable atmospheric water stress on photosynthesis, transpiration and water status of *Zea mays* L. and other species. *In* "Plant Responses to Climatic Factors (R. O. Slatyer, ed.), pp. 249–258. UNESCO, Paris.

Beardsell, M. F., Jarvis, P. G., and Davidson, B. (1972). A null-balance diffusion porometer suitable for use with leaves of many shapes. *J. Appl. Ecol.* **9**, 677–690.

Bierhuizen, J. F., and Slatyer, R. O. (1964). Photosynthesis of cotton leaves under a range of environmental conditions in relation to internal and external diffusive resistances. *Aust. J. Biol. Sci.* **17**, 348–359.

Bierhuizen, J. F., Slatyer, R. O., and Rose, C. W. (1965). A porometer for laboratory and field operation. *J. Exp. Bot.* **16**, 182–191.

Biscoe, P. V. (1972). The diffusion resistance and water status of leaves of *Beta vulgaris. J. Exp. Bot.* **23**, 930–940.

Biscoe, P. V., Littleton, E. J., and Scott, R. K. (1973a). Stomatal control of gas exchange in barley awns. *Ann. Appl. Biol.* **75**, 285–297.

Biscoe, P. V., Unsworth, M. H., and Pinckney, H. R. (1973b). The effects of low concentrations of sulphur dioxide on stomatal behaviour in *Vicia faba. New Phytol.* **72**, 1299–1306.

Brown, K. W., and Rosenberg, N. J. (1970). Effects of windbreaks and soil water potential on stomatal diffusion resistance and photosynthetic rate of sugar beets (*Beta vulgaris*). *Agron. J.* **62**, 4–8.

Brunner, U., and Eller, B. M. (1974). Öffnen der Stomata bei hoher Temperatur im Dunkeln. *Planta* **121**, 293–302.

Burrows, F. J. (1969). The diffusive conductivity of sugar beet and potato leaves. *Agr. Meteorol.* **6**, 211–226.

Burt, R. L. (1964). Carbohydrate utilization as a factor in plant growth. *Aust. J. Biol. Sci.* **17**, 867–877.

Byrne, G. F., Rose, C. W., and Slatyer, R. O. (1970). An aspirated diffusion porometer. *Agr. Meteorol.* **7**, 39–44.

Camacho-B, S. E., Hall, A. E., and Kaufmann, M. R. (1974). Efficiency and regulation of water transport in some woody and herbaceous species. *Plant Physiol.* **54**, 169–172.

Carman, P. C. (1956). "Flow of Gases through Porous Media." Butterworth, London.

Charles-Edwards, D. A., and Ludwig, L. J. (1975). The basis of expression of leaf photosynthetic activities. *In* "Environmental and Biological Control of Photosynthesis" (R. Marcelle, ed.). Junk, The Hague (in press).

Clough, B. F. (1971). "The Effects of Water Stress on Photosynthesis and Translocation in Tobacco." Ph.D. Thesis, University of Sydney.

Cowan, I. R. (1968). Mass, heat and momentum exchange between plant stands and their atmospheric environment. *Quart. J. Roy. Meteorol. Soc.* **94**, 523–544.

Cowan, I. R., and Troughton, J. H. (1971). The relative role of stomata in transpiration and assimilation. *Planta* **97**, 325–336.

Darwin, F., and Pertz, D. F. M. (1911). On a new method of estimating the aperture of stomata. *Proc. Roy. Soc., Ser. B* **84**, 136–154.

Davies, W. J., Kozlowski, T. T., and Pereira, J. (1974). Effect of wind on transpiration and stomatal aperture of woody plants. *Roy. Soc. N. Z., Bull.* **12**, 433–438.

DeMichele, D. W., and Sharpe, P. J. H. (1973). An analysis of the mechanics of guard cell motion. *J. Theor. Biol.* **41**, 77–96.

Denmead, O. T. (1970). Transfer processes between vegetation and air: Measurement, interpretation and modelling. *In* "Prediction and Measurement of Photosynthetic Productivity" (I. Šetlík, ed.), pp. 149–164. Pudoc, Wageningen.

Denmead, O. T., and McIlroy, I. C. (1971). Measurement of carbon dioxide exchange in the field. *In* "Plant Photosynthetic Production: Manual of Methods" (Z. Šesták, J. Čatský, and P. G. Jarvis, eds.), pp. 467–516. Junk, The Hague.

Drake, B. G., Raschke, K., and Salisbury, F. B. (1970). Temperature and transpiration resistances of *Xanthium* leaves as affected by air temperature, humidity, and wind speed. *Plant Physiol.* **46**, 324–330.

Duggar, W. M., and Ting, I. P. (1970). Air pollution oxidants—their effects on metabolic processes in plants. *Annu. Rev. Plant Physiol.* **21**, 215–224.

Eckardt, F. E. (1968). Techniques de mesure de la photosynthèse sur le terrain basées sur l'emploi d'encientes climatisées. *In* "Functioning of Terrestrial Ecosystems at the Primary Production Level" (F. E. Eckardt, ed.), pp. 289–319. UNESCO, Paris.

Evans, L. T., ed. (1975). "Crop Physiology—Some Case Histories." Cambridge Univ. Press, London and New York.

Flinn, A. M., and Pate, J. S. (1970). A quantitative study of carbon transfer from pod and subtending leaf to the ripening seeds of the field pea (*Pisum arvense* L.). *J. Exp. Bot.* **21**, 71–82.

Gale, J., and Poljakoff-Mayber, A. (1968). Resistances to the diffusion of gas and vapour in leaves. *Physiol. Plant.* **21**, 1170–1176.

Gifford, R. M. (1974). A comparison of potential photosynthesis, productivity and yield of plant species with differing photosynthetic metabolism. *Aust. J. Plant Physiol.* **1,** 107–117.

Gifford, R. M., and Musgrave, R. B. (1970). Diffusion and quasi-diffusion resistances in relation to the carboxylation kinetics of maize leaves. *Physiol. Plant.* **23,** 1048–1056.

Gifford, R. M., and Musgrave, R. B. (1972). Activation energy analysis and limiting factors in photosynthesis. *Aust. J. Biol. Sci.* **25,** 419–423.

Gifford, R. M., and Musgrave, R. B. (1973). Stomatal role in the variability of net CO_2 exchange rates by two maize inbreds. *Aust. J. Biol. Sci.* **26,** 35–44.

Goltz, S. M., and Tanner, C. B. (1972). Seed onion temperatures and their effects on stomata. *HortScience* **7,** 180–181.

Gregory, F. G., and Pearse, H. L. (1934). The resistance porometer and its application to the study of stomatal movement. *Proc. Roy. Soc., Ser. B* **114,** 477–493.

Hack, H. R. B. (1974). The selection of an infiltration technique for estimating the degree of stomatal opening in leaves of field crops in the Sudan and a discussion of the mechanism which controls the entry of test liquids. *Ann. Bot. (London)* [N.S.] **38,** 93–114.

Hall, A. J. (1975). "The Influence of Fruit Excision and Defloration on Photosynthesis in *Capsicum annuum* L." Ph.D. Thesis, Macquarie University, North Ryde, Australia.

Heath, O. V. S. (1941). Experimental studies of the relation between carbon assimilation and stomatal movement. II. The use of the resistance porometer in estimating stomatal aperture and diffusive resistance. *Ann. Bot. (London)* [N.S.] **5,** 455–500.

Heath, O. V. S. (1947). Role of starch in light-induced stomatal movement, and a new reagent for staining stomatal starch. *Nature (London)* **159,** 647–648.

Heath, O. V. S., and Orchard, B. (1956). Studies in stomatal behaviour. VII. Effects of anaerobic conditions upon stomatal movement—a test of William's hypothesis of stomatal mechanism. *J. Exp. Bot.* **7,** 313–325.

Heath, O. V. S., and Russell, J. (1951). The Wheatstone Bridge porometer. *J. Exp. Bot.* **2,** 111–116.

Hiron, R. W. P., and Wright, S. T. C. (1973). The role of endogenous abscisic acid in the response of plants to stress. *J. Exp. Bot.* **24,** 769–781.

Hodgkinson, K. C. (1974). Influence of partial defoliation on photosynthesis, photorespiration and transpiration by lucerne leaves of different ages. *Aust. J. Plant Physiol.* **1,** 561–578.

Holmgren, P., Jarvis, P. G., and Jarvis, S. J. (1965). Resistances to carbon dioxide and water vapour transfer in leaves of different plant species. *Physiol. Plant.* **18,** 557–573.

Hunt, L. A., Impens, I. I., and Lemon, E. R. (1968). Estimates of the diffusion resistance of some large sunflower leaves in the field. *Plant. Physiol.* **43,** 522–526.

Jarman, P. D. (1974). The diffusion of carbon dioxide and water vapour through stomata. *J. Exp. Bot.* **25,** 927–936.

Jarvis, P. G. (1971). The estimation of resistances to carbon dioxide transfer. In "Plant Photosynthetic Production: Manual of Methods" (Z. Šesták, J. Čatský, and P. G. Jarvis, eds.), pp. 566–631. Junk, The Hague.

Jarvis, P. G., and Slatyer, R. O. (1970). The role of the mesophyll cell wall in leaf transpiration. *Planta* **90,** 303–322.

Jarvis, P. G., Rose, C. W., and Begg, J. E. (1967). An experimental and theoretical comparison of viscous and diffusive resistances to gas flow through amphistomatous leaves. *Agr. Meteorol.* **4,** 103–117.

Jones, H. G., and Slatyer, R. O. (1972). Effects of intercellular resistances on estimates of the intracellular resistance to CO_2 uptake by plant leaves. *Aust. J. Biol. Sci.* **25,** 443–453.

Jordan, W. R., and Ritchie, J. T. (1971). Influence of soil water stress on evaporation, root absorption, and internal water status of cotton. *Plant Physiol.* **48,** 783–788.

Kanemasu, E. T., and Tanner, C. B. (1969a). Stomatal diffusion resistance of snap beans. I. Influence of leaf water potential. *Plant Physiol.* **44,** 1547–1552.

Kanemasu, E. T., and Tanner, C. B. (1969b). Stomatal diffusion resistance of snap beans. II. Effect of light. *Plant Physiol.* **44,** 1542–1546.

Kanemasu, E. T., Thurtell, G. W., and Tanner, C. B. (1969). Design, calibration and field use of a stomatal diffusion porometer. *Plant Physiol.* **44,** 881–885.

Kays, W. M. (1966). "Convective Heat and Mass Transfer." McGraw-Hill, New York.

King, R. W., Wardlaw, I. F., and Evans, L. T. (1967). Effect of assimilate utilization on photosynthetic rate in wheat. *Planta* **77,** 261–276.

Koller, D. (1970). The partitioning of resistances to photosynthetic CO_2 uptake in the leaf. *New Phytol.* **69,** 971–981.

Kriedemann, P. E. (1971). Photosynthesis and transpiration as a function of gaseous diffusive resistances in orange leaves. *Physiol. Plant.* **24,** 218–225.

Kriedemann, P. E., and Loveys, B. R. (1974). Hormonal mediation of plant responses to environmental stress. *Roy. Soc. N. Z., Bull.* **12,** 461–465.

Krizek, D. T., and Milthorpe, F. L. (1973). Effect of photoperiodic induction on the transpiration rate and stomatal behaviour of debudded *Xanthium* plants. *J. Exp. Bot.* **24,** 76–86.

Kuiper, P. J. C. (1961). The effects of environmental factors on the transpiration of leaves, with special reference to stomatal light response. *Meded. Landbouwhogesch. Wageningen* **61,** 1–49.

Kumura, A., and Naniwa, I. (1965). Studies on dry matter production of soybean plant. I. Ontogenetic changes in photosynthetic and respiratory capacity of soybean plant and its parts. *Proc. Crop Sci. Soc. Jap.* **33,** 467–472.

Loveys, B. R., and Kriedemann, P. E. (1974). Internal control of stomatal physiology and photosynthesis. I. Stomatal regulation and associated changes in endogenous levels of abscisic and phaseic acids. *Aust. J. Plant Physiol.* **1,** 407–415.

Ludlow, M. M. (1970). Effect of oxygen concentration on leaf photosynthesis and resistances to carbon dioxide diffusion. *Planta* **91,** 285–290.

Ludlow, M. M., and Jarvis, P. G. (1971). Photosynthesis in Sitka spruce (*Picea sitchensis* (Bong.) Carr.) I. General characteristics. *J. Appl. Ecol.* **8,** 925–953.

Ludlow, M. M., and Wilson, G. L. (1971a). Photosynthesis of tropical pasture plants. I. Illuminance, carbon dioxide concentration, leaf temperature and leaf-air vapour pressure difference. *Aust. J. Biol. Sci.* **24,** 449–470.

Ludlow, M. M., and Wilson, G. L. (1971b). Photosynthesis of tropical pasture plants. II. Temperature and illuminance history. *Aust. J. Biol. Sci.* **24,** 1065–1075.

Ludlow, M. M., and Wilson, G. L. (1971c). Photosynthesis of tropical pasture plants. III. Leaf age. *Aust. J. Biol. Sci.* **24,** 1077–1087.

McCree, K. J. (1974). Changes in the stomatal response characteristics of grain sorghum produced by water stress during growth. *Crop Sci.* **14,** 273–278.

McPherson, H. G., and Slatyer, R. O. (1973). Mechanisms regulating photosynthesis in *Pennisetum typhoides*. *Aust. J. Biol. Sci.* **26,** 329–339.

Meidner, H. (1967). The effect of kinetin on stomatal opening and the rate of intake of carbon dioxide in mature primary leaves of barley. *J. Exp. Bot.* **18**, 556–561.

Meidner, H. (1970a). Effects of photoperiodic induction and debudding in *Xanthium pennsylvanicum* and of partial defoliation in *Phaseolus vulgaris* on rates of net photosynthesis and stomatal conductances. *J. Exp. Bot.* **21**, 164–169.

Meidner, H. (1970b). A critical study of sensor element diffusion porometers. *J Exp. Bot.* **21**, 1060–1066.

Meidner, H., and Mansfield, T. A. (1968). "Physiology of Stomata." McGraw-Hill, New York.

Milborrow, B. V. (1974). The chemistry and physiology of abscisic acid. *Annu. Rev. Plant Physiol.* **25**, 259–307.

Milthorpe, F. L. (1955). The significance of the measurement made by the cobalt chloride paper method. *J. Exp. Bot.* **6**, 17–19.

Milthorpe, F. L., and Moorby, J. (1974). "An Introduction to Crop Physiology." Cambridge Univ. Press, London and New York.

Milthorpe, F. L., and Penman, H. L. (1967). The diffusive conductivity of the stomata of wheat leaves. *J. Exp. Bot.* **18**, 422–457.

Monteith, J. L. (1963). Gas exchange in plant communities. *In* "Environmental Control of Plant Growth" (L. T. Evans, ed.), pp. 95–112. Academic Press, New York.

Monteith, J. L. (1973). "Principles of Environmental Physics." Arnold, London.

Monteith, J. L., and Bull, T. A. (1970). A diffusion porometer for field use. II. Theory, calibration and performance. *J. Appl. Ecol.* **7**, 623–638.

Monteith, J. L., Biscoe, P. V., Cohen, Y., and Heine, R. (1974). "Porometry." School of Agriculture Report 1973–1974. University of Nottingham, Sutton Bonington.

Moreshet, S., Koller, D., and Stanhill, G. (1968). The partitioning of resistances to gaseous diffusion in the leaf epidermis and the boundary layer. *Ann. Bot. (London)* [N.S.] **32**, 695–701.

Morrow, P. A., and Slatyer, R. O. (1971a). Leaf temperature effects on measurements of diffusive resistance to water vapour transfer. *Plant Physiol.* **47**, 559–561.

Morrow, P. A., and Slatyer, R. O. (1971b). Leaf resistance measurements with diffusion porometers: Precautions in calibration and use. *Agr. Meteorol.* **8**, 223–233.

Moss, D. N. (1963). The effect of environment on gas exchange of leaves. *Conn. Agr. Exp. Sta., New Haven, Bull.* **664.**

Neales, T. F., and Incoll, L. D. (1968). The control of leaf photosynthesis rate by the level of assimilate concentration in the leaf: A review of the hypothesis. *Bot. Rev.* **34**, 107–125.

Osman, A. M., and Milthorpe, F. L. (1971). Photosynthesis of wheat leaves in relation to age, illuminance and nutrient supply. *Photosynthetica* **5**, 55–60 and 61–70.

Parkinson, K. J. (1968). Apparatus for the simultaneous measurement of water vapour and carbon dioxide exchanges of single leaves. *J. Exp. Bot.* **19**, 840–856.

Parkinson, K. J., and Legg, B. J. (1972). A continuous flow porometer. *J. Appl. Ecol.* **9**, 669–675.

Parkinson, K. J., and Penman, H. L. (1970). A possible source of error in the estimation of stomatal resistance. *J. Exp. Bot.* **21**, 405–409.

Parlange, J.-Y., and Waggoner, P. E. (1970). Stomatal dimensions and resistance to diffusion. *Plant Physiol.* **46**, 337–342.

Parlange, J.-Y., and Waggoner, P. E. (1973). Stomatal penetration by liquids. *Plant Physiol.* **51**, 596–597.

Penman, H. L. (1942). Theory of porometers used in the study of stomatal movements in leaves. *Proc. Roy. Soc., Ser. B* **130**, 416–434.

Penman, H. L., and Schofield, R. K. (1951). Some physical aspects of assimilation and transpiration. *Symp. Soc. Exp. Biol.* **5**, 115–129.

Pohlhausen, E. (1921). Der Wärmeaustauch zwischen festen Körpern und Flüssigkeiten mit kleiner Reibung und kleiner Wärmeleitung. *Z. Angew. Math. Mech.* **1**, 115–121.

Raschke, K. (1970). Stomatal responses to pressure changes and interruptions in the water supply of detached leaves of *Zea mays* L. *Plant Physiol.* **45**, 415–423.

Raschke, K. (1975). Stomatal action. *Annu. Rev. Plant Physiol.* **26**, 309–340.

Ryle, G. J. A., and Hesketh, J. D. (1969). Carbon dioxide uptake in nitrogen-deficient plants. *Crop Sci.* **9**, 451–454.

Salisbury, F. B., and Ross, C. (1969). "Plant Physiology." Wadsworth, Belmont.

Sambo, E. Y. (1974). "Leaf and Pod Photosynthesis of Soya Beans." M.Sc. Thesis, Macquarie University, North Ryde, Australia.

Schönherr, J., and Bukovac, M. J. (1972). Penetration of stomata by liquids: Dependance on surface tension, wettability, and stomatal morphology. *Plant Physiol.* **49**, 813–819.

Schulze, E.-D., Lange, O. L., Buschbom, U., Kappen, L., and Evenari, M. (1972). Stomatal responses to changes in humidity in plants growing in the desert. *Planta* **108**, 259–270.

Schulze, E.-D., Lange, O. L., Kappen, L., Buschbom, U., and Evenari, M. (1973). Stomatal responses to changes in temperature at increasing water stress. *Planta* **110**, 29–42.

Šetlík, I., ed. (1970). "Prediction and Measurement of Photosynthetic Productivity." Pudoc, Wageningen.

Shimshi, D. (1970a). The effect of nitrogen supply on transpiration and stomatal behaviour of beans (*Phaseolus vulgaris* L.). *New Phytol.* **69**, 405–412.

Shimshi, D. (1970b). The effect of nitrogen supply on some indices of plant-water relations of beans (*Phaseolus vulgaris* L.). *New Phytol.* **69**, 413–424.

Slatyer, R. O. (1969). Carbon dioxide and water vapour exchange in *Atriplex* leaves. *In* "The Biology of Atriplex" (R. Jones, ed.), pp. 23–29. CSIRO, Canberra.

Slatyer, R. O., and Shmueli, E. (1967). Measurement of internal water status and transpiration. *In* "Irrigation of Agricultural Lands" (R. M. Hagan, H. R. Haise, and T. E. Edminster, eds.), Agron. Ser. No 11, pp. 337–353. Amer. Soc. Agron. Madison, Wisconsin.

Slavík, B. (1973). Transpiration resistance in leaves of maize grown in humid and dry air. *In* "Plant Responses to Climatic Factors" (R. O. Slatyer, ed.), pp. 267–269. UNESCO, Paris.

Stålfelt, M. G. (1962). The effect of temperature on opening of the stomatal cells. *Physiol. Plant.* **15**, 772–779.

Stigter, C. J. (1972). Leaf diffusion resistance to water vapour and its direct measurement. I. Introduction and review concerning relevant factors and methods. *Meded. Landbouwhogesch. Wageningen* **72-3**, 1–47.

Stigter, C. J., and Lammers, B. (1974). Leaf diffusion resistance to water vapour and its direct measurement. III. Results regarding the improved diffusion porometer in growth rooms and fields of Indian corn (*Zea mays*). *Meded. Landbouwhogesch. Wageningen* **74-21**, 1–76.

Stigter, C. J., Birnie, J., and Lammers, B. (1973). Leaf diffusion resistance to water vapour and its direct measurement. II. Design, calibration and pertinent theory of an improved leaf diffusion resistance meter. *Meded. Landbouwhogesch. Wageningen* **73-15**, 1–55.

Stiles, W. (1970). A diffusion resistance porometer for field use. I. Construction. *J. Appl. Ecol.* **7**, 617–622.

Thom, A. S. (1968). The exchange of momentum, mass and heat between an artificial leaf and the air-flow in a wind-tunnel. *Quart. J. Roy. Meteorol. Soc.* **94**, 44–55.

Troughton, J. H. (1969). Plant water status and carbon dioxide exchange of cotton leaves. *Aust. J. Biol. Sci.* **22**, 289–302.

Turner, N. C. (1970). Response of adaxial and abaxial stomata to light. *New Phytol.* **69**, 647–653.

Turner, N. C. (1973). Illumination and stomatal resistance to transpiration in three field crops. *In* "Plant Responses to Climatic Factors" (R. O. Slatyer, ed.), pp. 63–68. UNESCO, Paris.

Turner, N. C. (1974a). Stomatal responses to light and water under field conditions. *Roy. Soc. N.Z., Bull.* **12**, 423–432.

Turner, N. C. (1974b). Stomatal behavior and water status of maize, sorghum, and tobacco under field conditions. II. At low soil water potential. *Plant Physiol.* **53**, 360–365.

Turner, N. C., and Begg, J. E. (1973). Stomatal behavior and water status of maize, sorghum, and tobacco under field conditions. I. At high soil water potential. *Plant Physiol.* **51**, 31–36.

Turner, N. C., and Parlange, J.-Y. (1970). Analysis of operation and calibration of a ventilated diffusion porometer. *Plant Physiol.* **46**, 175–177.

van Bavel, C. H. M. (1964). Measuring transpiration resistance of leaves. *U. S., Dep. Agr., Water Conserv. Lab., Rep.* No. 2.

van den Driessche, R., Connor, D. J., and Tunstall, B. R. (1971). Photosynthetic response of brigalow to irradiance, temperature and water potential. *Photosynthetica* **5**, 210–217.

Waggoner, P. E. (1965). Calibration of a porometer in terms of diffusive resistance. *Agr. Meteorol.* **2**, 317–329.

Wallihan, E. F. (1964). Modification and use of an electric hygrometer for estimating relative stomatal apertures. *Plant Physiol.* **39**, 86–90.

Whiteman, P. C., and Koller, D. (1964). Environmental control of photosynthesis and transpiration in *Pinus halepensis*. *Isr. J. Bot.* **13**, 166–176.

Whiteman, P. C., and Koller, D. (1967a). Species characteristics in whole plant resistances to water vapour and carbon dioxide diffusion. *J. Appl. Ecol.* **4**, 363–377.

Whiteman, P. C., and Koller, D. (1967b). Interactions of carbon dioxide concentration, light intensity and temperature on plant resistances to water vapour and carbon dioxide diffusion. *New Phytol.* **66**, 463–473.

Wilson, D. (1975). Stomatal diffusion resistances and leaf growth during droughting of *Lolium perenne* plants selected for contrasting epidermal ridging. *Ann. Appl. Biol.* **79**, 83–94.

Woledge, J. (1972). The effect of shading on the photosynthetic rate and longevity of grass leaves. *Ann. Bot. (London)* [N.S.] **36**, 551–561.

Wuenscher, J. E., and Kozlowski, T. T. (1971). The response of transpiration resistance to leaf temperature as a desiccation resistance mechanism in tree seedlings. *Physiol. Plant.* **24**, 254–259.

Zelitch, I. (1965). Environmental and biochemical control of stomatal movement in leaves. *Biol. Rev. Cambridge Phil. Soc.* **40**, 463–482.

CHAPTER 4

WATER DEFICITS AND PHOTOSYNTHESIS

J. S. Boyer

DEPARTMENTS OF BOTANY AND AGRONOMY
UNIVERSITY OF ILLINOIS, URBANA, ILLINOIS

A slight diminution of turgidity sufficient to close the stomata will render the absorption of carbon dioxide extremely difficult, and hence may markedly diminish the assimilation activity. . . .

W. Pfeffer (1900)

Since the earliest days of agriculture, man has recognized the inhibitory and sometimes devastating effects of drought on crops. Almost all aspects of plant growth are affected [see Hsiao (1973) for a recent review], among them the dry matter production that accounts for much of the production by agriculture. As the significance of photosynthesis began to emerge in the nineteenth century and it was found that the gases absorbed by this process carry most of the dry weight that comprises the plant, interest grew in studies of the factors that affect photosynthesis. Thus, Pfeffer (1900) knew that photosynthesis is reduced in wilted leaves and attributed the inhibition to stomatal closure. Soon thereafter studies began to appear that described the effects of water deficits on the photosyntheis process

153

in some detail. In the last 15 years, this effort has intensified and it now seems that some of the factors have been identified that limit photosynthesis during desiccation.

This chapter will attempt to summarize our knowledge of how desiccation inhibits photosynthesis. The discussion will be restricted to the effects of water deficits on photosynthesis in higher plants growing in soil. Allied subjects such as the effects of drought on lower plants or the effects of solutes, particularly in the form of salinity, will not be discussed.

In surveying the progress made in photosynthetic research on plants subjected to water deficits, several principles seem to have been established since the early 1900's. The first is that photosynthesis by crops is severely inhibited and may cease altogether as water deficits become severe. The second is that part of the photosynthetic limitation may be caused by reduction in leaf growth or by senescence of leaves, and part may be caused by inhibition of the photosynthetic activity of existing leaves. Since high turgidity is required for cell expansion, the effects on leaf area result initially from reduction in turgor. Inhibition of photosynthetic activity, on the other hand, results from stomatal closure or more directly from changes in chloroplast activity. The third principle that emerges from this survey is that drought-induced reductions in photosynthesis not only decrease the total dry matter accumulated by plants, but they also appear to be a major limitation for grain production. In other words, drought effects on photosynthesis may account for many drought effects on agricultural production.

In what follows, evidence for these assertions will be presented. As the discussion proceeds, however, it will become increasingly obvious that some of the primary events controlling the response of photosynthetic metabolism to water deficiency remain unknown, and our guesses today are little better then they were when Pfeffer (1900) made his comments at the turn of the century.

I. INHIBITION OF PHOTOSYNTHETIC ACTIVITY

The first observation of the inhibitory effects of water deficits on photosynthesis appears to have been made by Kreusler (1885). A few years later, Thoday (1910), in experiments designed for other purposes, noticed similar plant responses. Then, Iljin (1923), Brilliant (1924), Dastur (1924, 1925), and Wood (1929) each showed that a reduction in photosynthetic activity occurred upon water loss from leaves, and Dastur (1925) appears to have been the first to have expressed photosynthetic activity as a function of the measured water status of the tissue. His careful determinations provided evidence that photosynthesis responded linearly to leaf water content in several species. The relationship between transpiration and photosynthesis was subsequently noted (Heinicke and Childers, 1935), but the possible causal relationship between them was not discussed. It was in 1941 that work from the same group (Schneider and Childers, 1941) included preliminary observations of the stomata with measurements of net photosynthesis, transpiration, and dark respiration. They cautiously concluded that stomatal closure might be involved in the photosynthetic response to water deficiency. A similar comparison of transpiration and photosynthesis was soon made by Loustalot (1945), but once again the involvement of the stomata was only cautiously suggested.

Undoubtedly one of the reasons why stomata were not implicated by these early authors is that in all three comparative studies (Heinicke and Childers, 1935; Schneider and Childers, 1941; Loustalot, 1945), transpiration often was inhibited by desiccation without effects on photosynthesis. Here, the matter rested for several years while measurements of photosynthesis were made as a function of soil moisture content (Allmendinger et al., 1943; Simonis, 1947, 1952; Kozlowski, 1949; Bormann, 1953; Bourdeau, 1954; Upchurch et al., 1955; Ashton, 1956), or of leaf wilting (Verduin and Loomis, 1944; Scarth and Shaw, 1951). The latter two papers also included observations of stomatal aperture, which decreased when the leaves were severely wilted, although Scarth and Shaw (1951) considered stomatal aperture to be controlled by photosynthesis and not the reverse.

There was, therefore, little doubt that water deficits inhibited photosynthesis, but questions of physiological mechanisms and the agricultural importance of photosynthetic inhibition remained. Most evidence implicated the stomata as the cause, although the relationship was only roughly correlative. No data were available that related desiccation-inhibited photosynthesis to yield.

Recent work that measured photosynthesis as a function of leaf water status has confirmed the inhibition of the process, although stomatal measurements were not made (El-Sharkawy and Hesketh, 1964; Strain, 1970; Chen et al., 1971; Ghorashy et al., 1971; Bazzaz et al., 1972; Oechel et al., 1972; Beardsell et al., 1973; Bazzaz, 1974).

The possibility for evaluating the role of the stomata improved considerably when Gaastra (1959) published a penetrating analysis of the diffusion of CO_2 into leaves of several crop plants. In the species Gaastra (1959) used, photosynthesis appeared to be limited by the rate of CO_2 diffusion under high light, since rate was proportional to the external concentration of CO_2. It also was relatively insensitive to temperature, as would be expected for a diffusion-limited process. Therefore, he suggested that the rate of photosynthesis was determined by the size of the gradient in CO_2 between the CO_2-fixation sites in the leaves and the bulk air outside, as well as the resistance to CO_2 diffusion between these two points. With one or two assumptions, the resistances could be readily calculated (Gaastra, 1959). The equation governing photosynthesis was

$$P = (C_{air} - C_{chl})/r_a + r_s + r_m \qquad (1)$$

where P is the flux of CO_2 for net photosynthesis (gm cm^{-2} sec^{-1}, respiration was ignored), C_{air} is the CO_2 concentration of the bulk air (gm cm^{-3}), C_{chl} is the CO_2 concentration at the chloroplast expressed in an equivalent gas concentration (gm cm^{-3}, Gaastra assumed this to be virtually zero), r is the resistance to diffusion of CO_2 (sec cm^{-1}), and a, s, and m represent the segments of the diffusion path in the boundary layer, stomata, and mesophyll, respectively.

Subsequent to Gaastra's work, investigators have included the cuticular (r_c) and intercellular space resistance (r_{inter}) in Eq. (1) for completeness, and most conceive of the total of cuticular, stomatal, and intercellular space resistances as r_l (Slatyer, 1967). The cuticular and stomatal resistances are in parallel with each other whereas the resistance of the intercellular spaces is in series with the cuticular and stomatal paths. Thus

$$r_l = r_{inter} + r_c r_s/(r_c + r_s) \qquad (2)$$

In practice, r_l is usually measured as a combined resistance and its components are measured separately. Resistance r_c is generally so large that most changes in r_l are governed by r_s. Resistance r_{inter} is generally considered to be small and rarely measured separately (Slatyer, 1967), although it could change significantly if the geometry of the intercellular spaces changed.

After Gaastra's work, a number of studies attempted to quantify the importance of the stomata for photosynthetic activity during water deficits. Brix (1962) provided a thorough study of net photosynthesis, dark respiration, and transpiration in tomato (*Lycopersicon esculentum*) and loblolly pine (*Pinus taeda*). His was the first to measure these factors as a function of leaf water potential and the first, since the work of Dastur, to express metabolic response on the basis of tissue water status. From close correlations between photosynthesis and transpiration, Brix concluded that the stomata limited photosynthesis at low leaf water potentials. Shimshi (1963) followed photosynthesis and transpiration as a function of stomatal width, but calculated that, although the stomata decreased in aperture, there was a component of photosynthetic inhibition that could not be attributed to stomatal closure.

Gale *et al.* (1966) used the Gaastra concepts to partition the CO_2

resistance in bean subjected to water deficiency. Soon after a change in r_l was observed, r_m increased and, consequently, these investigators suggested that there is a component of the photosynthetic response that cannot be attributed to stomatal closure. Troughton (1969) showed a similar response in cotton (*Gossypium hirsutum*), Boyer (1970b) demonstrated an increase in r_m in maize (*Zea mays*) and soybean (*Glycine max*), Hansen (1971) showed an increase in r_m in beet (*Beta vulgaris*), and Slatyer (1973) found a similar increase in maize and cotton, but not in wheat (*Triticum*) and millet (*Pennisetum*).

Willis and Balasubramaniam (1968) measured photosynthesis, transpiration and r_l in *Pelargonium*, but did not partition resistances as Gale *et al.* (1966), Troughton (1969), Boyer (1970b), Hansen (1971), or Slatyer (1973) had done. *Pelargonium* photosynthesis was generally correlated with transpiration and changes in r_l. Willis and Balasubramaniam (1968) interpreted this to mean that stomatal aperture controlled photosynthesis, but a close look at their data shows that photosynthesis responded to desiccation and rehydration before changes in transpiration and r_l took place.

Figure 1 shows the type of correlation between r_l and photosynthetic inhibition that has been observed by a number of investigators (Baker and Musgrave, 1964; Boyer and Bowen, 1970; Boyer, 1970b; van den Driessche *et al.,* 1971; Kriedemann and Smart, 1971; Moldau, 1972; Frank *et al.,* 1973; Harris, 1973; Beadle *et al.,* 1973; Johnson *et al.,* 1974; Regehr *et al.,* 1975; O'Toole, 1975).

Barrs (1968) used a novel approach to the problem of stomatal control of photosynthesis by inducing stomatal cycling in leaves of sunflower (*Helianthus annuus*), cotton, and pepper (*Capsicum frutescens*). The cycling was associated with cycling in leaf water potential and he could observe photosynthesis through a series of desiccation-rehydration cycles. The cycles for photosynthesis and transpiration were in phase and a plot of rates of photosynthesis versus transpiration yielded a straight line for all but the most desiccated leaves. Barrs concluded that the stomata completely accounted for the photosynthetic response and that the effects of r_m were negligible in these species. Unfortunately, Barrs did not measure leaf temperature. Thus, it is possible that rates of photosynthesis followed changes in leaf temperature as latent heat transfer varied during the transpiration cycle and the correlation between photosynthesis and transpiration could have just as easily represented enzymatic control as stomatal control.

Regardless of this matter, however, the bulk of the measurements of photosynthesis and transpiration (or diffusive resistances) clearly show that the two are correlated in desiccated tissue. A few exceptions have been noted, especially during the early phases of desiccation, but across a wide

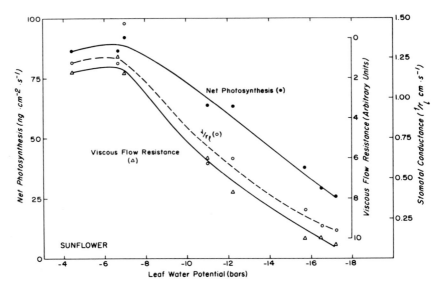

Fig. 1. Net photosynthesis at saturating irradiance, stomatal conductance $(1/r_l)$, and resistance to viscous flow of air through leaves in sunflower having various leaf water potentials. The plants were grown in soil from which water was withheld for various times. The viscous flow resistance was measured with a porometer that forced air through the stomata with a pressure that rose from -2 to -1 cm of water. The arbitrary units are the cube root of the time in seconds required for a standard pressure rise in the instrument (Boyer and Bowen, 1970). Since the measurement indicates leaf porosity, the determination provided a relative indication of stomatal aperture. Data for resistance to viscous flow and photosynthesis are from Boyer and Bowen (1970).

range of desiccation levels, the correlation has been amply documented. The relationship can be useful for those wishing to estimate the photosynthetic capacity of crops indirectly from measurements of transpiration or from the diffusive resistances of the stomata. Indeed, the data suggest that modelers and meteorologists who are able to estimate transpiration from physical inputs may also be able to estimate many effects of desiccation on photosynthesis, at least by taking certain precautions and using appropriate controls.

Nevertheless, in spite of this potentially useful relationship, it is invalid to conclude that stomatal closure controls photosynthesis on the basis of correlation alone. If photosynthesis is not limited by the availability of CO_2, it is possible for stomatal aperture to change without affecting the rate of photosynthesis. Analyses using the Gaastra (1959) approach, while they probably provide reliable estimates of leaf diffusive resistance, may have little bearing on photosynthesis unless CO_2 can be demonstrated to be limiting. With the exception of the initial experiments of Gaastra

(1959), none of the work cited heretofore has shown that CO_2 actually limited the rate of photosynthesis under all the conditions of the experiments.

As a result of this problem, there is a major ambiguity in most papers dealing with desiccation and photosynthesis. While it is true that there are frequent correlations between photosynthesis and transpiration (or diffusive resistances), it is not possible to say that the inhibition was caused by the restricted diffusion of CO_2 or by factors not involving diffusion. Rate-limiting processes have been hard to identify in photosynthesis and they can change as conditions change. Consequently, rather extensive studies and specific kinds of experiments are needed before a choice can be made.

A few papers have attempted to identify whether diffusion in fact limits photosynthesis at low water potentials, but before we consider them, it would be wise to consider some concepts of rate limitation that give rise to the ambiguity concerning the mechanism of the photosynthetic response to drought.

In any process that requires a number of completely independent factors acting simultaneously, the rate of the process is limited by the slowest. This basically was Blackman's concept of photosynthesis (Blackman, 1905). The limitation of photosynthesis by low irradiances probably comes closest to being an example of limitation by an independent factor in that changes in enzyme activity, CO_2 availability, and other factors often have little effect on the rate of photosynthesis in this situation. As many authors have pointed out, however, photosynthesis often is not controlled by independent factors [see Rabinowitch (1951) for a description of several cases]. For example, the rate of enzyme reactions at high irradiances (most notably the carboxylases) and the rate of CO_2 supply are highly interdependent, and become relatively independent only in extreme situations. By extreme situations, the author means that if the CO_2 supply vastly exceeds the ability of ribulose-1,5-diphosphate carboxylase to fix CO_2 (that is, the enzyme is operating at maximum velocity), the system is limited by carboxylase activity and changes in stomatal aperture would have little effect on the rate of carboxylation until the changes became large. On the other hand, if the CO_2 supply is exceptionally low (perhaps because the diffusive resistance of the stomata is large), CO_2 diffusion would limit the process and changes in enzyme activity would have little effect on the rate of CO_2 fixation.

In the first case, an increase in CO_2 would not bring about an increase in fixation because the enzyme was already operating at a maximum rate, but CO_2 fixation would probably respond to temperature since the maximum velocity of enzymes is temperature dependent. In the second case, an increase in CO_2 would cause a proportional increase in CO_2 fixation, but temperature would have little effect since gas diffusion is only slightly responsive to temperature.

In practice, however, it is rare to find a plant that exhibits either of these extremes. Usually, both an increase in CO_2 above the level found in the atmosphere (300 $\mu l\, l^{-1}$) and a change in temperature (from approximately room temperature) cause changes in photosynthesis at high irradiances. The intermediate situation is to be expected, since it would be a gross waste of metabolic energy for the plant to build large amounts of enzymatic protein that was unused or to synthesize so little carboxylase that it was unable to utilize the CO_2 that was available.

But regardless of this latter point, the intermediate situation generates a problem when attempts are made to distinguish the stomatal and enzymatic contributions to a particular rate of photosynthesis. This is caused by the compensating character of the enzymatic and diffusive components under these conditions. If, for example, the stomata close slightly, the CO_2 concentration inside the leaf momentarily falls. However, the drop in CO_2 concentration causes CO_2 fixation by the carboxylase to decrease, which reduces the rate of CO_2 utilization by the leaf. With the flux through the stomatal resistance thus reduced, the gradient in CO_2 required between the internal and external air is less and the CO_2 concentration in the leaf rises. The end result is that although the diffusive resistance of the stomata increased, it did not cause as great a reduction in CO_2 fixation as expected, because of the compensatory property of the carboxylase. Of course, the adjustment process would be continuous rather than several steps, but the end result would be the same.

In this situation, one would have to conclude that both enzyme activity and CO_2 diffusion contributed to the limitation of photosynthesis, but the exact contribution of each would be unknown. Jones (1973a) concerned himself with the evaluation of each contribution in a situation such as this and suggests two or three methods of analyzing the components, but he points out that each may give a different answer and is subject to certain assumptions. Ultimately, the CO_2 response and temperature response of photosynthesis probably provide the most unambiguous tests for carboxylation and stomatal limitation, and they can identify intermediate situations. However, there remains a need for methods to quantify the components of limitation in the intermediate cases and it is hoped that a satisfactory procedure might appear soon.

Clearly, however, it is important to understand what causes the reduction in photosynthesis at low water potentials and the simple tests for limitation should be conducted. Some investigators have recently recognized this problem and have done the required measurements. A few have used ingenious methods of circumventing it.

Wardlaw (1967) studied the CO_2 response of photosynthesis in wheat that had been desiccated. He did not find the linear response expected if CO_2 diffusion alone limited photosynthesis, although the complicating effects of direct stomatal responses to CO_2 were not measured. More importantly, photosynthesis remained depressed in the desiccated plants at low light levels, where CO_2 diffusion should not have limited the process. This effect can only be explained by some alteration at the chloroplast level.

Troughton and Slatyer (1969) recognized the complicating problem of stomatal diffusion and circumvented it by forcing air directly through the leaves of cotton. By also working at low O_2 concentrations, possible changes in photorespiration did not affect their measurements. At low CO_2 concentrations (100 $\mu l\ l^{-1}$ and less), they found only small effects of desiccation on photosynthetic rates. They infer from this that neither diffusive resistances in the liquid phase of the leaf nor biochemical reactions were affected by desiccation. However, since their experiments were done at limiting CO_2, it is possible that changes in the carboxylating steps of photosynthesis or in other aspects of chloroplast activity would have gone unobserved.

Redshaw and Meidner (1972) did a similar through-flow experiment with tobacco (*Nicotiana*) at an average CO_2 concentration of 240 $\mu l\ l^{-1}$, but observed a change in the activity of photosynthesis with desiccation. They concluded that the activity of a biochemical step or photorespiration had changed (they did not use low O_2 in their experiments). Since they used higher CO_2 concentrations than Troughton and Slatyer (1969), they would have been more likely to detect alterations in chloroplast activity.

In a somewhat similar vein, Boyer (1971b) studied the response of photosynthesis when CO_2 concentrations were varied in the air outside sunflower leaves. In this situation, the rate of CO_2 diffusion through the stomata could be varied and the CO_2 concentration inside the leaf would then change. Simultaneous measurements of r_l were made and showed that, under the conditions of the experiment, r_l remained constant at the CO_2 concentrations used (200 to 400 $\mu l\ l^{-1}$). While photosynthesis was affected by the concentration of external CO_2 at high leaf water potentials, it was progressively less so as leaf water potentials declined. At leaf water potentials of -18 to -20 bars, there was no response to external CO_2. Thus, photosynthesis must have been limited by something other than CO_2 availability.

Another approach to avoiding the stomatal problem has been to strip the epidermis from leaf segments (Graziani and Livne, 1971). In tobacco, an inhibition of photosynthesis could be observed when the tissue was dehydrated, but only if dehydration was fairly severe.

Since these experiments indicate that there were nonstomatal changes in photosynthesis under dry conditions, attempts have been made to measure the activity of chloroplasts in desiccated leaves. The work of Todd and Basler (1965) suggests that Hill activity in isolated chloroplasts was affected by desiccation, when activity is recalculated on a chlorophyll basis. Others have shown that Hill activity (Nir and Poljakoff-Mayber, 1967; Fry, 1970) and cyclic photophosphorylation (Nir and Poljakoff-Mayber, 1967) were inhibited when chloroplasts were isolated from leaves that had previously been severely desiccated. Chloroplasts desiccated *in vitro* also showed reduced Hill activity (Santarius and Heber, 1967). Santarius (1967) demonstrated that the NADPH and ATP content of leaves of *Beta vulgaris* decreased during severe desiccation. Wilson and Huffaker (1964) described changes in amounts of phosphorylated intermediates. Since each of these studies involved severely desiccated tissue, however, it is not clear whether the metabolic alterations were significant for a range of tissue water potentials or only reflected metabolic changes that accompany tissue death.

In an effort to answer this question, Boyer and Bowen (1970) assayed for oxygen evolution by chloroplasts isolated from sunflower leaf tissue having a wide range of water potentials. They reported that oxygen evolution in the presence of dichloroindophenol [which later was shown to measure the activity of Photosystem 2 (Potter and Boyer, 1973; Keck and Boyer, 1974)] decreased as soon as photosynthesis in the whole leaves began to be affected. There was good correlation between chloroplast activity and the photosynthetic activity of the intact leaves over the entire range of leaf water potentials (Fig. 2). Since Photosystem 2 activity is one of the partial reactions of the light-activated portion of photosynthesis, these authors also made *in vivo* measurements of the rate of photosynthesis under light-limiting conditions. The measurements should have indicated changes in the activity of the light-activated portion of photosynthesis *in vivo* and indeed show a correlation with chloroplast activity *in vitro* (Fig. 2). When measurements of transpiration and stomatal aperture were included, these also were correlated with leaf photosynthesis (see Fig. 1).

It is clear that whole leaf photosynthesis could have been controlled either by stomatal aperture or by chloroplast activity in the Boyer and Bowen (1970) work, but the CO_2 experiment conducted with whole sunflower leaves [described above (Boyer, 1971b)] showed that the limitation was at the chloroplast level and was not caused by CO_2 diffusion. In view of the effects on Photosystem 2, it appears that sunflower photosynthesis was limited by an inhibition of chloroplast electron transport that remained effective at both high irradiances and low leaf water potentials (Boyer, 1971b).

In a sequel to this work, Keck and Boyer (1974) explored the extent to which a number of partial reactions of photosynthesis were affected in

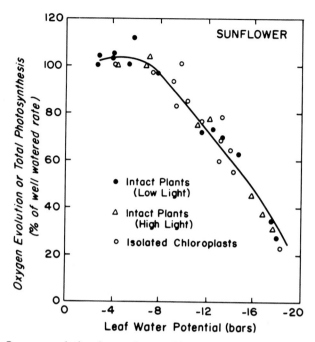

Fig. 2. Oxygen evolution by sunflower chloroplasts and total photosynthesis by sunflower leaves at various leaf water potentials. Oxygen evolution was measured in the presence of the electron acceptor dichloroindophenol (Photosystem 2). The chloroplasts were exposed to low leaf water potentials *in vivo*, then isolated and assayed. Total photosynthesis represents net photosynthesis plus dark respiration *in vivo* and was measured in similar leaves. The chloroplast assays and some of the measurements with intact leaves were conducted at saturating irradiances (high light), whereas others with intact leaves were conducted at limiting irradiances (low light). Adapted from Boyer and Bowen (1970) and Boyer (1971b).

sunflower. Photosystem 1 and 2, as well as whole-chain electron transport, were inhibited at leaf water potentials that corresponded to those causing effects in intact plants. Photosystem 2 activity appeared to be most affected. Cyclic and noncyclic photophosphorylation responded at leaf water potentials that were somewhat lower than those that influenced electron transport, but the response then became very large. At severe desiccation levels, inhibition of photophosphorylation exceeded inhibition of electron transport and Keck and Boyer (1974) suggest that the limitation of photosynthesis may shift from electron transport to photophosphorylation when sunflower becomes severely desiccated (leaf water potentials of −17 bars and below).

   In addition to effects on the "light" reactions of photosynthesis, leaf desiccation alters the activity of some of the "dark" reactions. The activity of ribulose-1,5-diphosphate carboxylase is reduced when assays are performed on extracts of desiccated leaves (Huffaker *et al.,* 1970; Jones, 1973b; Lee *et al.,* 1974; Johnson *et al.,* 1974; O'Toole, 1975). In those studies that made comparisons with photosynthesis in the intact leaf, none demonstrated an inhibition of enzyme activity large enough to fully account for the inhibition of photosynthesis *in vivo,* however (Jones, 1973b; Lee *et al.,* 1974; Johnson *et al.,* 1974; O'Toole, 1975). Phosphoenolpyruvate carboxylase activity and ribulose-5-phosphate kinase activity also decrease, but only slightly (Huffaker *et al.,* 1970; Shearman *et al.,* 1972). Plaut (1971) found little effect of high concentrations of sorbitol on extracts of ribulose-1,5-diphosphate carboxylase or ribulose-5-phosphate kinase activities, although he demonstrated an inhibition of the activities in isolated, but intact, chloroplasts. Unfortunately, as Plaut (1971) points out, the *in situ* assays in the intact plastids were dependent on endogenous ATP and, consequently, a decrease in photophosphorylation (Nir and Poljakoff-Mayber, 1967; Keck and Boyer, 1974) could have accounted for his results. Plaut and Bravdo (1973) demonstrated that CO_2 fixation was inhibited in intact chloroplasts isolated from desiccated leaves.

The bulk of the evidence suggests, therefore, that appreciable changes occur in the partial reactions of the light-activated phase of photosynthesis, but that the changes in the dark reactions, while quite detectable, are not large enough to completely account for the photosynthetic behavior of desiccated leaves. Since the chloroplast changes take place at the same leaf water potentials that cause stomatal closure, the rate limitation could be caused by either factor and might change as the conditions change. At low light, for example, photosynthesis would almost certainly be rate limited by reduced chloroplast activity. At higher radiation levels, stomatal limitation would usually be more effective [although examples of chloroplast limitation have been cited (Boyer, 1971b; Redshaw and Meidner, 1972)].

As is readily apparent, most of the chloroplast and stomatal effects of low leaf water potentials have been measured during the dehydration of plant tissue. Although far fewer studies have been concerned with the characteristics of photosynthetic recovery, they tend to confirm the desiccation experiments. Brix (1962) showed that recovery of transpiration and photosynthesis followed similar kinetics, which could be shortened by excising the roots under water. Willis and Balasubramaniam (1968) demonstrated a similar correlation between transpiration, r_l, and recovery of photosynthesis. Boyer (1971a) also showed this effect, but illustrated that severely desiccated plants may not recover completely for at least several days, in spite of a complete recovery of the water potentials of the leaves. This aftereffect could not be attributed to chloroplast photochemical activity, since it recovered almost completely (Boyer, 1971a; Potter and Boyer, 1973). However, stomatal apertures remained reduced (Fischer *et al.,* 1970; Fischer, 1970; Boyer, 1971a), so at least part of the incomplete recovery could have been associated with stomatal behavior.

As these data on desiccation and recovery accumulate, they permit comparison of the photosynthetic response of different species. There are simple methods now available for measuring leaf water potentials and, consequently, it is possible to rank photosynthetic response of a number of species on the basis of a method that has the advantage of having a physically defined reference. Table I shows that the sensitivity of photosynthesis varies between species. Since the data are based solely on desiccation experiments, the comparison may not reflect ecologically significant aftereffects of desiccation. Nevertheless, it is clear that mesic crops, native herbs, and some trees show a similar behavior and exhibit the largest inhibition at leaf water potentials between -5 and -20 bars. Small grains such as barley and wheat appear somewhat less sensitive, and species from semiarid environments (*Atriplex, Acacia, Larrea, Prosopis*) are capable of photosynthesis at leaf water potentials of -30 to -60 bars. In wheat and maize, the plants become less subject to inhibition after grain-filling has commenced, but soybean does not appear to show this tendency.

II. CONTROL OF STOMATAL APERTURE, CHLOROPLAST ACTIVITY, AND RESPIRATION

A. STOMATAL APERTURE

There are at least three theories that might account for the dramatic closure of stomata at low leaf water potentials. The first, and by far the most widely accepted, is that the guard cells lose turgor as a natural consequence of the general decrease in the water potential of the leaves. This theory rests on the idea that guard-cell bending depends on the creation of a higher turgor in the guard cells than in the accessory cells, which then deform as the guard cells bend. The turgor difference occurs because of a difference in osmotic potential between the two types of cells; the guard cells contain the larger amount of osmotically active solute. The low osmotic potential of the guard cells is thought to result from a rapid uptake of potassium (Fischer, 1968, 1971; Fischer and Hsiao, 1968; Humble and Hsiao, 1969, 1970; Sawhney and Zelitch, 1969; Humble and Raschke, 1971; Raschke and Fellows, 1971; Allaway and Hsiao, 1973; Hsiao *et al.,* 1973) accompanied by an exchange of protons (Raschke and Humble, 1973) or chloride (Raschke and Fellows, 1971) and the production of an organic acid [largely malate (Allaway, 1973)] from guard cell starch

J. S. Boyer

TABLE I
Net Photosynthesis at Various Leaf Water Potentials for a Number of Plant Species

Species	ψ_w of Initial inhibition	ψ_w at 50% inhibition	ψ_w at 100% inhibition	Technique for measuring ψ_w	Source	Remarks
Zea mays	−3 bars	−12 bars	—	Isopiestic thermocouple psychrometer	Boyer (1970b)	Young vegetative plants
Zea mays	−3	−11	−12 bars	Dewpoint thermocouple psychrometer	Beadle *et al.* (1973)	Vegetative plants
Zea mays	−8	−17	−20	Isopiestic thermocouple psychrometer	H. G. McPherson and J. S. Boyer (unpublished)	Reproductive plants
Vitis vinifera	−5	−9	−13	Pressure chamber	Kriedemann and Smart (1971)	Shadehouse grown
Vitis vinifera	−13	—	−15	Pressure chamber	Kriedemann and Smart (1971)	Field grown
Pinus taeda	−5	−7	−11	Thermocouple psychrometer	Brix (1962)	Young plants
Phaseolus vulgaris	−6	−7	−10	Dewpoint thermocouple psychrometer	O'Toole (1975)	Young plants
Helianthus annuus	−7	−14	−22	Isopiestic thermocouple psychrometer	Boyer and Bowen (1970)	Young plants
Lycopersicon esculentum	−7	−9	−14	Thermocouple psychrometer	Brix (1962)	Young plants
Populus deltoides	−8	−10	−11	Pressure chamber	Regehr *et al.* (1975)	Young plants
Ambrosia artemisiifolia	−9	−12	—	Pressure chamber	Bazzaz (1974)	Young plants
Sorghum bicolor	−10	−14	—	Dewpoint thermocouple psychrometer	Beadle *et al.* (1973)	Young plants
Glycine max	−11	−16	—	Isopiestic thermocouple psychrometer	Boyer (1970b)	Young plants, laboratory grown

Species				Method	Reference	Conditions
Glycine max	−10 to −12	−18	−25	Pressure chamber	Ghorashy et al. (1971)	Reproductive plants, field grown
Hordeum vulgare	−10	−15	−30	Thermocouple psychrometer	Johnson et al. (1974)	Reproductive plants, glasshouse grown
Triticum aestivum	−10	−15	−30	Thermocouple psychrometer	Johnson et al. (1974)	Reproductive plants, glasshouse grown
Triticum aestivum	−10	−15	—	Thermocouple psychrometer	Frank et al. (1973)	Young vegetative plants, laboratory grown
Triticum aestivum	−20	−30	—	Thermocouple psychrometer	Frank et al. (1973)	Reproductive plants, laboratory grown
Sassafras albidum	−12	−17	−22	Pressure chamber	Bazzaz et al. (1972)	Young plants
Atriplex hastata		—	−27	—	Slatyer (1973)	Young plants
Atriplex spongiosa		—	−35	—	Slatyer (1973)	Young plants
Prosopis juliflora		—	Below −45	Dye method	Strain (1970)	Field grown
Larrea divaricata		—	Below −60	Dye method	Strain (1970)	Field grown
Larrea divaricata	−25	−35	Below −55	Pressure chamber	Oechel et al. (1972)	Field grown
Beta vulgaris	−20	−30	−50	Gravimetric vapor equilibration	Hansen (1971)	Field grown
Acacia harpophylla	−30	−45	Below −60	Pressure chamber	van den Driessche et al. (1971)	Young plants, laboratory grown

(Mansfield and Jones, 1971). The anion and the potassium form a highly dissociated salt having a large effect on the osmotic potential of the guard cells.

Since the water potential of the guard cells should be close to that of the bulk leaf, at least over times of several minutes, the components of the water potential (ψ_w) of the guard cells (ψ^{guard}) and accessory cells (ψ^{acc}) would be

$$\psi_w^{guard} = \psi_s^{guard} + \psi_p^{guard} \tag{3}$$

$$\psi_w^{acc} = \psi_s^{acc} + \psi_p^{acc} \tag{4}$$

and

$$\psi_s^{guard} + \psi_p^{guard} = \psi_s^{acc} + \psi_p^{acc} \tag{5}$$

so that

$$\psi_p^{guard} - \psi_p^{acc} = -(\psi_s^{guard} - \psi_s^{acc}) \tag{6}$$

$$\Delta\psi_p = \Delta\psi_s \tag{7}$$

The subscripts s and p represent solutes and pressure and the cell membranes are considered to be ideally semipermeable. A number of studies have established that $\Delta\psi_s$ is 8 to 35 bars, depending on the species (Bearce and Kohl, 1970; Mansfield and Jones, 1971; Humble and Raschke, 1971; Allaway, 1973; Allaway and Hsaio, 1973).

Equation 6 indicates that, when ψ_p^{acc} becomes zero, a further reduction in leaf water potential will result in a diminution of $\Delta\psi_p$ and the stomata will begin to close. This sequence of events suggests that guard cell closure should begin just as turgor begins to disappear in the bulk tissue of the leaf (assuming that the elastic modulus behaves similarly for all cells at a given ψ_p and that ψ_s^{acc} is similar to that of the bulk leaf). In spite of the assumptions inherent in the analysis, Fig. 3 shows that stomatal closure does begin as the turgor of the bulk leaf is reduced below 1 bar in sunflower leaves, as expected from Eq. (6). It is important to note that chloroplast activity also shows this behavior (cf. Figs. 2 and 3B). Thus chloroplast activity and stomatal closure appear to be correlated, as pointed out earlier (Boyer and Bowen, 1970).

The observation that chloroplast activity and stomatal closure occur at similar leaf water potentials leads to the second theory of stomatal closure—that there could be a role for the guard cell chloroplasts in stomatal closure during leaf desiccation. Humble and Hsiao (1970) provided evidence that cyclic photophosphorylation can provide energy for the accumulation of potassium by guard cells, and action spectra for the opening of stomata also suggest that photosynthesis can be involved (Hsiao et al.,

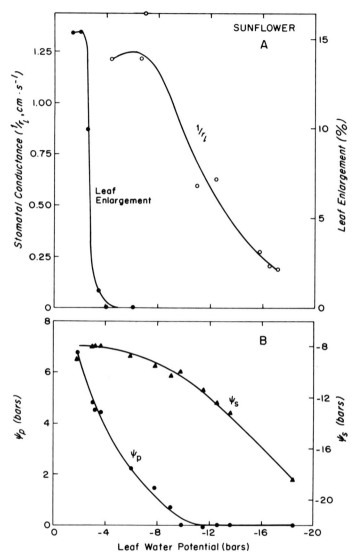

Fig. 3 (A) Leaf enlargement and stomatal conductance and (B) leaf osmotic potential (ψ_s) and leaf turgor (ψ_p) in sunflower having various leaf water potentials. The data for ψ_s and ψ_p were obtained with a thermocouple psychrometer and those for ψ_s were corrected for dilution by water in the cell walls. Adapted from Boyer (1968) and Boyer and Potter (1973).

1973). Thus, stomatal closure could be mediated by changes in chloroplast activity, quite apart from any direct effects of turgor on the stomatal apparatus. Unfortunately, the photosynthetic activity of the guard cells has not been measured separately during leaf desiccation, and so it cannot be said with certainty that guard cell chloroplasts change in activity as leaf water potentials decrease.

The third or hormonal theory of stomatal closure at low leaf water potentials is based on the recent observation that abscisic acid increases in desiccated leaves (Wright and Hiron, 1969; Zeevaart, 1971) and has a marked ability to cause stomatal closure (Little and Eidt, 1968; Mittelheuser and van Steveninck, 1969, 1971; Jones and Mansfield, 1970; Mansfield and Jones, 1971; Cummins et al., 1971; Kriedemann, et al., 1972; Cummins, 1973). The effects are rapid (Cummins et al., 1971; Kriedemann et al., 1972) and result in less potassium accumulation by the guard cells (Mansfield and Jones, 1971). It is presumed that the rise in abscisic acid concentration in leaf tissue during desiccation could cause loss of potassium from the guard cells and result in stomatal closure. Unfortunately, however, no data are available to test this idea in desiccated plants.

B. CHLOROPLAST ACTIVITY

The large decreases in electron transport that occur in chloroplasts of desiccated leaves do not depend on the particular means of desiccation, provided desiccation occurs *in vivo*. Fry (1972) showed that desiccation by exposing leaf tissue to light or expressing sap under pressure resulted in similar Hill activities of the chloroplasts after isolation. The rapidity of desiccation made little difference, since experiments requiring only a few minutes (excised tissue) and those requiring several days (whole plants) gave similar results (Fry, 1972; Potter and Boyer, 1973).

When isolated chloroplasts were exposed to osmotic potentials *in vitro* that corresponded to those *in vivo,* however, the inhibition of chloroplast activity was considerably less than occurred *in vivo* (Fig. 4). Potter and Boyer (1973) showed that the small inhibition of Photosystem 2 activity by sorbitol solution was reversible, whereas the inhibition *in vivo* could not be reversed after the chloroplasts had been isolated. Nevertheless, it was readily reversed if the chloroplasts were rehydrated *in vivo* before isolation (Potter and Boyer, 1973). This suggests that, although chloroplast activity may be correlated with tissue ψ_s (cf. Figs. 2 and 3B), the Gibbs free energy of water represented by ψ_s does not cause the changes in chloroplast activity that were observed. It remains possible, however, that solute-chloroplast interactions specific to the cell environment could bring about changes in chloroplast activity, since the sorbitol solutions obviously reproduced only the free energy environment of the cell.

This demonstrates that efforts to use solute systems as models for desiccation must be approached with caution. Plaut (1971) and Plaut and

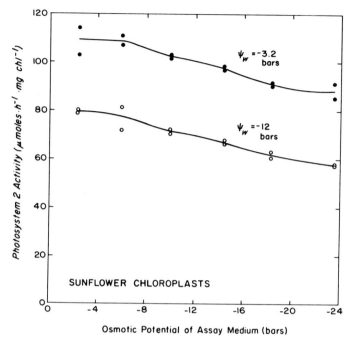

Fig. 4. Photosystem 2 activity measured as the photoreduction of dichloroindophenol by chloroplasts assayed in media containing sorbitol to give various osmotic potentials. Chloroplasts were isolated from half of one sunflower leaf that was well watered (leaf water potential = —3.2 bars) and from the other half of the same leaf after desiccation (leaf water potential = —12 bars). Data from Potter and Boyer (1973).

Bravdo (1973) assumed that high sorbitol concentrations caused effects on spinach chloroplasts that were similar to those occurring in intact leaves. However, chloroplasts isolated from desiccated tissue exhibited a response to desiccation that differed from the response to sorbitol solutions (Plaut and Bravdo, 1973). Fry (1972) also measured Hill activity in sorbitol solutions. Although he states that they caused a chloroplast response that was similar in magnitude to that of chloroplasts desiccated *in vivo*, inspection of the data indicates that sorbitol was only about half as effective. Santarius and Ernst (1967) showed that high concentrations of solutes differed in their effects on chloroplasts according to the solute used, which would not be expected at isosmotic concentrations if only the free energy of the solution was required to duplicate the cell environment. As a result, the interpretation of these experiments remains difficult until we more fully understand the effects of various solutes on chloroplast systems.

Regardless of the molecular character of the cell environment that

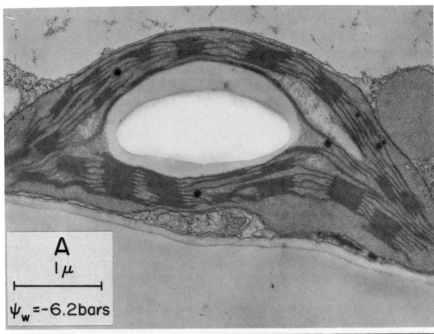

A
1μ

$\psi_w = -6.2\,\mathrm{bars}$

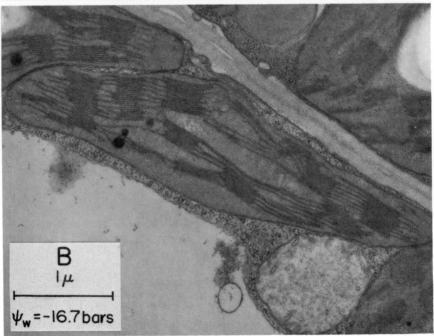

B
1μ

$\psi_w = -16.7\,\mathrm{bars}$

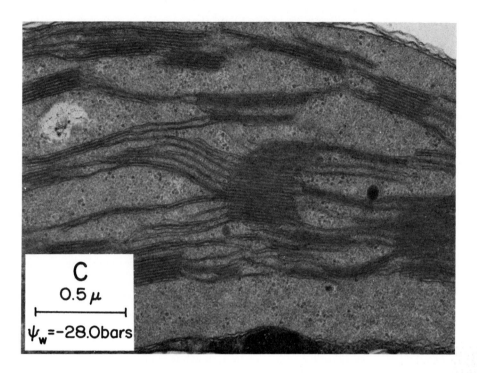

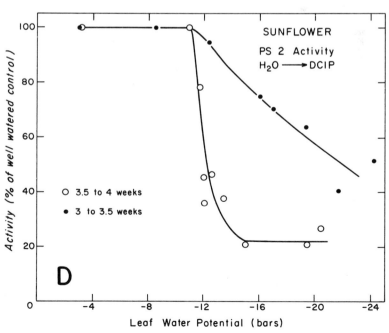

might cause these events, it seems likely that a major part of the response of chloroplasts to low leaf water potentials is associated with some aspect of the lamellar membrane systems. Electron transport and photophosphorylation, both membrane-associated phenomena, show large inhibitions of activity, whereas activities of the stroma enzymes are less affected. However, electron micrographs show that chloroplast lamellae appear normal in cells that were desiccated to an extent that would cause considerable loss in Photosystem 2 activity (Fig. 5). Only in tissue that was desiccated irreversibly (leaf water potentials below —25 bars, Fig. 5C) did lamellar membranes begin to show evidence of an alteration of structure characterized by indistinctness of the membranes. Thus, if any ultrastructural changes accompany the alterations in activity, they must involve rather subtle aspects of lamellar conformation. Giles *et al.* (1974) have shown that certain chloroplasts of maize, notably the "C_3" type (which are similar to those shown in Fig. 5), exhibit little evidence of structural changes after the leaves were severely desiccated, but that "C_4" chloroplasts were altered.

Although there are no obvious ultrastructural changes in sunflower chloroplasts at low leaf water potentials (Fig. 5), the large decrease in Photosystem 2 activity and other aspects of electron transport (Keck and Boyer, 1974) reflect fundamental changes in the primary light harvesting process of photosynthesis. Figure 6 shows that the quantum yield of desiccated leaves decreased as leaf water potentials decreased. Upon rehydration, the quantum yield returned to a level close to that of the leaf before desiccation. Chloroplasts isolated from similar leaves showed a similar reduction in quantum yield Photosystem 2 when the leaf tissue had previously been desiccated (Fig. 6B). The large magnitude of the reduction in quantum yield clearly would cause significant reductions in photosynthesis at irradiances below saturation. Furthermore, the similarity between the behavior of the whole leaf and the chloroplasts suggests that observa-

---

Fig. 5. Ultrastructure and Photosystem 2 activity of chloroplasts of sunflower leaves having various water potentials. (A) Chloroplast of well-watered leaf of 3.5-wk plant. Grana and lamellae appear normal, and a large starch grain is visible in the center of the plastid. Leaf water potential = —6.2 bars. Magnification = 23,300×. (B) Chloroplast of leaf desiccated to water potential of —16.7 bars. Plant, 3.5 wk old. Grana and lamellae appear normal. Magnification = 26,700×. (C) Chloroplast of leaf desiccated to water potential of —28.0 bars. Plant, 3.5 wk old. Much of the leaf was irreversibly desiccated and death was occurring, as indicated by severe inhibition of dark respiration. Grana appear normal but lamellae are beginning to become slightly indistinct. Magnification = 48,200×. (D) Photosystem 2 activity measured as the photoreduction of dichloroindophenol in sunflower chloroplasts isolated from leaves having various water potentials. The plants were 3 to 3.5 wk (●) and 3.5 to 4 wk (○) old. Electron micrographs from R. J. Fellows and J. S. Boyer (unpublished) and Photosystem 2 data from Keck and Boyer (1974).

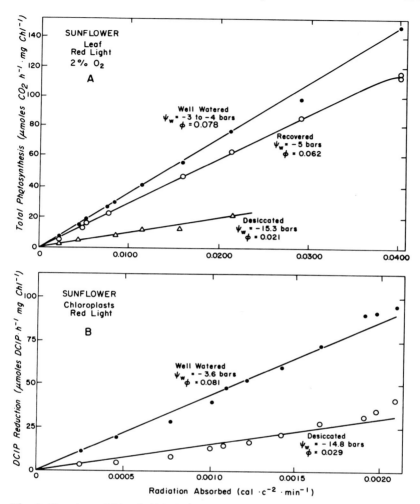

Fig. 6. Quantum yields of (A) desiccated and recovered leaves and (B) chloroplasts isolated from desiccated leaves in sunflower. Photosynthetic measurements with whole leaves were conducted at 2% O_2 and total photosynthesis was considered to be net photosynthesis plus dark respiration under these conditions. The chloroplast assays measured primarily Photosystem 2 activity. The quantum yields are calculated from the slope of the photosynthesis-irradiance relationship. Data from P. Mohanty and J. S. Boyer (unpublished).

tions of altered electron transport *in vitro* represent similar alterations *in vivo*.

In principle, certain aspects of the inhibition of chloroplast activity might be caused by a change in the plastid pigments. The synthesis of chlorophyll can be highly sensitive to low leaf water potentials. Virgin

TABLE II

CHLOROPHYLL CONTENT OF AN ATTACHED SUNFLOWER LEAF
THAT HAD BEEN DESICCATED 3 DAYS AND
PERMITTED TO RECOVER[a]

Treatment	Leaf water potential (bars)	Chlorophyll a + b (mg/gm dry weight)
Well watered	−3 to −4	11.4
Desiccated	−20.2	11.2
Recovered	−4.5	11.2

[a] P. Mohanty and J. S. Boyer, unpublished.

(1965) and Bourque and Naylor (1971) showed that water deficits caused significantly slower accumulation of chlorophyll in greening leaves. In the Bourque and Naylor (1971) experiments, the small water deficit induced by low humidities was enough to cause a significant response.

In green leaves, however, chlorophyll content often shows no significant changes during desiccation of a few days (Table II). The measurements were made on the same leaves during both desiccation and recovery. There were no significant changes in the absorption spectrum of chloroplasts isolated from sunflower leaves having different water potentials, which suggests that chloroplast pigments other than chlorophyll were unaffected as well (Fig. 7).

On the other hand, it is a common observation that leaf yellowing can occur when leaves have had low water potentials for a considerable time. Frequently, this loss in chlorophyll is associated with reduction in

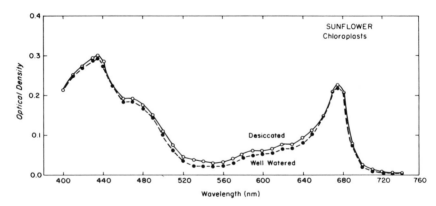

Fig. 7. Optical density at various wavelengths for sunflower chloroplasts isolated from well-watered (leaf water potential = −4 bars) and desiccated (leaf water potential = −15.1 bars) leaf halves. Optical density was measured in an integrating sphere. Data from P. Mohanty and J. S. Boyer (unpublished).

the flux of nitrogen into the tissue as well as alterations in the activity of enzyme systems such as nitrate reductase (Huffaker *et al.,* 1970; Morilla *et al.,* 1973) or nitrogenase in legumes (Engin and Sprent, 1973). A loss of chlorophyll may also arise in situations where desiccation causes leaf senescence. However, most studies of the effects of low leaf water potentials on photosynthesis have been made for short enough times that nutrient and senescence effects would be small.

C. Respiration

Extensive measurements show that desiccation affects the rate of dark respiration, but that considerable respiration occurs even when no net photosynthesis is detectable (Schneider and Childers, 1941; Upchurch *et al.,* 1955; Brix, 1962; Boyer, 1970a). In the early phases of desiccation, Schneider and Childers (1941), Upchurch *et al.* (1955), Brix (1962), and Kaul (1966) demonstrated that dark respiration increased, but then decreased as desiccation became more severe. Others showed that dark respiration decreased whenever leaf water potentials decreased (Flowers and Hanson, 1969; Boyer, 1970a; Bell *et al.,* 1971; Miller *et al.,* 1971; Koeppe *et al.,* 1973). The study by Brix (1962) is particularly revealing because it shows that respiration initially increased in pine but decreased in tomato. This suggests that the early response of respiration is probably a characteristic of the species. In any case, the ultimate effect of severe desiccation is always to decrease dark respiration.

Respiration in the light (or photorespiration, which involves both dark respiration and light-induced respiration) also shows a decrease in sunflower leaves having low leaf water potentials (Fig. 8). The decrease is steeper than that for dark respiration (Fig. 8), which indicates that the light-induced component is more inhibited than the dark component at low leaf water potentials.

In principle, a rise in dark respiration or light-induced respiration could account for decreases in net photosynthesis during desiccation. However, since the increases are modest and in many species simply not present, it is unlikely that increases in respiration can account for the decreases in photosynthesis associated with low leaf water potentials.

It has been reported that the CO_2 compensation point of leaves rises as desiccation becomes severe (Heath and Meidner, 1961; Fischer, 1970; Glinka and Katchansky, 1970). Since the CO_2 compensation point represents the external CO_2 concentration at which CO_2 influx for photosynthesis just equals CO_2 efflux for respiration in the light, low water potentials could be affecting either process. Since photosynthesis is more sensitive than dark respiration to low leaf water potentials (Schneider and Childers,

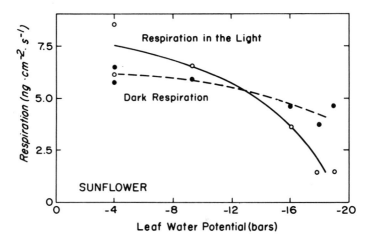

Fig. 8. Dark respiration and respiration in the light in sunflower leaves having various leaf water potentials. Respiration in the light was measured as CO_2 evolution into CO_2-free air. Although it estimates any contribution by dark respiration and light-induced respiration, the measurements underestimate respiration because of internal recycling of CO_2 by photosynthesis. Data from Boyer (1971b).

1941; Upchurch *et al.*, 1955; Brix, 1962; Boyer, 1970a), it seems likely that the increased compensation point is caused by a reduction in CO_2 influx at a time when CO_2 efflux remains reasonably substantial (of course, light-induced CO_2 loss would have declined and the remaining respiration would be mostly of the dark type). Recent measurements of the O_2 sensitivity of the CO_2 compensation point confirmed this idea, since the compensation point became less sensitive to 2% O_2 as leaf water potentials decreased (J. S. Boyer, unpublished data). This would be expected if photosynthesis and light-induced CO_2 loss were severely reduced. Then the remaining dark respiration (which is insensitive to 2% O_2) would cause the CO_2 compensation point to rise.

III. INHIBITION OF LEAF GROWTH

The production of leaves by crops represents the production of photosynthetic surface. Low leaf water potentials influence this process through their effects on leaf initiation in meristems and the subsequent enlargement of leaves. The rate of leaf initiation may become slower or even cease as desiccation proceeds (Husain and Aspinall, 1970), and there is evidence that cell division itself may be reduced (Terry *et al.*, 1971; Kirkham *et al.*, 1972; Meyer and Boyer, 1972; McCree and Davis, 1974). In general, cell enlargement appears to be more sensitive than cell division (Meyer and Boyer, 1972), although Kirkham *et al.* (1972) found an early effect of osmotic solutions on cell division.

The exceptional sensitivity of leaf enlargement was first demonstrated by Boyer (1968, 1970a), who showed that leaf enlargement was reduced to 25% of the controls or less when leaf water potentials decreased to —4 bars in maize, soybean, and sunflower. Acevedo *et al.* (1971) have shown a similar response in maize except that the leaf water potentials appeared to be a little lower. Photosynthesis was unaffected by these leaf water potentials (Boyer, 1970a). High irradiances and low humidities caused enough desiccation to significantly restrict leaf elongation in sunflower (Boyer, 1968) and maize (Acevedo *et al.*, 1971). Short exposures permitted rapid recovery of original growth rates when the plants were returned to more moderate conditions (Acevedo *et al.*, 1971). On the other hand, Boyer (1970a) showed that there was a reduced rate of growth in the recovered leaf if growth-inhibiting desiccation lasted for several days. This suggests that the normal diurnal changes in leaf desiccation are likely to cause reversible effects on leaf growth, but that desiccation for prolonged periods can cause partially irreversible inhibition.

It is important to note that in certain situations such as plants in saline soils or the leaves at the tops of tall trees, leaves grow even though leaf water potentials may be continuously more negative than —4 bars. It therefore seems certain that leaves are capable of adjusting in some way so that enlargement is less affected than in those cases described by Boyer (1970a) and Acevedo *et al.* (1971). The investigation of how these adaptations occur should be highly worthwhile.

Some investigations have shown that these adaptations may take the form of adjustments in the osmotic potentials of cell contents. Thus, Meyer and Boyer (1972) described an internal osmotic compensation for low tissue water potentials that caused the tissue osmotic potential to change by the same amount as the water potential in soybean hypocotyls. The overall effect of the compensation was to keep turgor high, which resulted in less inhibition of cell enlargement than occurred when compensation did not take place. Greacen and Oh (1972) have shown a similar phenomenon in roots, and Goode and Higgs (1973) and Biscoe (1972) report that it may occur in leaves.

The mechanism by which a reduction in leaf enlargement occurs at low cell water potentials seems to involve the characteristics of the viscoelastic "creep" of the cell walls under the action of turgor (Cleland, 1971). Turgor above a certain threshold is required before enlargement begins (Cleland, 1959; Probine and Preston 1962; Boyer, 1968; Green *et al.*, 1971) and, since that threshold appears to be relatively high in certain cases (Boyer, 1968, 1970a), the turgor-response of the system is quite

large. Figure 3 shows that, for sunflower, a turgor of approximately 3.5
bars is required before leaf enlargement begins. Above this threshold level,
the entire range of growth rates occurs within a turgor interval of just 2
bars. Obviously, wilt symptoms are of little use when one is concerned
with drought effects on leaf enlargement!

In addition to the effects of low leaf water potentials on the production of new leaf area, low water potentials may cause the loss of existing
leaf area if desiccation is prolonged and severe. The loss is caused by a
hastened form of leaf senescence, which is particularly rapid in grasses.
Figure 9 shows the change in viable leaf area of maize that was subjected
to leaf water potentials of −18 to −20 bars during the grain-fill period.
The plants, which had only mature leaves during this time, lost all their
leaf area within 30 days after leaf water potentials first reached −18 to
−20 bars. The lowest leaves senesced first.

The early senescence of leaf area in maize represents an irreversible
loss of photosynthetic capability by the crop. The fact that it occurs in
the lowest part of the canopy first is fortunate, because this effectively saves
the most important part of the photosynthetic surface until last.

It is interesting that the metabolic changes that bring about hastened
senescence are set in motion rapidly after the onset of low leaf water potentials in some plants. Ribonuclease is an enzyme that rises in activity
during senescence and the increase is one of the first signs of the onset
of leaf senescence (Sacher, 1973). It increases markedly in maize leaves
4 to 5 hr after low leaf water potentials occur (Morilla et al., 1973). Furthermore, nitrate reductase activity, which is characteristic of active
growth, falls dramatically within this time (Morilla et al., 1973).

IV. SIGNIFICANCE OF INHIBITION OF PHOTOSYNTHESIS FOR GRAIN PRODUCTION

It is well accepted that production of dry matter is reduced when
plants are grown at low leaf water potentials [see Salter and Goode (1967)
for a thorough review]. Although this suggests a reduction in photosynthesis, the exact relationship of decreased photosynthesis to decreased yield,
particularly of grain, is unknown. A recent experiment conducted by the
author and Dr. H. G. McPherson at the Climate Laboratory in Palmerston
North, New Zealand, attempted to provide some information about this
question. The intent of the experiment was to expose maize to low leaf
water potentials during the early grain-filling period (after flowering and
pollination had been completed), and then to continue the exposure to
low leaf water potentials for the remainder of the growing season. Since
maize must accumulate 40 to 50% of the total plant dry weight during

this time, the treatment provided an opportunity to study the effects of photosynthate deprivation on the grain-filling process when grain development was virtually the only development that was occurring.

Figure 9 shows that net photosynthesis ceased when leaf water potentials decreased to -18 to -20 bars. The plants were inactive photosynthetically for the remainder of the grain-filling period. In spite of this situation, the desiccated plants produced 4900 kg ha^{-1} (62.2 gm per plant) and the controls produced 10,500 kg ha^{-1} (133 gm per plant) of grain, an incredibly high production when it is considered that the desiccated plants produced only 17% of the photosynthate of the controls during grain-fill.

Calculations of the integrated net photosynthesis that occurred during the grain-filling period bore little relationship to grain yield (Table III). However, integrations of the net photosynthesis during the whole growing season were closely related to grain yield (Table III). Indeed, grain yield was a constant fraction of the total photosynthate accumulated during the growing season (Table III). This result indicates that photosynthesis must have been an important determinant of grain yield, but that it was the total season's photosynthate rather than the grain-fill photosynthate that was significant.

These results also demonstrate another point. The dry weight accumu-

TABLE III

GRAIN YIELD, INTEGRATED NET PHOTOSYNTHESIS DURING GRAIN-FILLING, AND INTEGRATED NET PHOTOSYNTHESIS DURING THE WHOLE SEASON FOR MAIZE THAT HAD BEEN DESICCATED THROUGHOUT MOST OF THE GRAIN-FILLING PERIOD[a]

	Observed dry weight of grain (gm/plant)	Integrated[c] photosynthesis, grain-fill (gm/plant)	Ratio of grain wt. to photosynthesis during grain-fill	Integrated[c] photosynthesis, whole season (gm/plant)	Ratio of grain wt. to photosynthesis, whole season
Well watered[b]	133	199	0.67	311	0.43
Desiccated	62.2	32.8	1.90	144	0.43

[a] H. G. McPherson and J. S. Boyer, unpublished.
[b] Leaf water potentials were -4 to -5 bars for well watered plants, -18 to -20 bars for desiccated plants (see Fig. 9A).
[c] Photosynthetic integrals were based on measurements of net photosynthesis made on the fourth leaf from the top of the plant during grain fill (see Fig. 9) and assumed that (1) the fourth leaf reflected the photosynthetic behavior of the whole plant (corrected for percent viable leaf area) and (2) photosynthesis per unit dry weight increase was the same for the well-watered and desiccated plants.

J. S. Boyer

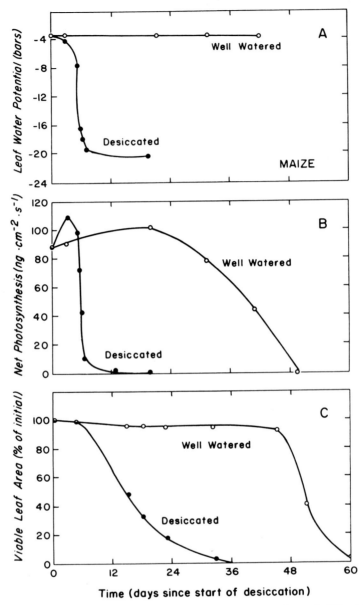

Fig. 9. (A) Leaf water potential, (B) net photosynthesis, and (C) viable leaf area of maize during desiccation in the grain-filling period. Desiccation was initiated by withholding water from the soil shortly after pollination had been completed, and low leaf water potentials were maintained by adding small amounts of water (1/7 that used for the controls) for the remainder of the growing season. Data from H. G. McPherson and J. S. Boyer (unpublished).

lated by the ear comes almost entirely from translocation in maize, since the leaves of the ear contribute little to the grain. Thus, the constant fraction of grain weight/total season's photosynthesis indicates that translocation continued in spite of the virtual cessation of photosynthesis (Table III). This suggests that translocation was less limiting than photosynthesis to grain yield at low leaf water potentials. Furthermore, it shows that plant reserves rather than current photosynthesis must have contributed dry weight to the grain, an observation also noted by Wardlaw (1967) in desiccated wheat. The contribution of plant reserves appeared as a loss in dry weight of the vegetative parts of the plant during the desiccation period (H. G. McPherson and J. S. Boyer, unpublished).

Since the photosynthate that was produced throughout the growing season seemed to limit grain yield in maize, some measure of the integrated photosynthetic activity of crops ought to be useful for predicting yield reductions by drought. The measurements could be important both for those crops having their economic yield in the form of total shoot dry matter, as well as for those in which yield is represented only by the reproductive structures, since the season's photosynthesis rather than the capacity for translocation appears to limit reproductive development after flowering. For grain crops, however, desiccation during floral development and the flowering period itself can cause irreversible damage to floral structures and result in a diminished amount of pollination. In this situation, grain production would be curtailed in spite of adequate dry matter production for the season as a whole.

V. SUMMARY

There is abundant evidence that photosynthesis is inhibited at water deficits that are commonly met by higher plants under natural conditions. The inhibition is accompanied by stomatal closure and losses in chloroplast activity, but it is not caused by increases in dark respiration or light-induced respiration nor by changes in chloroplast pigment content, at least for periods of 2 or 3 days. Stomatal closure and the restriction of CO_2 diffusion which results may account for the loss in photosynthetic activity under some conditions, particularly high irradiances. Decreased chloroplast activity probably is the cause of depressed photosynthesis under other conditions, notably low irradiances (although there are also examples at high irradiances). The loss in chloroplast activity appears to be largest in the "light reactions" and is not directly caused by changes in the free energy of water or by alterations in membrane ultrastructure visible in thin sections under the electron microscope. Decreased development of leaf area also contributes to a decrease in the photosynthetic productivity of water deficient plants and may be the earliest sign of a water deficit. Leaf senescence is accelerated and often results in irreversible loss of photosynthetic surface. The loss in leaf area and photosynthetic activity, when taken together, represent a potentially large loss of photosynthate for crops. In the cereal grains, photosynthesis is more sensitive than translocation to a water deficiency. Therefore, the production of grain is limited more by deprivation of photosynthate than by loss in transport capability for photosynthate. Grain yield is determined more by the total photosynthesis taking place for the growing season as a whole than by the photosynthesis occurring during the grain filling period alone. Consequently, water deficiencies decrease grain yield by decreasing the photosynthate accumulated for the season.

REFERENCES

Acevedo, E., Hsiao, T. C., and Henderson, D. W. (1971). Immediate and subsequent growth responses of maize leaves to changes in water status. *Plant Physiol.* **48**, 631–636.

Allaway, W. G. (1973). Accumulation of malate in guard cells of *Vicia faba* during stomatal opening. *Planta* **110**, 63–70.

Allaway, W. G., and Hsiao, T. C. (1973). Preparation of rolled epidermis of *Vicia faba* L. so that stomata are the only viable cells: Analysis of guard cell potassium by flame photometry. *Aust. J. Biol. Sci.* **26**, 309–318.

Allmendinger, D. F., Kenworthy, A. L., and Overholser, E. L. (1943). The carbon dioxide intake of apple leaves as affected by reducing the available soil water to different levels. *Proc. Amer. Soc. Hort. Sci.* **42**, 133–140.

Ashton, F. M. (1956). Effects of a series of cycles of alternating low and high soil water contents on the rate of apparent photosynthesis in sugar cane. *Plant Physiol.* **31**, 266–274.

Baker, D. N., and Musgrave, R. B. (1964). The effects of low level moisture stresses on the rate of apparent photosynthesis in corn. *Crop Sci.* **4**, 249–253.

Barrs, H. D. (1968). Effect of cyclic variations in gas exchange under constant environmental conditions on the ratio of transpiration to net photosynthesis. *Physiol. Plant.* **21**, 918–929.

Bazzaz, F. A. (1974). Ecophysiology of *Ambrosia artemisiifolia:* A successional dominant. *Ecology* **55**, 112–119.

Bazzaz, F. A., Paape, V., and Boggess, W. R. (1972). Photosynthetic and respiratory rates of *Sassafras albidum*. *Forest Sci.* **18**, 218–222.

Beadle, C. L., Stevenson, K. R., Neumann, H. H., Thurtell, G. W., and King, K. M. (1973). Diffusive resistance, transpiration, and photosynthesis in single leaves of corn and sorghum in relation to leaf water potential. *Can. J. Plant Sci.* **53**, 537–544.

Bearce, B. C., and Kohl, H. C., Jr. (1970). Measuring osmotic pressure of sap within live cells by means of a visual melting point apparatus. *Plant Physiol.* **46**, 515–519.

Beardsell, M. F., Mitchell, K. J., and Thomas, R. G. (1973). Effects of water stress under contrasting environmental conditions on transpiration and photosynthesis in soybean. *J. Exp. Bot.* **24**, 579–586.

Bell, D. T., Koeppe, D. E., and Miller, R. J. (1971). The effects of drought stress on respiration of isolated corn mitochondria. *Plant Physiol.* **48,** 413–415.

Biscoe, P. V. (1972). The diffusion resistance and water status of leaves of *Beta vulgaris. J. Exp. Bot.* **23,** 930–940.

Blackman, F. F. (1905). Optima and limiting factors. *Ann. Bot. (London)* **19,** 281–295.

Bormann, F. H. (1953). Factors determining the role of loblolly pine and sweetgum in early old-field succession in the Piedmont of North Carolina. *Ecol. Monogr.* **23,** 339–358.

Bourdeau, P. (1954). Oak seedling ecology determining segregation of species in Piedmont oak-hickory forests. *Ecol. Monogr.* **24,** 297–320.

Bourque, D. P., and Naylor, A. W. (1971). Large effects of small water deficits on chlorophyll accumulation and ribonucleic acid synthesis in etiolated leaves of jack bean (*Canavalia ensiformis* (L.) DC.). *Plant Physiol.* **47,** 591–594.

Boyer, J. S. (1968). Relationship of water potential to growth of leaves. *Plant Physiol.* **43,** 1056–1062.

Boyer, J. S. (1970a). Leaf enlargement and metabolic rates in corn, soybean, and sunflower at various leaf water potentials. *Plant Physiol.* **46,** 233–235.

Boyer, J. S. (1970b). Differing sensitivity of photosynthesis to low leaf water potentials in corn and soybean. *Plant Physiol.* **46,** 236–239.

Boyer, J. S. (1971a). Recovery of photosynthesis in sunflower after a period of low leaf water potential. *Plant Physiol.* **47,** 816–820.

Boyer, J. S. (1971b). Nonstomatal inhibition of photosynthesis in sunflower at low leaf water potentials and high light intensities. *Plant Physiol.* **48,** 532–536.

Boyer, J. S., and Bowen, B. L. (1970). Inhibition of oxygen evolution in chloroplasts isolated from leaves with low water potentials. *Plant Physiol.* **45,** 612–615.

Boyer, J. S., and Potter, J. R. (1973). Chloroplast response to low leaf water potentials. I. Role of turgor. *Plant Physiol.* **51,** 989–992.

Brilliant, B. (1924). Le teneur en eau dans les feuilles et l'energie assimilatrice. *C. R. Acad. Sci.* **178,** 2122–2125.

Brix, H. (1962). The effect of water stress on the rates of photosynthesis and respiration in tomato plants and loblolly pine seedlings. *Physiol. Plant.* **15,** 10–20.

Chen, L. H., Mederski, H. J., and Curry, R. B. (1971). Water stress effects on photosynthesis and stem diameter in soybean plants. *Crop Sci.* **11,** 428–431.

Cleland, R. (1959). Effect of osmotic concentration on auxin-action and on irreversible and reversible extension of *Avena* coleoptile. *Physiol. Plant.* **12,** 809–825.

Cleland, R. (1971). Cell wall extension. *Annu. Rev. Plant. Physiol.* **22,** 197–222.

Cummins, W. R. (1973). The metabolism of abscisic acid in relation to its reversible action on stomata in leaves of *Hordeum vulgare* L. *Planta* **114,** 159–167.

Cummins, W. R., Kende, H., and Raschke, K. (1971). Specificity and reversibility of the rapid stomatal response to abscisic acid. *Planta* **99,** 347–351.

Dastur, R. H. (1924). Water content, a factor in photosynthesis. *Ann. Bot. (London)* **38,** 779–788.

Dastur, R. H. (1925). The relation between water content and photosynthesis. *Ann. Bot. (London)* **39,** 769–786.

El-Sharkawy, M. A., and Hesketh, J. D. (1964). Effects of temperature and water deficit on leaf photosynthetic rates of different species. *Crop Sci.* **4,** 514–518.

Engin, M., and Sprent, J. I. (1973). Effects of water stress on growth and nitrogen-fixing activity of *Trifolium repens. New Phytol.* **72,** 117–126.

Fischer, R. A. (1968). Stomatal opening: Role of potassium uptake by guard cells. *Science* **168**, 784–785.

Fischer, R. A. (1970). After-effect of water stress on stomatal opening potential. II. Possible causes. *J. Exp. Bot.* **21**, 386–404.

Fischer, R. A. (1971). Role of potassium in stomatal opening in the leaf of *Vicia faba. Plant Physiol.* **47**, 555–558.

Fischer, R. A., and Hsiao, T. C. (1968). Stomatal opening in isolated epidermal strips of *Vicia faba.* II. Responses to KCl concentration and the role of potassium absorption. *Plant Physiol.* **43**, 1953–1958.

Fischer, R. A., Hsiao, T. C., and Hagan, R. M. (1970). After-effect of water stress on stomatal opening potential. I. Techniques and magnitude. *J. Exp. Bot.* **21**, 371–385.

Flowers, T. J., and Hanson, J. B. (1969). The effect of reduced water potential on soybean mitochondria. *Plant Physiol.* **44**, 939–945.

Frank, A. B., Power, J. F., and Willis, W. O. (1973). Effect of temperature and plant water stress on photosynthesis, diffusion resistance, and leaf water potential in spring wheat. *Agron. J.* **65**, 777–780.

Fry, K. E. (1970). Some factors affecting the Hill reaction activity in cotton chloroplasts. *Plant Physiol.* **45**, 465–469.

Fry, K. E. (1972). Inhibition of ferricyanide reduction in chloroplasts prepared from water-stressed cotton leaves. *Crop Sci.* **12**, 698–701.

Gaastra, P. (1959). Photosynthesis of crop plants as influenced by light, carbon dioxide, temperature, and stomatal diffusion resistance. *Meded. Landbouwhogesch. Wageningen* **59**, 1–68.

Gale, J., Kohl, H. C., and Hagan, R. M. (1966). Mesophyll and stomatal resistances affecting photosynthesis under varying conditions of soil water and evaporation demand. *Isr. J. Bot.* **15**, 64–71.

Ghorashy, S. R., Pendleton, J. W., Peters, D. B., Boyer, J. S., and Beuerlein, J. E. (1971). Internal water stress and apparent photosynthesis with soybeans differing in pubescence. *Agron. J.* **63**, 674–676.

Giles, K. L., Beardsell, M. F., and Cohen, D. (1974). Cellular and ultrastructural changes in mesophyll and bundle sheath cells of maize in response to water stress. *Plant Physiol.* **54**, 208–212.

Glinka, Z., and Katchansky, M. Y. (1970). The effect of water potential on the CO_2 compensation point of maize and sunflower leaf tissue. *Isr. J. Bot.* **19**, 533–541.

Goode, J. E., and Higgs, K. H. (1973). Water, osmotic and pressure potential relationships in apple leaves. *J. Hort. Sci.* **48**, 203–215.

Graziani, Y., and Livne, A. (1971). Dehydration, water fluxes, and permeability of tobacco leaf tissue. *Plant. Physiol.* **48**, 575–579.

Greacen, E. L., and Oh, J. S. (1972). Physics of root growth. *Nature (London), New Biol.* **235**, 24–25.

Green, P. B., Erickson, R. O., and Buggy, J. (1971). Metabolic and physical control of cell elongation rate. *Plant Physiol.* **47**, 423–430.

Hansen, G. K. (1971). Photosynthesis, transpiration and diffusion resistance in relation to water potential in leaves during water stress. *Acta Agr. Scand.* **21**, 163–171.

Harris, D. G. (1973). Photosynthesis, diffusion resistance and relative water content of cotton as influenced by induced water stress. *Crop Sci.* **13**, 570–572.

Heath, O. V. S., and Meidner, H. (1961). The influence of water strain on the minimum intercellular space CO_2 concentration, Γ, and stomatal movement in wheat leaves. *J. Exp. Bot.* **12**, 226–242.

Heinicke, A. J., and Childers, N. F. (1935). The influence of water deficiency in photosynthesis and transpiration of apple leaves. *Proc. Amer. Soc. Hort. Sci.* **33**, 155–159.

Hsiao, T. C. (1973). Plant responses to water stress. *Annu. Rev. Plant Physiol.* **24**, 519–570.

Hsiao, T. C., Allaway, W. G., and Evans, L. T. (1973). Action spectra for guard cell Rb⁺ uptake and stomatal opening in *Vicia faba*. *Plant Physiol.* **51**, 82–88.

Huffaker, R. C., Radin, T., Kleinkopf, G. E., and Cox, E. L. (1970). Effects of mild water stress on enzymes of nitrate assimilation and of the carboxylative phase of photosynthesis in barley. *Crop Sci.* **10**, 471–474.

Humble, G. D., and Hsiao, T. C. (1969). Specific requirement of potassium for light-activated opening of stomata in epidermal strips. *Plant Physiol.* **44**, 230–234.

Humble, G. D., and Hsiao, T. C. (1970). Light-dependent influx and efflux of potassium of guard cells during stomatal opening and closing. *Plant Physiol.* **46**, 483–487.

Humble, G. D., and Raschke, K. (1971). Stomatal opening quantitatively related to potassium transport. *Plant Physiol.* **48**, 447–453.

Husain, I., and Aspinall, D. (1970). Water stress and apical morphogenesis in barley. *Ann. Bot. (London)* [N.S.] **34**, 393–408.

Iljin, W. S. (1923). Der Einfluss des Wassermangels auf die Kohlenstoff-assimilation durch die Pflanzen. *Flora (Jena)* **116**, 360–378.

Johnson, R. R., Frey, N. M., and Moss, D. N. (1974). Effect of water stress on photosynthesis and transpiration of flag leaves and spikes of barley and wheat. *Crop Sci.* **14**, 728–731.

Jones, H. G. (1973a). Limiting factors in photosynthesis. *New Phytol.* **72**, 1089–1094.

Jones, H. G. (1973b). Moderate-term water stresses and associated changes in some photosynthetic parameters in cotton. *New Phytol.* **72**, 1095–1105.

Jones, R. J., and Mansfield, T. A. (1970). Suppression of stomatal opening in leaves treated with abscisic acid. *J. Exp. Bot.* **21**, 714–719.

Kaul, R. (1966). Effect of water stress on respiration by wheat. *Can. J. Bot.* **44**, 623–632.

Keck, R. W., and Boyer, J. S. (1974). Chloroplast response to low leaf water potentials. III. Differing inhibition of electron transport and photophosphorylation. *Plant Physiol.* **53**, 474–479.

Kirkham, M. B., Gardner, W. R., and Gerloff, G. C. (1972). Regulation of cell division and cell enlargement by turgor pressure. *Plant Physiol.* **49**, 961–962.

Koeppe, D. E., Miller, R. J., and Bell, D. T. (1973). Drought-affected mitochondrial processes as related to tissue and whole plant responses. *Agron. J.* **65**, 566–569.

Kozlowski, T. T. (1949). Light and water in relation to growth and competition of Piedmont forest tree species. *Ecol. Monogr.* **19**, 207–231.

Kreusler, U. (1885). Ueber eine Methode zur Beobachtung der Assimilation und Athmung der Pflanzen und über einige diese Vorgäng beeinflussende Momente. *Landwirt. Jahrb.* **14**, 913–965.

Kriedemann, P. E., and Smart, R. E. (1971). Effects of irradiance, temperature, and leaf water potential on photosynthesis of vine leaves. *Photosynthetica* **5**, 6–15.

Kriedemann, P. E., Loveys, B. R., Fuller, G. L., and Leopold, A. C. (1972). Abscisic acid and stomatal regulation. *Plant Physiol.* **49**, 842–847.

Lee, K. C., Campbell, R. W., and Paulsen, G. M. (1974). Effects of drought stress and succinic acid-2,2-dimethylhydrazide treatment on water relations and photosynthesis in pea seedlings. *Crop Sci.* **14**, 279–282.

Little, C. H. A., and Eidt, D. C. (1968). Effect of abscisic acid on budbreak and transpiration in woody species. *Nature (London)* **220**, 498–499.

Loustalot, A. J. (1945). Influence of soil moisture conditions on apparent photosynthesis and transpiration of pecan leaves. *J. Agr. Res.* **71**, 519–532.

McCree, K. J., and Davis, S. D. (1974). Effect of water stress and temperature on leaf size and on size and number of epidermal cells in grain sorghum. *Crop Sci.* **14**, 751–755.

Mansfield, T. A., and Jones, R. J. (1971). Effects of abscisic acid on potassium uptake and starch content of stomatal guard cells. *Planta* **101**, 147–158.

Meyer, R. F., and Boyer, J. S. (1972). Sensitivity of cell division and cell elongation to low water potentials in soybean hypocotyls. *Planta* **108**, 77–87.

Miller, R. J., Bell, D. T., and Koeppe, D. E. (1971). The effects of water stress on some membrane characteristics of corn mitochondria. *Plant Physiol.* **48**, 229–231.

Mittelheuser, C. J., and van Steveninck, R. F. M. (1969). Stomatal closure and inhibition of transpiration induced by (RS)-abscisic acid. *Nature (London)* **221**, 281–282.

Mittelheuser, C. J., and van Steveninck, R. F. M. (1971). Rapid action of abscisic acid on photosynthesis and stomatal resistance. *Planta* **97**, 83–86.

Moldau, K. H. (1972). Effects of a water deficit and the light regime on photosynthetic activity of leaves. *Sov. Plant Physiol.* **19**, 970–975.

Morilla, C. A., Boyer, J. S., and Hageman, R. H. (1973). Nitrate reductase activity and polyribosomal content of corn (*Zea mays* L.) having low leaf water potentials. *Plant Physiol.* **51**, 817–824.

Nir, I., and Poljakoff-Mayber, A. (1967). Effect of water stress on the photochemical activity of chloroplasts. *Nature (London)* **213**, 418–419.

Oechel, W. C., Strain, B. R., and Odening, W. R. (1972). Tissue water potential, photosynthesis, ^{14}C-labeled photosynthate utilization, and growth in the desert shrub *Larrea divaricata* Cav. *Ecol. Monogr.* **42**, 127–141.

O'Toole, J. C. (1975). "Photosynthetic Response to Water Stress in *Phaeolus vulgaris* L." Ph.D. Dissertation, Cornell University, Ithaca, New York.

Pfeffer, W. (1900). "The Physiology of Plants," Vol. 1. Oxford Univ. Press (Clarendon), London and New York (English translation).

Plaut, Z. (1971). Inhibition of photosynthetic carbon dioxide fixation in isolated spinach chloroplasts exposed to reduced osmotic potentials. *Plant Physiol.* **48**, 591–595.

Plaut, Z., and Bravdo, B. (1973). Response of carbon dioxide fixation to water stress. *Plant Physiol.* **52**, 28–32.

Potter, J. R., and Boyer, J. S. (1973). Chloroplast response to low leaf water potentials. II. Role of osmotic potential. *Plant Physiol.* **51**, 993–997.

Probine, M. C., and Preston, R. D. (1962). Cell growth and the structure and mechanical properties of the wall of internodal cells of *Nitella opaca*. *J. Exp. Bot.* **13**, 111–127.

Rabinowitch, E. I. (1951.) "Photosynthesis and Related Processes," Vol. II, Part 1, pp. 858–880. Wiley (Interscience), New York.

Raschke, K., and Fellows, M. P. (1971). Stomatal movement in *Zea mays:* Shuttle of potassium and chloride between guard cells and subsidiary cells. *Planta* **101**, 296–316.

Raschke, K., and Humble, G. D. (1973). No uptake of anions required by opening stomata of *Vicia faba:* Guard cells release hydrogen ions. *Planta* **115**, 47–57.

Redshaw, A. J., and Meidner, H. (1972). Effects of water stress on the resistance to uptake of carbon dioxide in tobacco. *J. Exp. Bot.* **23**, 229–240

Regehr, D. L., Bazzaz, F. A., and Boggess, W. R. (1975). Photosynthesis, transpiration, and leaf conductance of *Populus deltoides* in relation to flooding and drought. *Photosynthetica* **9**, 52–61.

Sacher, J. A. (1973). Senescence and postharvest physiology. *Annu. Rev. Plant Physiol.* **24**, 197–224.

Salter, P. J., and Goode, J. E. (1967). "Crop Responses to Water at Different Stages of Growth." Commonwealth Agricultural Bureaux, Farnham Royal, Bucks, England.

Santarius, K. A. (1967). Das Verhalten von CO_2-Assimilation, NADP- und PGS-Reduktion und ATP-Synthese intakter Blattzellen in Abhängigkeit vom Wassergehalt. *Planta* **73**, 228–242.

Santarius, K. A., and Ernst, R. (1967). Das Verhalten von Hill-Reaktion und Photophosphorylierung isolierter Chloroplasten in Abhängigkeit vom Wassergehalt. I. Wasserentzug mittels konzentrierter Lösungen. *Planta* **73**, 91–108.

Santarius, K. A., and Heber, U. (1967). Das Verhalten von Hill-Reaktion und Photophosphorylierung isolierter Chloroplasten in Abhängigkeit vom Wassergehalt. II. Wasserentzug über $CaCl_2$. *Planta* **73**, 109–137.

Sawhney, B. L., and Zelitch, I. (1969). Direct determination of potassium ion accumulation in guard cells in relation to stomatal opening in the light. *Plant Physiol.* **44**, 1350–1354.

Scarth, G. W., and Shaw, M. (1951). Stomatal movement and photosynthesis in *Pelargonium.* II. Effects of water deficit and of chloroform: photosynthesis in guard cells. *Plant Physiol.* **26**, 581–597.

Schneider, W. G., and Childers, N. F. (1941). Influence of soil moisture on photosynthesis, respiration, and transpiration of apple leaves. *Plant Physiol.* **16**, 565–583.

Shearman, L. L., Eastin, J. D., Sullivan, C. Y., and Kinbacher, E. J. (1972). Carbon dioxide exchange in water stressed sorghum. *Crop Sci.* **12**, 406–409.

Shimshi, D. (1963). Effect of soil moisture and phenylmercuric acetate upon stomatal aperture, transpiration, and photosynthesis. *Plant Physiol.* **38**, 713–721.

Simonis, W. (1947). CO_2-Assimilation und Stoffproduction trocken gezogener Pflanzen. *Planta* **35**, 188–224.

Simonis, W. (1952). Untersuchungen zum Dürreeffekt. I. Mitteilung. Morphologische Struktur, Wasserhaushalt, Atmung und Photosynthese feucht und trocken gezogener Pflanzen. *Planta* **40**, 313–332.

Slatyer, R. O. (1967). "Plant-Water Relationships," p. 256. Academic Press, London.

Slatyer, R. O. (1973). Effects of short period of water stress on leaf photosynthesis. *In* "Plant Response to Climatic Factors" (R. O. Slatyer, ed.), Proc. Uppsala Symp., pp. 271–276. UNESCO, Paris.

Strain, B. R. (1970). Field measurements of tissue water potential and carbon dioxide exchange in the desert shrub *Prosopis julifera* and *Larrea divaricata. Photosynthetica* **4**, 118–122.

Terry, N., Waldron, L. J., and Ulrich, A. (1971). Effects of moisture stress on the multiplication and expansion of cells in leaves of sugar beet. *Planta* **97**, 281–289.

Thoday, D. (1910). Experimental researches on vegetable assimilation and respiration. VI. Some experiments on assimilation in the open air. *Proc. Roy. Soc., Ser. B* **82,** 421–450.

Todd, G. W., and Basler, E. (1965). Fate of various protoplasmic constituents in droughted wheat plants. *Phyton* **22,** 79–85.

Troughton, J. H. (1969). Plant water status and carbon dioxide exchange of cotton leaves. *Aust. J. Biol. Sci.* **22,** 289–302.

Troughton, J. H., and Slatyer, R. O. (1969). Plant water status, leaf temperature, and the calculated mesophyll resistance to carbon dioxide of cotton leaves. *Aust. J. Biol. Sci.* **22,** 815–827.

Upchurch, R. P., Peterson, M. L., and Hagan, R. M. (1955). Effect of soil-moisture content on the rate of photosynthesis and respiration in ladino clover (*Trifolium repens* L.). *Plant Physiol.* **30,** 297–303.

van den Driessche, R., Conner, D. J., and Tunstall, B. R. (1971). Photosynthetic response of brigalow to irradiance, temperature and water potential. *Photosynthetica* **5,** 210–217.

Verduin, J., and Loomis. W. E. (1944). Absorption of carbon dioxide by maize. *Plant Physiol.* **19,** 278–293.

Virgin, H. I. (1965). Chlorophyll formation and water deficit. *Physiol. Plant.* **18,** 994–1000.

Wardlaw, I. F. (1967). The effect of water stress on translocation in relation to photosynthesis and growth. I. Effect during grain development in wheat. *Austr. J. Biol. Sci.* **20,** 25–39.

Willis, A. J., and Balasubramaniam, S. (1968). Stomatal behavior in relation to rates of photosynthesis and transpiration in *Pelargonium. New Phytol.* **67,** 265–285.

Wilson, A. M., and Huffaker, R. C. (1964). Effects of moisture stress on acid-soluble phosphorus compounds in *Trifolium subterraneum. Plant Physiol.* **39,** 555–560.

Wood, J. G. (1929). The relation between water content and amount of photosynthesis. *Aust. J. Exp. Biol. Med. Sci.* **6,** 127–131.

Wright, S. T. C., and Hiron, R. W. P. (1969). (+)-Abscisic acid, the growth inhibitor induced in detached wheat leaves by a period of wilting. *Nature* (*London*) **224,** 719–720.

Zeevaart, J. A. D. (1971). (+)-Abscisic acid content of spinach in relation to photoperiod and water stress. *Plant Physiol.* **48,** 86–90.

CHAPTER 5

WATER SUPPLY AND LEAF SHEDDING

T. T. Kozlowski

DEPARTMENT OF FORESTRY, UNIVERSITY OF WISCONSIN, MADISON, WISCONSIN

I. INTRODUCTION

Premature shedding of leaves often is associated with water deficits in many herbaceous and woody plants. In some species, the loss of leaves as a result of drought may involve true abscission. In this case, cytolysis weakens cells of a separation layer until the subtended parts fall of their own weight or because some external force, such as wind, is applied. In other species, the leaves merely wither and rapidly decay (Kozlowski, 1973).

191

When water deficits induce abscission, the critical internal changes include (1) alteration in amounts and balances of several hormones involved in abscission: auxin, abscisic acid, gibberellin, cytokinin, and ethylene, and (2) synthesis and activity of enzymes that finally hydrolyze pectins of the middle lamella between cells of the abscission layer.

The auxin gradient across the abscission zone exerts a strong regulatory effect on leaf separation. Auxin moves from the leaf to the abscission zone and there maintains physiological processes so as to delay abscission. Auxin also has a role in mobilizing nutrients from weak to vigorous organs and thereby promotes abscission of the weaker organs. Abscisic acid (ABA) appears to accelerate preabscission senescent changes in tissue distal to the abscission zone (Addicott, 1970). However, ABA does not appear to be closely involved in late phases of abscission. Although applied ABA stimulates abscission in petiole explants containing presumptive abscission zones and senescent leaves, abnormally high concentrations are needed (Milborrow, 1974). Gibberellin tends to inhibit abscission by promoting growth of the subtending organ. However, when it is applied to abscission zones proximal to them it induces abscission. In part, gibberellin may also act by promoting synthesis of auxins or hydrolytic enzymes. The influences of cytokinins are similar to those of auxin in maintaining biochemical processes and directing flow of nutrients. These effects may either retard or promote abscission (Addicott, 1970).

In recent years, much emphasis has been placed on the central role of ethylene in the chain of factors controlling abscission (Abeles, 1973; Osborne, 1973). Addicott (1970) suggested that the principal role of ethylene in abscission may be the stimulation of synthesis of energy-rich compounds required for production of enzymes involved in hydrolysis of cell-wall components. Ethylene also tends to reduce the amount of auxin reaching the abscission zone.

Ethylene is a powerful inducer of abscission, with concentrations as low as 0.1 μl/liter capable of initiating the process. In most species where ethylene production was monitored, tissues distal to abscission zones produced large quantities of ethylene during senescence of the tissue and prior to abscission (Jackson and Osborne, 1970). In some herbaceous plants (e.g., *Phaseolus, Xanthium*), no more ethylene was produced by senescing than by green leaves. Nevertheless, in these plants the tissue of the pulvinus or petiole that was distal to and adjoining the abscission zone produced large amounts of ethylene at senescence. In *Phaseolus,* the initiation of abscission was attributed to such localized ethylene production. Applied ethylene or large amounts of ethylene produced by leaves of evergreens that had been treated with auxin also induced senescence and abscission of leaf blades and petioles (Hallaway and Osborne, 1969; Abeles *et al.,*

1971). As emphasized by Abeles (1973), the capacity of ethylene to induce and accelerate abscission depends on the level of auxin at or near the abscission layer.

In normal abscission, blade senescence involving loss of chlorophyll precedes the actual shedding of leaves. Following a severe drought, however, leaves often are shed while still somewhat green. When this happens, the leaf blades may exhibit symptoms similar to those encountered in senescence, with the abscission region reacting similarly to that during normal abscission. Hence, whereas senescence progresses normally from the distal to the proximal region of a leaf, with senescence taking place last in the cells that abut the point of separation, the sequence may be reversed under conditions of severe water deficit so the cells close above an abscission zone may senesce first and the leaf may be shed while green (Osborne, 1973).

Sometimes leaf shedding of water-deficient plants may not occur until after rehydration following drought damage that is not immediately apparent. Then leaves fall rapidly, suggesting that abscission is initiated by a response to water stress injury that cannot be completed without adequate water. This is in accord with the observation that water is needed for normal functioning of cells of the abscission zone. It thus appears that, in some species undergoing water deficits, the leaf abscission zone cannot compete successfully with the rest of the plant for water or obtain enough water for hydrolysis of the middle lamella and cell walls. Also, rehydration allows stems to regain turgor and, thereby, to facilitate shedding of partly abscised leaves (Addicott and Lynch, 1955).

II. SUMMER DROUGHT AND LEAF SHEDDING

Addicott and Lyon (1973) divided woody plants showing summer leaf fall into (1) summer deciduous species and (2) species that shed only some of their leaves during a period of water deficit. Although many tropical trees are classified as summer deciduous, some of them tend to shed their leaves during dry periods whether these occur during the actual summer months or not. Leaf shedding in tropical rain forests is similarly not restricted to a particular time of year. On the other hand, in semievergreen and tropical deciduous forests the loss of leaves is a distinct seasonal characteristic.

Summer deciduous woody plants include many temperate zone species (Marsden, 1950; Kozlowski, 1958, 1971, 1972, 1973) as well as various Mediterrean and desert species (Mooney and Dunn, 1970; Orshan, 1972). Species that shed only some of their leaves during summer droughts include some broad-leaved evergreens (e.g., *Eucalyptus* sp., *Citrus* sp.) and many tropical and subtropical evergreen plants (Addicott and Lyon, 1973).

A. TEMPERATE ZONE SPECIES

Shedding of leaves during dry summers occurs commonly, with the degree of reduction of the total leaf surface varying with moisture conditions from year to year. When herbaceous plants shed leaves by abscission, they do so sequentially in order of their development, with the oldest leaves at the base of the plant dropping first.

Weaver *et al.* (1935) described responses of prairie grasses and forbs to a severe drought in 1934. Grasses, especially *Andropogon scoparius, Koeleria cristata,* and *Bouteloua curtipendula* were completely dried by midsummer. In *Andropogon furcatus,* the two or three basal leaves of a plant turned brown and died. Simultaneously or a little later, the remaining leaves died at the tip and then progressively toward the base as drought increased. *Aster sericeus* was defoliated to near the tip, but species of *Acerates* appeared unharmed. *Euphorbia corollata* retained only a few of its upper leaves. In some cases, all leaves had fallen, but the dwarfed plants continued to bloom. According to Weaver and Albertson (1936), *Andropogon furcatus* was injured less than *Andropogon scoparius* because of the deeper root system of the former.

In broad-leaved woody plants of the temperate zone, leaf yellowing, "scorching," and early leaf fall in response to drought are well known and only a few examples will be given.

Early leaf fall of deciduous street trees occurred in the United States during the severe 1913 drought. In some cases, almost all leaves were shed by the end of July. Near Lincoln, Nebraska, *Celtis occidentalis, Ulmus americana,* and *Populus deltoides* were conspicuously affected. Toward the end of the summer a number of trees, which had been defoliated by drought, put out a second crop of small leaves from previously dormant buds. Most conspicuous examples were *Celtis* and *Gymnocladus dioicus* (Pool, 1913).

During the severe drought of 1913, at least one fourth of the *Acer platanoides* trees in the District of Columbia were visibly affected (Hartley and Merrill, 1915). Practically all injury occurred along leaf margins. When injury was severe, almost the entire leaf was killed and the leaf dropped soon thereafter. *Acer saccharum* and *Acer saccharinum* were injured somewhat less. *Tilia americana* and *Ulmus americana* also were severely injured. Most of the injured trees were nearly defoliated. The injury was localized within trees, with some branches almost wholly defoliated and adjacent ones nearly uninjured. Injury often was most severe on south and southwestern sides of crowns. Leaves in the extreme crown periphery, not shaded by other leaves, were injured more than those in the crown interior.

In the late 1920's and early 1930's the central states were undergoing severe droughts. The 12-month period following June, 1933, was the driest recorded for the Dakotas, Minnesota, Nebraska, Iowa, Illinois, and Missouri. Kansas, Oklahoma, and Colorado also were undergoing extreme drought conditions (Kincer, 1934). In Nebraska, many trees were partially defoliated by drought early in the summer of 1934. Among the species that showed extensive leaf scorching and defoliation were *Prunus virginiana, Salix interior, Ulmus americana, Quercus borealis maxima,* and *Tilia americana.* Trees on hilltops and upper slopes were affected most. In general, injury to foliage and defoliation were most apparent in portions of the crown that were in full sun. By the end of August, leaf rolling, folding, curling, and shedding were intensified. An early outward sign of prolonged drought was reduction in size and number of leaves of deciduous trees. Defoliation of outer portions of the crown was a common sight (Albertson and Weaver, 1945).

Often the shedding of leaves is associated with hot, dry winds that have been variously called foehns (Switzerland, Formosa, U. S.), chinooks (northwestern U. S.), siroccos (Palestine, N. Africa), levantos (Canary Islands), Santa Anas or northers (southern California), and sukhoveys (southern Russia). In mountainous regions, such as the east slope of the northern and central Rocky Mountains of the United States, warm, dry chinook winds appear suddenly (Henson, 1952). In southern California, hot, dry northers from the interior deserts cause extensive desiccation injury to citrus groves. Similar injury to *Citrus* trees by hot, dry winds has been reported in North Africa (Boeuf and Genet, 1906), the West Indies (Chalot and Deslandes, 1914), and Palestine (Georgii, 1920).

In California, leaf desiccation of *Citrus* by dry winds was followed by two types of injury described as "windburn" and "scorch" (Reed and Bartholomew, 1930). Leaves killed by windburn first wilted, then dried out rapidly, and became brittle within a day. If the wind ceased after a few hours, the wilted leaves sometimes recovered. However, if the leaves were desiccated beyond a critical threshold, they did not recover. Old leaves were killed faster than young ones. In a few days, the windburn-killed leaves were shed. Scorch, the less common type of injury, resulted from hot winds. Blades of scorched leaves generally turned brown and became brittle within a few hours, without an intermediate wilting stage. Scorched leaves remained attached to twigs for several weeks, even though exposed to strong winds.

In parts of Australia (New South Wales and southern Queensland), the first half of 1965 was unusually dry. Precipitation from January 1 to

June 30 was less than one-fourth of normal and by early April plants were undergoing severe water stress. Foliage of several species of *Eucalyptus* first became dull, turned yellow, and wilted. Leaves then turned brown and brittle, but were retained on the trees except in windy situations. Leaf injury was common on trees growing in light-textured, stony, or shallow soils, whereas leaves of trees on adjacent, heavy soils were not injured. In addition to desiccation of foliage, the occurrence of bark fissuring and separation of wood from bark indicated a state of severe dehydration (Pook *et al.,* 1966).

Much attention has been given to premature abscission of leaves and bolls of cotton (*Gossypium hirsutum*) resulting from water stress (Bruce and Romkens, 1965; Bloodworth *et al.,* 1956; Stockton *et al.,* 1961). Actual separation does not occur during the period of water stress, but follows relief from the deficit and rehydration of the abscission zone (Mc-Michael *et al.,* 1972; Jordan *et al.,* 1972). Figure 1 shows that leaf abscission of cotton was a linear function of the degree of water deficit. At plant water potential recovery values (ψ_R) of -5 to -6 bars little abscission occurred. Appreciable leaf fall occurred only after ψ_R reached -8 to -10

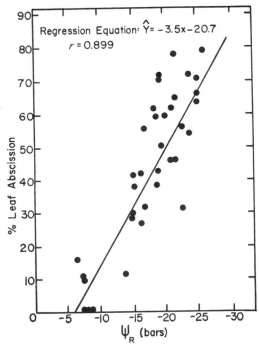

Fig. 1. Relation of leaf abscission in cotton following relief from water deficits to severity of water deficit (Ψ_R). From McMichael *et al.* (1973).

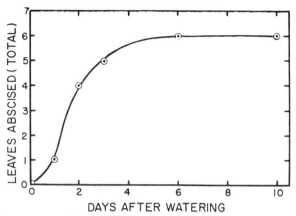

Fig. 2. Time course of leaf abscission in cotton following relief from a severe plant water deficit. Nine leaves were present on the plant at the time of rewatering. From McMichael *et al.* (1973).

bars. As shown in Fig. 2, the leaves abscised only after plant water deficits were relieved by irrigation. Then abscission occurred rapidly, with the oldest leaves being shed first. Abscission layers of leaves at the first or second nodes generally formed within 12 hours after rewatering, whereas several days were needed for formation of abscission layers of upper, more juvenile leaves (McMichael *et al.,* 1973).

Increased production of ethylene during drought appears to be important in abscission of cotton leaves (Jordan *et al.,* 1972). As shown in Fig. 3, there was little abscission during the first 4 days after water deficits were relieved by irrigation in plants that were not treated with ethylene. Water deficits alone, within the water potential range of −6.7 to −12.3 bars, did not cause abscission. Abscission was stimulated by exposing plants to ethylene, however, and the amount of abscission was related to the degree of water deficit. Results of a more intensive experiment on the relation between plant water potential and ethylene-induced leaf abscission of cotton are shown in Fig. 4. Leaf abscission was measured 24 hr after treatment with ethylene and relief from water deficits. Treatment with ethylene reduced the plant water deficit required to induce abscission from −17 to −7 bars, which indicated that water deficits caused the tissue to become predisposed to ethylene action.

B. TROPICAL SPECIES

In the tropics, the availability of water often determines whether a species is classified as evergreen, deciduous, or partly deciduous. For example, *Gossampinus malabarica* and *Tectona grandis* are deciduous in areas

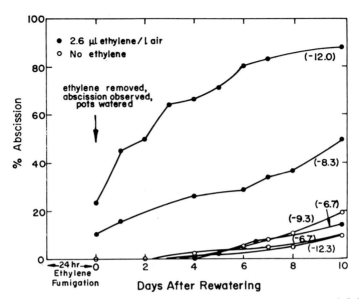

Fig. 3. Cotyledonary leaf abscission of cotton resulting from water deficits alone (○) or water deficits in combination with exogenous ethylene (●). Values in parentheses are plant water potential in bars before treatment with ethylene. From Jordan *et al.* (1972).

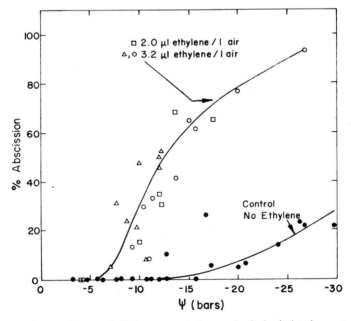

Fig. 4. Effects of water deficits on cotyledonary leaf abscission in control cotton plants receiving no ethylene (●) and plants treated with ethylene at concentrations of 2.0 (□) and 3.2 μl ethylene/liter air (△, ○). From Jordan *et al.* (1972).

with alternating wet and dry seasons, but nondeciduous in areas that are constantly wet (Merrill, 1945).

In tropical areas with a distinct dry season, as in the "caatinga" and "cerrado" of Brazil, much leaf shedding occurs at the height of the dry season (Ferri, 1961; Eiten, 1972). In cerrado, when the upper layers of soil dry out, usually below the wilting point, the aerial parts of most of the herbaceous plants dry up, and the dead leaves of many forbs disintegrate and disappear. The woody plants are only partially drought evading. Individuals of some species lose all their leaves and those of other species shed only some of their leaves before new ones grow. Hence, the mass of woody plants has only a fraction of the living leaves that were present in the wet season. In many species, new leaves emerge even before the rains start again. The leaves that are retained during the dry season lose considerable water by stomatal transpiration. Such transpiration is possible because the plants have deep roots that obtain water from moist soil above a deep water table (Eiten, 1972).

In West Africa, many trees lose all their leaves with onset of the dry season, but some of them produce a new leaf crop before the dry season ends (Richards, 1952). In Trinidad, some species are leafless for only a few days or weeks at the beginning of the dry season and leaf out completely during the height of the dry season (Beard, 1946). Growth of a leaf crop before the dry season ends also is characteristic of trees of Ceylon and Java (Wright, 1905; Richards, 1952).

Many trees of tropical rain forests readily lose all their leaves in response to even mild droughts, with their pattern of leaf shedding not necessarily tied to an annual cycle. Richards (1952) considered variations among rain forest trees to be so great that he could not distinguish clearly between evergreen and deciduous trees. For convenience, he regarded a species as evergreen if its members carried a substantial number of leaves throughout the year. Deciduous species were considered to be those that lost all or almost all of their leaves, even if only for a few days. Deciduous trees, as defined above, are numerous in rain forests in all tropical areas, even in climates with very evenly distributed rainfall.

C. Leaf Shedding and Drought Resistance

Reduction of the transpiring surface is an effective drought evading adaptation of many species of desert and Mediterranean woody plants (Table I). Orshansky (1954) considered, for example, that reduction of leaf surface was the most important factor in the water economy and survival of plants in the Near East. Evenari *et al.* (1971) also emphasized that low transpiration rates of desert plants during the dry season were associ-

TABLE I

EFFECT OF SHEDDING OF LEAVES AND SHOOTS OF SHRUBS
OF ARID REGIONS ON SHOOT WEIGHT[a]

Climatic zone and species	Shoot weight (% of spring weight)	
	1957	1958
Mediterranean: Judean Hills		
Phlomis viscosa	51	52
Cistus villosus	61	59
Teucrium polium	45	43
Scrophurlaria xanthoglossa	50	44
Thymus capitatus	36	50
Trano-Turanian, Semidesertic		
Artemisia herba-alba	28	26
Noea mucronata	2	0
Saharo-Sindian, Desertic		
Haloxylon articulatum	28	33
Zygophyllum dumosum	14	4
Anabasis spp.	45	33
Anabasis spp. (drier locality)	27	33
Artemisia monosperma	45–52	63

[a] Data Represent Weight of Shoots at End of Dry Summer as % of Weight of Shoots in the Spring. From Oppenheimer (1960).

ated with marked reduction in transpiring surface by shedding of leaves and branches.

Reduction of the transpiring surface of arid-zone plants of the Near East occurs in a number of patterns which may involve shedding of leaflets, death and shedding of the green sheath of some plants, and acropetal leaf fall in others (Orshan, 1972). *Retama raetam* and other broomlike xerophytes produce green shoots with small leaves. The latter are shed during droughts, with the shoots remaining as photosynthetic organs. In *Noea mucronata*, the shedding of leaves is combined with shedding of branches. *Artemisia herba-alba* has large leaves at the beginning of the rainy season. These are shed early in the dry season and leaves subsequently produced during the dry season are increasingly smaller, until the last-formed ones are very small scales. *Zygophyllum dumosum* sheds leaf blades, but not petioles. During shedding of blades, the petioles undergo structural changes that restrict the capacity of stomata to open fully.

In southern California, *Encelia farinosa* is leafless during the dry season, but rain causes production of new leaves. Their structure depends on the amount of water available to induce cell expansion. As soil moisture is depleted, the leaves that develop on the higher nodes have a denser structure. Further depletion of soil moisture causes leaves of lower density thickness to develop progressively higher water deficits until they are shed. The leaves of greater density thickness, higher on a branch, do not develop water deficits as severe as those in the leaves of less-dense structure. These small leaves therefore are retained longer into the dry period. Such seasonal leaf variability undoubtedly is a drought evading mechanism (Cunningham and Strain, 1969).

Ocotillo (*Fouquieria splendens*) of the arid southwestern United States sheds its leaves during drought and grows a new crop several days after a rain. The number of leaf crops varies, but as many as four or five may be produced annually (McCleary, 1968). Desert plants of Central Asia that shed leaves or shoots at the beginning or middle of the dry season include species of *Calligonum, Ephedra, Kochia, Artemisia, Salsola,* and *Chondrilla* (Sveshnikova and Zalensky, 1956).

The Mediterranean climatic type (summer drought, winter rain) occurs in California, South Africa, Central Chile, southern Australia and Mediterranean region. In these areas, most of the dominant shrubs evade drought and survive by shedding all or a portion of their leaves during the dry summer (Mooney and Dunn, 1970). An example is the California coastal sage plant community, which grows at elevations up to 300 m. The most common species of this community include *Artemisia californica, Encelia californica, Lotus scoparius, Ergiogonum fasciculatum, Viguera laciniata,* and species of *Eriodictyon* and *Salvia.* The soft leaves of these species readily wither and die as summer drought progresses (Harrison *et al.,* 1971). The leaf shedding habit is also shown by a comparable group of plants in Chile which has a similar climate (Mooney *et al.,* 1970).

The advantage of leaf shedding during the dry season in some tropical areas has been questioned. Richards (1952) pointed out that leaf shedding sometimes was poorly correlated with occurrence of the dry season. For example, in evergreen forests of West Africa, some species of trees expand their leaves long before the dry season ends, when dangers of desiccation are greatest. This also is true with some species of the Middle Niger Valley (Roberty, 1946), Trinidad (Beard, 1946), and Ceylon (Coster, 1923). Richards (1952) concluded that tropical rain forest trees could retain their leaves throughout the year without risking injury.

III. WINTER DESICCATION AND LEAF SHEDDING

Considerable winter and early-spring leaf scorching and defoliation occur in evergreens of the temperate zone because of leaf desiccation. Such injury is well known in ornamental woody plants (Felt, 1943; Havis, 1971) and forest trees (Kozlowski, 1955, 1958, 1964, 1967, 1968a, 1970, 1971; Kramer and Kozlowski, 1960). When the injury is severe, all the leaves and buds (and often the trees) are killed. More commonly, however, the leaves are killed but the buds are not, and the trees survive. Eventually the dead leaves are shed, with the most obvious symptoms occurring on the windward sides of trees (Boyce, 1961). Such winter desiccation injury often has been erroneously attributed to direct thermal effects.

A. Extent of Winter Desiccation Injury

Winter desiccation damage to gymnosperms is one of the most important factors limiting the range of forest trees in several countries including Japan (Sakai, 1968, 1970), parts of the United States (Parker, 1951, 1965; Kozlowski, 1958, 1968a,b; White and Weiser, 1964), Austria (Pisek and Winkler, 1953; Pisek and Larcher, 1954; Larcher, 1972), and Germany (Michael, 1963, 1966).

On northeastern slopes of windswept mountainous areas of Japan at elevations above 600 m, the soil often is frozen and considerable desiccation injury occurs to several species (e.g., *Cryptomeria japonica; Picea glehnii, Abies sachalinensis*). Trees growing on the lee side of windbreaks escape injury. Practically no desiccation injury occurs on southern and eastern slopes in middle and eastern Japan where the soil does not freeze during the winter (Sakai, 1968, 1970; Okonoue and Sasaki, 1960). In Hokkaido, where the soil usually freezes all winter, even in the lowlands, winter desiccation injury occurs commonly in all windswept areas regardless of direction of slope (Sakai, 1970). The time of appearance of desiccation injury varies from year to year and is influenced by time of soil freezing, depth of frozen soil, depth of snowfall, depth of snowcover, humidity, and wind velocity (Sakai, 1968, 1970).

Many alpine plants rarely undergo severe water deficits during the summer. In winter, however, evergreen shrubs and small trees on exposed sites without snow at or above timberline undergo severe drying of tissues (Tranquillini, 1963, 1964; Larcher, 1963; Lindsay, 1967). If shrubs (e.g., *Rhododendron ferrugineum*) are covered with snow, they tend to rehydrate; but if the snow is removed, such plants desiccate rapidly.

Winter desiccation of gymnosperms in North America has been well documented. Lindsay (1967) found that leaves of *Picea engelmannii* in the windy timberline regions of Wyoming (3300 m) had water potentials near or below −30 bars from September to early June. By midsummer the values rose to −15 bars, but during late-summer drought these values decreased again to the low winter values. By comparison, leaf water potentials of this species in a closed forest 250 m lower in altitude were higher throughout the year. There was a deep snowcover in the forest and leaf water potential generally did not drop below —20 bars, but at timberline it was —34 bars. Similar results were obtained for *Abies lasiocarpa.*

In the Adirondack Mountains of New York State, winter desiccation injury of gymnosperms appeared in early March of 1948 with whole mountainsides becoming covered with discolored trees (Curry and Church, 1952). Some of the more heavily damaged stands appeared as if scorched by fire. Within a few weeks, extensive defoliation occurred. Injury varied among species in the following descending order of severity: *Picea rubens, Tsuga canadensis, Pinus strobus,* and *Abies balsamea.* Relatively little damage was noted on *Picea mariana, Picea glauca,* and *Pinus resinosa.* In general, damage occurred through the length of the crown, but injury was largely confined to the south and southwest sides of trees.

During the winter of 1947–1948 in Wisconsin, gymnosperms in forest stands, plantations, and nurseries, as well as ornamentals, were severely damaged by drought. Nearly all species of gymnosperms were affected, with injury most prominent on *Pinus silvestris, P. resinosa,* and *P. banksiana,* in order of decreasing severity. The injury was preceded by an extended period of autumn drought and high temperature. In January, when injury occurred, several bright, unseasonably warm days with strong southerly winds were followed by very cold nights. The injury was common on dry uplands, but not on lowlands with an accessible groundwater table (Voigt, 1951).

Winter desiccation injury often occurs when warm winds sweep over a region, especially following a cold period. In the mountainous regions of the western United States, *Pinus ponderosa, P. contorta, Pseudotsuga menziesii,* and *Picea pungens* are frequently injured by desiccating winds. Winter drying injury has been called red belt when it occurs in horizontal stripes on mountainsides. Such zonation may result when the soil is frozen at certain elevations and not others, or when warm chinook winds follow belts (Hedgcock, 1912; Melrose, 1919). The bands of red belt may be regular to irregular and they often vary in width. When lower branches of trees are buried by snow, they usually do not sustain injury and have a full set of healthy green needles. In typical red-belt injury, there is a tendency for the older needles to be shed first and for the youngest needles to remain uninjured and attached to the tree (Hubert, 1930).

MacHattie (1963) described red-belt injury in the Rocky Mountain forests of western Alberta. The injury affected *Picea glauca, Pseudotsuga menziesii,* and *Pinus contorta,* but was less obvious in the first two species because they shed their needles soon after being injured. By comparison, the reddish discoloration of desiccated *Pinus contorta* needles was noticeable for a distance of 5 miles. Snow-covered foliage was not injured and the snow even protected portions of needles. When snow had been lying on top of a branch, the upper needles remained green, whereas those growing on the underside of the branch were injured. Windward sides of trees were injured appreciably more than the protected sides. The injury also was more evident on slopes exposed toward the south than on those exposed to the north.

Winter drying injury of conifers in western Oregon and Washington, called parch blight, occurs when dry winds from the east suddenly cross the Cascade Mountains (Boyce, 1961). Generally, the soil is not frozen when injury occurs. Nevertheless, the rate of absorption of water lags behind transpiration to such a degree that the foliage becomes desiccated.

B. MECHANISM OF WINTER DESICCATION INJURY

Winter desiccation injury occurs when absorption of water cannot keep up with transpirational losses. In many parts of the temperate zone, appreciable transpiration occurs as the air warms up sufficiently during sunny winter or spring days and increases the vapor pressure gradient between the leaves and surrounding air. Since the soil is cold or frozen, water cannot be absorbed through the roots rapidly enough to replace transpirational losses (Kramer, 1942; Kozlowski, 1943) and the shoots become desiccated.

It is well known that low soil temperatures decrease the rate of absorption of water. Because greenhouse plants sometimes wilt if irrigated with cold water, growers sometimes heat the water. Usually low temperatures decrease absorption of water more in plants that normally grow in warm soil than in those that grow in cooler soil. For example, absorption of water of such southern species as cotton, watermelon (*Citrullus vulgaris*), *Pinus taeda,* and *P. elliottii* was reduced more by cooling the soil than was absorption of water of such northern species as collard (*Brassica oleracea acephala*), *Pinus strobus, P. resinosa,* and *Ulmus americana* (Kramer, 1942; Kozlowski, 1943). The rate of absorption is reduced more by rapid cooling than by slow cooling of soil (Böhning and Lusanandana, 1952).

Among the important causes of decreased absorption of water at low temperatures are physical effects on root resistance of increased water viscosity and decreased permeability of root cell membranes. Other factors, such as decrease in rate of root growth, hydraulic conductivity, and root metabolism, may also contribute to decreased absorption, but these appear to be much less important (Kramer, 1940, 1969).

Although some investigators have questioned whether xylem water freezes (Handley, 1939), it has been shown that at low air temperatures the supply of water to the leaves is reduced in stems chilled to temperatures approaching the freezing point (Johnston, 1959). Furthermore, when stems are chilled slightly below 0°C, the water in xylem elements freezes. Zimmermann (1964) determined the freezing point of xylem sap in *Acer rubrum, A. saccharum, Betula papyrifera, Fraxinus americana, Prunus serotina, Quercus rubra,* and *Ulmus americana* trees. The lowest supercooling was −3°C, but in most cases freezing began at about −1.5°C. Once freezing started, the temperature rose briefly to between 0° and 1°C and then decreased again to the value previously reached. Wilting of trees never occurred above −1°C. However, as soon as the temperature was lowered appreciably below −1°C, wilting occurred in all species investigated. These experiments generally corroborated those of Johnston (1959) who determined the freezing point of stem water of *Pinus radiata.* When stem temperatures were kept above −2°C, the rate of transpiration was not decreased. However, when stem temperature was kept slightly below −2°C for 7 days, leaf water deficits of shoots increased greatly, but returned to normal levels within 4 days after freezing was discontinued.

Winslow and Havis (1967) chilled *Ilex opaca* stems to temperatures below 0°C and determined their capacity to conduct water. The freezing points of 8 stem sections varied from −0.7° to −1.2°C, with an average of −1.0°C. These freezing points agreed closely with those of Zimmermann (1964) for other species. As temperature decreased, water transport was not decreased down to a temperature of 7.2°C. However, capacity for water movement was reduced by about half at 0.6°C. Reduced flow continued down to −1.0°C and stopped in all plants when temperatures reached −1.1° to −1.4°C.

After freezing, a relatively high stem temperature was required for resumption of water movement. For example, water did not move in stems held between 0.0° and 0.3°C for 2 hr. However, water movement resumed immediately when the temperature was raised to 0.5°C. The rate of transport at between 0.5°C and 1.0°C was about half that above 1°C. The rate above 1°C after chilling was about the same as the rate at 7.0°C or higher before chilling.

Havis (1971) demonstrated that when water froze in vessels of *Kalmia latifolia* and *Ilex opaca,* some water continued to move upward, presumably through the micropores in cell walls. On further freezing, however, the water in the micropores also froze and upward water transport then ceased.

IV. FLOODING AND LEAF SHEDDING

Saturation of soil with water for extended periods leads to a number of sequential physiological disturbances in plants, with reduced absorption of water and leaf water deficits among the earliest responses. Later responses to flooding include reduced mineral uptake, leaf epinasty, stem thickening, formation of callus and adventitious roots, leaf senescence, and leaf abscission, usually beginning with the lower leaves. The extent of injury by flooding varies with species, soil factors, timing and duration of flooding, and physicochemical conditions of the floodwater (Gill, 1970; Rowe and Beardsell, 1973).

A. Variations in Response to Flooding

Wide inter- and intraspecific differences in response to inundation have been demonstrated. Among forest trees, the gymnosperms as a group generally are injured more than angiosperm trees (Gill, 1970). However, flooding is tolerated more by a few gymnosperms (e.g., *Taxodium distichum*) than by many angiosperms. There often is wide variation in flooding tolerance of closely related species. For example, *Nyssa aquatica* tolerates flooding much better than *Nyssa silvatica* or *Nyssa sylvatica* var. *biflora* (Hall and Smith, 1955). Marked interspecific variations in tolerance to flooding are evident in zonation of species along margins of lakes, reservoirs, or river floodplains, as well as in differences in injury caused by occasional flooding of species growing in normally unflooded communities.

Roots of species such as *Taxodium distichum, Nyssa* spp., and *Salix* spp. have low oxygen requirements and are very tolerant to flooding. Parker (1950) noted that transpiration of *Taxodium distichum* seedlings was not decreased by flooding whereas that of *Juniperus virginiana, Quercus alba, Q. rubra, Q. lyrata, Q. prinus,* and *Cornus florida* was markedly reduced. When the soil was drained after flooding, the rate of transpiration usually increased. Pirone (1972) classified susceptibility of shade trees to poor soil aeration (caused by flooding, heavy soils, compaction, etc.) as shown in the following tabulation.

Yelenosky (1963) noted variations among species in time of leaf abscission in response to flooding. Seedlings of *Liriodendron tulipifera* lost all their leaves within 2 wk after the soil was flooded. *Quercus alba* and

Most severely injured	Less severely injured	Least injured
Acer saccharum	*Betula* spp.	*Salix* spp.
Fagus grandifolia	*Carya* spp.	*Ulmus americana*
Cornus spp.	*Tsuga* spp.	*Populus* spp.
Quercus spp.		*Platanus* spp.
Liriodendron tulipifera		*Quercus palustris*
Pinus spp.		*Robinia pseudoacacia*
Picea spp.		

Acer saccharinum lost their leaves within 3 wk, *Gleditsia triacanthos* within 4 wk, and *Ulmus americana* after 8 wk or more. For further information on species variations in tolerance of forest trees to flooding, the reader is referred to McDermott (1954), Ahlgren and Hansen (1957), Hosner (1960), Hosner and Boyce (1962), Hermann and Lavender (1967), and Gill (1970).

Rowe and Beardsell (1973) pointed up wide differences in sensitivity of orchard trees to waterlogging, with species varying in the following order of decreasing sensitivity: olive > almond = peach = apricot > cherry > plum = citrus > apple > pear > quince. Among citrus rootstocks, sensitivity to flooding varied as follows: sour orange = sweet lime > Cleopatra = mandarin = sweet orange > rough lemon = trifoliata.

As emphasized by Gill (1970), the specific rankings of flooding tolerance of species as determined by different investigators should be viewed cautiously because of variations in experimental procedures. For example, *Fraxinus pennsylvanica* was considered more tolerant than *Populus deltoides* to flooding following complete inundation. However, when soil saturation tests were used, the ranking of these species was reversed (Hosner, 1958, 1959). Another problem with species rankings is that most studies are based on work with seedlings. Yet it is known that mature vigorous trees are damaged less by flooding than are seedlings or overmature trees of the same species. Gill (1970) also called attention to intraspecific variation in flooding tolerance. When genetic variation in flooding tolerance existed within a species, cyclic flooding exerted selective effects so that populations of regularly flooded plants became genotypically more resistant to flooding than did populations of normally unflooded plants.

B. Timing and Duration of Flooding

For a wide variety of angiosperms and gymnosperms, flooding is much more harmful during the growing season than during the dormant season. For example, Childers and White (1942) observed that flooding during the summer injured apple trees and Heinicke (1932) noted that winter flooding did not injure apple trees if the excess water was drained before the growing season started. According to Alben (1958), free water in the lower root zone of Stuart pecan trees in September and October caused scorching of leaves and premature defoliation from the tops downward. By comparison, waterlogged subsoil conditions in April and May did not cause visible injury. McAlpine (1961) showed that *Liriodendron tulipifera* seedlings were not adversely affected by flooding during the dormant season. Yet seedlings were killed after 4 days of flooding in the early growing season (May) and after 3 days in early summer (June). Wilting and death of shoots increased as duration of flooding increased. According to Hall and Smith (1955), even the most flood-tolerant species must be free of flooding for at least half of the growing season.

Greater flooding injury during the growing season than during the dormant season may be expected because of higher oxygen requirements of growing roots and higher water requirements of plants. Root respiration increases as temperature rises, which results in increase in oxygen requirements that induces more rapid root injury than would occur when the oxygen requirement is low. Also, at high temperatures, the increased activity of microorganisms leads to rapid depletion of oxygen and accumulation of CO_2. Increased severity of injury to root systems is also important at high temperatures. For example, Kramer and Jackson (1954) reported that at 20°C root systems of flooded tobacco plants showed little damage, whereas at 34°C roots and stem bases were dead or dying.

C. CAUSES OF FLOODING INJURY

There has been vigorous debate about causes of injury to plants in flooded soils. In addition to causing water deficits in leaves, prolonged inundation of soils by standing water is associated with a decrease of soil oxygen, increase in CO_2, and accumulation of a variety of toxic substances including nitrites, reduced forms of iron and manganese, hydrogen sulfide, aliphatic carboxylic acids, and ethylene (Rowe and Beardsell, 1973). Flooding by stagnant water is much more injurious than by flowing water (Hunt, 1951).

The causes of flooding injury are complex as indicated by occurrence of sequential plant responses to prolonged inundation. Yelenosky (1963) noted that a very early symptom of flooding injury was pronounced water stress and wilting of leaves. Much later responses included leaf senescence and abscission. It appears likely that the very early effects of flooding (e.g., reduced absorption of water and minerals) are responses to poor soil aeration. However, reduction in the rate of water and mineral uptake cannot explain a number of later responses to flooding such as epinastic curvatures, formation of callus and adventitious roots, and leaf abscission. Using approach-grafted *Lycopersicon* and *Nicotiana* plants with split-root systems, Jackson (1956) showed that growing shoots depended on functional root systems for growth factors other than water and minerals. He also showed that chlorosis or reduced growth of lower leaves was not due to root toxins. Such observations suggest that the environmental and internal factors that induce early flooding responses set in motion a complex series of physiological disturbances which, in later stages, are influenced by additional internal factors, to finally induce leaf abscission.

Flooding reduces water absorption directly by decreasing the permeability of roots to water and indirectly by reducing the size of the root system. Childers and White (1942) found that absorption of water by roots of Stayman Winesap apple trees was reduced within 2 to 7 days after the roots were flooded. In some cases, the rate of transpiration became so low under continuous flooding that it could not be measured. Development of new roots and formation of root hairs also were inhibited by continuous flooding. Chang and Loomis (1945) observed that absorption of water by wheat, maize, and rice in nutrient solution was reduced by 14 to 50% when CO_2 was bubbled through the solution for 10 min out of each hr. When air was then bubbled through the solution, the rate of absorption of water increased appreciably.

Flooding of soil caused rapid decrease in permeability of tobacco roots to water (Kramer and Jackson, 1954). The rate of water absorption after only an hour of flooding decreased to 60% of the rate after 15 min of flooding. There was a steady decrease in water intake until the rate leveled off, after 3 hr of flooding, to about one-fourth the rate after 15 min of flooding. After 24 hr, the rate of water absorption had increased, and, at 43 hr, it was about 5 times as high. The data indicated that roots were injured after a day or two to increase their permeability, but they were not sufficiently damaged to decrease their absorbing capacity. Kramer and Jackson (1954) concluded that the usual course of events under sustained flooding was the following: Deficient aeration produced rapid but temporary decrease in permeability of roots, probably because of increased viscosity of protoplasm. Continued deficient aeration caused injury and ultimately death of some cells, which resulted in increased permeability. Plugging of xylem vessels then occurred, decreasing permeability, with decay finally resulting in permeability or total destruction of the root system.

There has not been general agreement on whether a lack of oxygen or excess of CO_2 is most important in causing early injury to flooded plants. Data of Kramer and Jackson (1954) and Harris and van Bavel (1957) were interpreted as showing that excessive CO_2 was more detrimental to growth of tobacco than was a low level of O_2. However, Williamson and Splinter (1968) showed that either a lack of O_2 or excessive CO_2 caused injury to tobacco. Oxygen levels of 1% caused wilting and reduced growth, and pure nitrogen treatments caused a typical "drowning" reaction (e.g., severely wilted leaves and senescence of lower leaves). Addition of 20% CO_2 to 1% O_2 caused additional wilting over that obtained with 1% O_2, but it also caused leaf mottling, an injury not normally associated with "drowning." These studies indicated very little effect of adding 18.5% CO_2 to an atmosphere of 2.5% O_2. Because the pure nitrogen treatment was considerably more injurious to tobacco than the 1% O_2–20% CO_2 treatment, and because it killed the roots only slightly less quickly than did 21% CO_2 without O_2, Williamson and Splinter (1968) concluded that lack of O_2 was more injurious than the concentration of CO_2 per se in flooded soil. Willey (1970) also emphasized the importance of oxygen deficiency in flooded soil.

1. Hormone Relations

Much interest has been shown in hormone balances in flooded plants in relation to abscission. Epinastic curvatures and formation of callus and adventitious roots in flooded plants generally have been correlated with high auxin levels. Kramer (1951) suggested that oxygen deficiency around the stem base caused a decrease in polar transport through that part of the stem, leading to a high auxin content in shoots. Phillips (1964a) envisaged an early effect of flooding to raise auxin levels in shoots (either by preventing entry of auxin from the stem, or by inhibiting oxidative catabolism of shoot auxin in the roots) and also to stop synthesis of a nonauxin shoot growth hormone in the roots. Phillips (1964b) found that ethanol extracts of shoots and roots of *Helianthus annuus* contained an auxin similiar to indole-3-acetic acid. After 14 days of flooding, the concentration of this auxin tripled in shoots. There was a subsequent decrease in auxin content to the control level. The effect of flooding on shoot auxin content was correlated with characteristic morphological changes. Flooding also delayed the normal decrease in auxin content of roots of control plants.

Both gibberellin and cytokinin levels are influenced by flooding. It is known that gibberellins that are synthesized in roots are translocated in the xylem sap to the shoots (Burrows and Carr, 1969). Since waterlogging reduces gibberellin levels in roots, it also decreases the amount available for shoots. Furthermore, applications of GA_3 stimulated growth of flooded plants more than growth of unflooded plants (Reid *et al.,* 1969; Reid and Crozier, 1971). Flooding also reduced activity and production of cytokinin in roots. Chlorosis of lower leaves of flooded plants has been attributed, at least in part, to decrease in cytokinin synthesis in roots (Burrows and Carr, 1969).

Ethylene has a central role in inducing injury and eventual leaf shedding in flooded plants. This view is based on three lines of evidence: (1) ethylene produces responses in unflooded plants that are similar to those associated with flooding, such as leaf epinasty (Denny and Miller, 1935), stem thickening (Zimmermann *et al.,* 1939), and leaf senescence and abscission (Abeles, 1973; Osborne, 1973); (2) ethylene production is stimulated in poorly aerated or flooded soils and plants (Smith and Russell, 1969; Smith and Restall, 1971); and (3) application to plants of ethylene-releasing chemicals induces leaf abscission (Abeles, 1973).

Smith and Russell (1969) compared concentrations of ethylene produced in soil at field capacity under aerobic and anaerobic conditions. Cylinders of soil were either sealed or left open to the atmosphere. After 7 days, the anaerobic soil had ethylene concentrations of 9.3 to 10.6 ppm, whereas soil exposed to the atmosphere had concentrations of only 0.07 ppm at a depth of 15 cm to 0.14 ppm at a depth of 60 cm. Smith and Restall (1971) showed that, under anaerobic conditions, if no loss of ethylene occurred, its concentration in the soil atmosphere reached or exceeded 20 ppm in widely different soil types. Microorganisms appeared to be responsible for occurrence of ethylene in anaerobic soils. Lynch (1972) attributed high ethylene production in poorly aerated soils to *Mucor hiemalis* and two yeasts.

Flooding increased ethylene concentrations in softwood cuttings of *Malus robusta, Ligustrum obtusifolium,* and *Chrysanthemum morifolium* (Table II). Flooding the bases of intact plants of *Chrysanthemum morifolium, Helianthus annuus, Lycopersicon esculentum,* and *Raphanus sativus* for 24 hr increased ethylene concentration in submerged stems and roots, while it did not change ethylene concentration in controls (Kawase, 1972). When *Helianthus annuus* plants were flooded, ethylene in roots and stems below the water line began to increase. This coincided with beginning of hypocotyl hypertrophy and new root formation in hypocotyls. Correlations were strong between ethylene concentration and chlorophyll

TABLE II

EFFECT OF SUBMERSION ON ETHYLENE CONCENTRATION IN *Malus robusta,*
Chrysanthemum morifolium, AND *Ligustrum obtusifolium*[a]

| | Ethylene concentration (ppm) | | | |
| | | | LSD | |
Cutting and treatment	Before treatment	After treatment	5%	1%
Malus robusta				
Control	0.07	0.32	0.13	0.23
Submerged	0.06	1.66		
Chrysanthemum morifolium				
Control	0.06	0.37	0.30	0.55
Submerged	0.06	0.67		
Ligustrum obtusifolium				
Control	0.06	0.50	0.40	0.74
Submerged	0.06	1.20		

[a] Cuttings were completely submerged in water for 20 hr and controls were steeped upright in water 2.5 cm deep. From Kawase (1972).

breakdown and epinasty. Plant responses caused by soil application of Ethephon, which was absorbed and decomposed to ethylene in plants, were almost identical with those caused by flooding. The data indicated that the increase in ethylene concentration in flooded plants was, largely, but not exclusively, responsible for flooding damage symptoms (Kawase, 1974).

2. *Tolerance to Flooding*

Both anatomical and physiological adaptations are important in tolerance of flooding. Formation of aerenchyma is important in some species and production of adventitious roots appears to be one of the most important adaptations. In some flooded species the amount of anaerobic respiration occurring in roots is supplemented by transport of oxygen from the atmosphere to the roots; for example, in rice (Barber *et al.*, 1962), bog species, *Salix* and *Myrica* (Armstrong, 1964, 1968), *Salix atrocinerea* (Leyton and Rousseau, 1958), *Betula pubescens* (Huikari, 1954), and *Nyssa sylvatica* var. *biflora* (Hook *et al.*, 1971).

In many studies of tolerance to flooding, species that formed adventitious roots at or below the water level survived best (Armstrong, 1968). The adventitious roots of flooded plants comprise a supplementary absorbing system in the somewhat aerobic zone, while the original root system does not function normally because of low oxygen tension (Gill, 1970). Although adventitious roots did not prevent injury to shoots of tomato when the original roots were flooded, epinasty of leaves was less and shoot growth greater than in flooded plants without adventitious roots. Shoots resumed growing if adventitious roots were allowed to develop. If adventitious roots were removed and the cut stumps sealed, shoot growth was prevented (Jackson, 1955).

In some plants (e.g., *Juncus effusus, Senecio aquaticus*) metabolic adaptations occur in which respiratory metabolism produces reduced amounts of toxic end products (e.g., malate) (Crawford, 1967; Crawford and McManmon, 1968; Crawford and Tyler, 1969). *Nyssa sylvatica* var. *biflora* seedlings developed tolerance to flooding through sequential anatomical and physiological adaptations. An early response to flooding was formation of water roots. These appeared to be beneficial because they occurred in the upper flood water where the oxygen content was higher and toxic compounds were lower than in the soil. The water roots also increased the absorbing surface for water and minerals. Under flooding, the newly initiated roots oxidized the rhizosphere whereas unflooded roots did not. Oxygen entered the stem through lenticels and appeared to be transported through the cortex or phloem. The combined adaptations of increased anaerobic respiration, oxidation of the rhizosphere, and tolerance of CO_2 by the new roots appeared to account for flood tolerance of this species (Hook *et al.,* 1970, 1971, 1972).

V. DISEASE-INDUCED WATER DEFICITS AND LEAF SHEDDING

By impeding absorption of water, blocking water transport in plants, increasing leaf-water loss, or by various combinations of these, a number of plant diseases induce desiccation and premature shedding of leaves (Kozlowski, 1964, 1968a, Kozlowski *et al.,* 1962; Duniway, 1971).

A. ROOT ROTS AND CANKERS

Reduction in water-absorbing efficiency of roots is associated with several plant diseases including root rots, viruses, and rusts. Destruction of the root system often occurs before aboveground symptoms are evident. The effects of root infections on water uptake of host plants often are similar to those traceable to mechanical injury to roots (Subramanian and Saraswathi-Devi, 1959).

Plant water potentials decreased with time after root infection of *Verticicladiella wagenerii* in pine (Helms *et al.,* 1971). Black shank, a root disease of tobacco, also caused water deficits in shoots, promoted wilting, and lowered transpiration (Schramm and Wolf, 1954). According to Powers (1954), black shank of tobacco caused wilting by localized obstruction of water transport in the area of lesions. This conclusion was reached because (1) diseased plants, even after prolonged wilting, regained turgor when water was introduced into stems above the lesions, (2) bridge grafts around lesions delayed wilting until the advancing lesions reached the grafted areas, and (3) dyes moved up to, but not through, the lesions. Desiccation and wilting in rhododendron also were caused by a root pathogen suspected to be *Phytophthora cinnamomi* (De Roo, 1969). Accurate diagnosis of root diseases often is difficult because the symptoms on aerial portions may resemble those caused by other diseases, particularly vascular wilts. For example, infection with *Armillaria mellea,* which causes root rot, is followed by yellowing of leaves, premature leaf fall, and branch dieback, all symptoms that are characteristic also of plants infected with vascular wilt diseases.

B. VASCULAR DISEASES

The water economy of both herbaceous and woody plants is often disrupted by vascular wilt disease leading to wilting, yellowing, and shedding of leaves. In herbaceous plants, the oldest leaves are dropped first, and the shedding sometimes is attributed to ethylene formed by the pathogen and host (Dimond and Waggoner, 1953). The disease spreads upward (or inward in rosette plants). Examples of vascular wilt diseases in herbaceous plants include Verticillium wilt of tomato, hop, lupin, sunflower, cotton, and potato, and Fusarium wilt of tomato and banana. Leaf abscission does not occur in monocotyledons infected with wilt diseases (Talboys, 1968).

The response of potato plants to infection by *Verticillium albo-atrum* were described by Isaac and Harrison (1968) and Harrison (1970). Infected plants exhibited wilting, chlorosis, and necrosis of leaves, with disease symptoms progressing upward from the lowermost leaves. The necrotic leaves were shed prematurely, and only a few shriveled ones remained attached to the plant, so that the lower portion of the stem was rapidly defoliated. The water content of infected plants, whether they showed disease symptoms or not, was lower than in healthy plants. Water content of diseased plants decreased throughout the growing season as the disease progressed, and it fluctuated more during the day than it did in healthy plants.

In perennial woody plants, the acute forms of vascular wilt disease syndrome are characterized by rapid wilting of the youngest leaves in the crown periphery. The leaves may die while green or after becoming chlorotic and/or necrotic. Leaves generally are shed but, if they are killed rapidly, shedding may be delayed since normal abscission processes are arrested. Some or all branches may be affected initially, but eventually the disease spreads throughout the crown and the plant usually dies within a year. In the mild disease syndrome, leaves yellow on one side of the tree or in a few branches, followed by acropetal defoliation of affected branches. Among the important vascular wilt diseases of woody plants are Verticillium wilt of maple and stone fruits, Dutch elm disease caused by *Ceratocystis ulmi,* oak wilt caused by *Ceratocystis fagacearum,* and persimmon wilt caused by *Cephalosporium diospyri.*

As Talboys (1968) emphasized, it should not be assumed that abscission is an invariable feature of the vascular wilt syndrome. Some diseases produce typical symptoms of water deficits leading to chlorosis and leaf shedding. These generally are associated with occlusion of vessels by gums, tyloses, pathogen-produced polysaccharides, or debris of enzymatic degradation of host tissues. Other diseases, which are characterized by foliar chlorosis and necrosis, appear to be associated largely with accumulation of pathogen-produced or pathogen-induced toxins. Furthermore, the same vascular disease may or may not induce leaf shedding. For example, the acute form of hop verticillium wilt is characterized by extensive mycelial growth in vessels, negligible formation of tyloses, marginal and interveinal leaf chlorosis, with green color along the major veins. In the mild syndrome, however, there is limited mycelial growth in vessels, but many tyloses are formed. Leaf shedding may occur following leaf chlorosis and necrosis (Talboys, 1958b).

Transpiration rates of plants with vascular diseases decrease greatly as wilt symptoms develop (Duniway, 1973). For example, water loss per unit of leaf surface of tomato plants decreased after inoculation with *Verticillium albo-atrum*. When diseased plants began to wilt, water loss temporarily increased, but then rapidly decreased to become less than that of healthy plants with adequate or restricted water supply. Resistance to water flow in diseased plants was high and was correlated with vessel blockage (Threlfall, 1959). Similarly potato plants infected with *V. albo-atrum* or *V. dahliae* had lower transpiration rates than healthy plants. The difference increased as the disease progressed. Diurnal fluctuations in transpirations were smaller in infected plants than in controls because infection reduced water loss during midday. Transpiration at night was not affected by infection (Harrison, 1971).

As mentioned previously, water deficits in diseased plants may be associated with occlusion of vessels by the pathogen, host responses induced by the pathogen, products of degradation of the host tissue, or by combinations of these. In some vascular wilt diseases, the vessels may be blocked by mycelium or bacteria. However, blockage by pathogens generally is not a primary cause of wilting, except in bacterial wilt (Dimond, 1955). In most vascular wilts, the occlusion of water-transporting xylem elements by fungus mycelium accounts for only little reduction in water translocation.

Major causes of vascular plugging and dysfunction in diseased plants are tyloses and gums. Tyloses, which result from metabolic activity of xylem parenchyma and ray cells, occlude water-conducting vessels in diseases caused by *Ceratocystis fagacearum* in oak (Struckmeyer et al., 1954); *Verticillium albo-atrum* in hops (Talboys, 1958a), tomato (Blackhurst and Wood, 1963), and cherry (Van der Meer, 1925); *Fusarium oxysporum* in oil palm (Kovachich, 1948), *F. oxysporum f. cubense* in banana (Beckman et al., 1961), and *F. oxysporum f. batatas* in sweet potato (McClure, 1950).

Most major vascular pathogens produce pectolytic enzymes. These may act on the middle lamella of vascular elements to produce fragments that block water movement and thereby cause wilting. Among products so released are large molecules consisting of polysaccharides, glycopeptides, and hydrolytic enzymes. Polysaccharides can impede water transport by accumulating in pit membranes of vessels, which act as ultrafilters (Dimond, 1967), or by disrupting water transport in veins of leaves. The evidence

for a role of enzymes in pathogenesis includes (1) their presence in diseased plants, (2) capacity of pathogens to produce them on host tissues,
(3) disappearance of their substrates in diseased plants, (4) degradation
of middle lamellae and cell walls, (5) presence in diseased vessels and
in pits of molecules that represent cleavage products from cell walls, and
(6) capacity of such products to produce vascular dysfunction (Dimond,
1970).

In diseased tomato plants, various polysaccharides of plant and micro-
bial origin induce wilting by mechanically interfering with water transport
(Hodgson *et al.,* 1949). The middle lamella of cells of diseased plants
may be acted on by pectolytic enzymes to produce fragments causing vascular plugging. Pierson *et al.* (1955) found that vascular plugging was
similar in tomato plants infected with *Fusarium* wilt and those treated with
pectin-digesting enzymes. The enzymes split pectin compounds of xylem
parenchyma cells to form soluble fragments that moved into vessels and
formed pectin gel plugs.

Leaf abscission has long been recognized as a characteristic symptom
of oak wilt. Te Beest *et al.* (1973) compared oak wilt-induced abscission
with artificially induced abscission and natural leaf abscission in mature
and juvenile *Quercus rubra* trees. Disease-induced abscission was examined
in leaves of greenhouse-grown seedlings inoculated with a conidial suspension of *Ceratocystis fagacearum.* Abscission of debladed petioles was also
studied. Artificially induced abscission was attempted by deep-stem girdles
in small trees or by withholding water.

Leaves on untreated mature and juvenile trees naturally abscised in
the autumn. The abscission process involved lysis of cell walls within a
separation layer. Abscission of leaves of inoculated trees usually occurred
within 3 to 5 days after appearance of oak-wilt symptoms. On individual
trees the leaves took longer to abscise from the first branch showing symptoms than from branches that developed symptoms later. The abscission
process induced by oak wilt was indistinguishable from that in natural abscission. Lignification of cells distal to the separation layer developed first
on the side of the petiole opposite the bud and gradually extended across
the whole width of the petiole. It stopped at the separation layer. Tyloses
were extensive even before leaf symptoms developed, in contrast to the
situation in natural abscission. As in natural abscission, the separation layer
was discontinuous and developed around tracheal bundles. Lysis of cell
walls developed at first on the abaxial side of the petiole and then across
the entire petiole.

Abscission of debladed petioles of healthy trees developed from a
layer that was not found in natural abscission or oak wilt-induced abscis-

sion. Leaf abscission was not induced in drought-killed or girdled trees.

Tyloses appear to play a major role in inducing symptoms of oak wilt in some oaks, with gums also contributing to vascular occlusion in severely diseased trees. Nair *et al.* (1966) demonstrated that 3 to 5 days before incipient wilting of foliage, tyloses formed in xylem vessels of *Quercus rubra,* the rate of water transport decreased greatly, and diurnal shrinkage and swelling of stems decreased. In infected *Quercus rubra* trees, tyloses developed profusely in stems and branches, but in *Quercus alba,* only in narrow infected areas. Tyloses were limited in roots.

In *Quercus macrocarpa* inoculated with conidial suspension of *Ceratocystis fagacearum,* fungus distribution and disease development were confined to vertical sectors of the trunk and vascularly connected branches. Eight days after inoculation, no foliar symptoms were evident, no tyloses were found in the current xylem except near the wound, and no vascular discoloration had developed.

Struckmeyer *et al.* (1954) found that in twigs with early oak-wilt symptoms, up to three-fourths of the large vessels were blocked. Tyloses were not found in tracheids, but gummosis was evident in both tracheids and smaller vessels. Plugging of vessels occurred prior to wilting. Marked resistance to water movement in vessels was shown by drooping of leaves during the day followed by recovery of turgor during the night. According to Beckman *et al.* (1953), the rate of transport of radioactive rubidium in trees with oak wilt was normal up to 5 days before wilting symptoms developed. However, during the fourth and third days before initial symptoms of permanent wilt developed, water transport (as indicated by rubidium transport) was decreased by about 90%. Hence, wilting symptoms were preceded by vascular plugging. Wilt symptoms could also be induced by mechanically interrupting the water supply.

The oak-wilt fungus not only induces formation of tyloses and gums, but also degrades the walls of infected sapwood cells, including the middle lamella. Sachs *et al.* (1970) found that hyphae of the oak-wilt fungus invaded xylem vessels of the outer sapwood of *Quercus rubra* and penetrated adjacent xylem parenchyma cells through pits. Lysis of vessels progressed from the lumen outward, with both cellulose and lignin affected. Cavitation and degradation of cell-wall layers were distinct. In many areas, the S_3 layer was etched and corroded. In other areas, etched zones apparently due to fungal enzymes also occurred in the S_2 and S_1 layers. Lysis of the middle lamella often accompanied intercellular penetration. In addition to providing cellulase and pectinase enzymes (Geary and Kuntz, 1962), the oak-wilt fungus apparently produced enzymes that acted on lignin.

There has been some controversy about the cause of the imbalance of water economy of trees having Dutch elm disease caused by *Ceratocystis ulmi*. Wilting and leaf shedding could occur if there were (1) reduction of water available to the leaves, or (2) an increase in transpirational water loss (Roberts, 1966).

Ouellette (1962) described the infection process of *Ceratocystis ulmi* in *Ulmus americana* trees. The pathogen grew extensively in the xylem and passed from cell to cell through pits and by penetrating cell walls. As infection developed, both bordered pits and cell walls disintegrated. Acute symptoms of Dutch elm disease were attributed mainly to plugging of vessels of small branches by the fungus and disintegration products. Gagnon (1967) found lignin and pectin-positive materials in plugged vessels of infected *Ulmus americana* trees. He suggested that the alterations of lignin and pectins resulted from action on cell walls of pectolytic enzymes produced by the pathogen.

MacHardy and Beckman (1973) showed that, following inoculation of *Ulmus americana* trees with a spore suspension of *Ceratocystis ulmi,* Dutch elm disease symptoms appeared resulting from desiccation because of a reduction in water-carrying capacity of invaded xylem vessels. The appearance of foliar wilt and chlorosis followed or accompanied a decrease in transpiration.

In uninfected branches, the rate of transpirational water loss remained constant or increased gradually in the light and in the dark, which indicating no vascular blockages (Fig. 5A). In infected branches, however, water loss was impeded to varying degrees. For example, in one branch (Fig. 5B) water loss was relatively constant before wilting. The wilting occurred at about the time transpiration suddenly decreased by almost 90%. Wilting was followed by leaf necrosis and abscission, but little chlorosis. In two other branches (Figs. 5C,D) the appearance of symptoms occurred after 3 to 8 days of a gradual decline in transpiration. Although little wilting occurred, the leaves became chlorotic and necrotic. Finally they abscised. Further evidence of reduction in water availability to leaves of infected plants was shown in an undulating pattern of transpiration in inoculated plants following a dark period (Fig. 6). MacHardy and Beckman explained this pattern as follows: Stomata at first opened normally because water was readily available after a dark period. Stomatal closure began as soon as the rate of transpiration exceeded the reduced supply of water to the leaves. When an appreciable number of stomata closed, leaf hydration began to increase. When guard-cell turgor increased sufficiently, the stomata reopened. Hence, the undulating transpiration curve was the result of such periodic stomatal opening and closing in unison. Gradually, however, stomatal opening and closing became more random, resulting in a reduced "equilibrium" curve. The undulating pattern of transpiration was

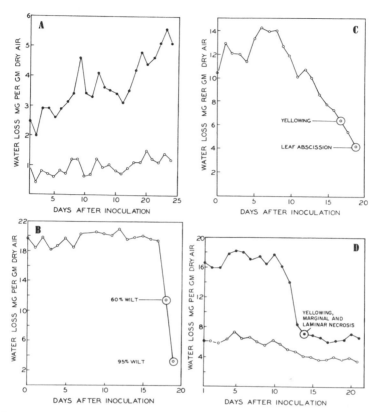

Fig. 5. Water loss from *Ulmus americana* trees inoculated with a spore suspension of *Ceratocystis ulmi* and noninoculated trees. (A) Daily water loss from noninoculated control branch during illumination (●———●) and darkness (○———○). (B) and (C) Daily water loss from 2 different branches during illumination in relation to appearance of foliar symptoms of Dutch elm disease. (D) Daily water loss from an inoculated branch during illumination (●———●) and darkness (○———○) in relation to appearance of foliar symptoms. From MacHardy and Beckman (1973).

similar to that of banana leaves infected with *Pseudomonas solanacearum* (Beckman *et al.,* 1962).

Distribution of dye in infected branches showed that the decrease in water loss was related to blocking of infected water conducting vessels. In noninfected twigs, the distribution of dye within xylem elements was uniform. Blocking of xylem elements in the main stem and large branches did not cause symptoms of Dutch elm disease because of alternate channels of water flow. However, blockage of xylem elements in twigs and green shoots readily caused impeded dye transport and induced wilt symptoms because of absence of alternate channels of water transport.

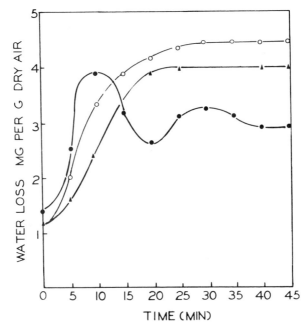

Fig. 6. Course of water loss from (*Ulmus americana*) branches exposed to continuous light after 9 hr of darkness. The data are from a noninoculated control (○——○), and an inoculated branch 1 day (▲——▲) and 23 days (●——●) after inoculation. From MacHardy and Beckman, (1973).

VI. CHEMICALLY INDUCED LEAF SHEDDING

Induction of leaf shedding with chemicals is an important agricultural and horticultural practice that facilitates harvesting of important crop plants. This may be accomplished with (1) chemical desiccants which by destroying selective permeability of membranes allow for rapid leaf dehydration, or (2) chemical defoliants which accelerate physiological processes leading to leaf separation at an abscission layer. These two classes of inducers of leaf shedding will be discussed separately.

A. DESICCANTS

Chemical desiccation is useful only on plants harvested for reproductive tissues or underground organs (e.g., tubers) or when leaf survival following harvest is not important. Among the potential advantages of desiccation with chemicals are control of insects and diseases, permitting scheduling of harvests, controlling weeds that interfere with harvesting, facilitating mechanical harvesting, attracting hand harvesters, improving quality of the crop, reducing moisture in seeds, and increasing yields (Addicott and Lynch, 1957). Although the advantages of chemical desiccation vary greatly for different crops, they are substantial for some. For example, killing of potato vines is a widely used preharvest practice for one or more of the following reasons: (1) to prevent oversized, knobby, and second-growth tubers, (2) to prevent spread of virus diseases, (3) to kill late-blight spores on foliage and surface of soil, (4) to reduce late-blight tuber-rot infection, (5) to facilitate hand and machine harvesting by eliminating vines and weeds, (6) to terminate growth of the potato crop whenever desirable, (7) to hasten plant maturity, (8) to permit harvesting at a predetermined time, and (9) to facilitate early harvesting of seed potatoes (Murphy, 1968).

As mentioned, chemical desiccants cause rapid drying and killing of the tissues covered. The desiccants apparently kill or denature protoplasm and destroy selective permeability of cell membranes, which thereby permits rapid dehydration. Among commonly used desiccants are ammonium nitrate, arsenic acid, sodium chlorate, and the bipyridiliums, paraquat and diquat. The latter two compounds, in particular, have been found useful on many crops. Annuals and seedling perennials are killed, but woody plants and perennials that can grow from underground rooting systems will regrow. Paraquat is inactivated by soil contact, so there is no hazard. The bipyridiliums do not cause any of the abnormal growth responses associated with hormone-type herbicides.

The bipyridiliums have been found useful in aquatic weed control (Darter and Wright, 1963); drying of hay (Atzmon and Volcani, 1968); desiccation of sugar cane (Chen and Liu, 1965; Cochrane and Procter, 1970; Arvier, 1962); preharvest desiccation of sugar beets (Austin and Longden, 1968); killing of potato vines (Scudder, 1966; Murphy, 1968; Nelson and Nylund, 1969); preharvest desiccation of fiber crops, such as hemp, flax, and cotton (Cegna, 1965); preharvest desiccation of lucerne (Jacarossi, 1965); and retarding maturation of cattle feed grasses which thereby prevents normal rapid decline in nutritional value of the grasses (Sneva, 1967; Freyman, 1970).

Plants treated with bipyridiliums show rapid wilting. This is followed by necrosis and ultimate death of the entire leaf. At the cellular level, these compounds cause loss of integrity of cell membranes and chloroplasts. The mechanism of action of bipyridiliums involves formation of the free radical by reduction of the ion and subsequent autoxidation to yield the original ion. The free radical itself does not appear to be the primary toxicant, but rather the OH radical or H_2O_2 which is formed during the autoxidation of the free radical to the ion. The photosynthetic apparatus, light, and molecular oxygen are required cofactors for these reactions (Mees, 1960; Homer *et al.,* 1960; Moreland, 1967; Ashton and Crafts, 1973).

B. DEFOLIANTS

Chemical defoliants generally are nonspecific and injure leaves without injuring the abscission zone. By such leaf injury, the physiological events that effect cytolysis and abscission are accelerated. An early loss of leaf moisture commonly is an associated response to the application of chemical defoliants (F. T. Addicott, personal communication). Defoliants are more effective on turgid than on flaccid leaves (Cooper and Henry, 1973).

Among important dust defoliants are sodium chlorate, tributyl phosphorotrithioate, and tributyl phosphorotrithioite. Important spray defoliants include ammonium nitrate, anhydrous ammonia, sodium chlorate, tributyl phosphorotrithioate, tributyl phosphorotrithioite, and Ethrel (2-chloroethylphosphonic acid, CEPA, ethephon).

In recent years much interest has been shown in Ethrel (ethephon), which appears to be unusually effective in forming ethylene which, in turn, is importantly involved in inducing abscission (Warner and Leopold, 1967, 1969). Most Ethrel that is applied to plants eventually is converted to ethylene (Abeles, 1973).

Ethrel accelerated leaf senescence and caused premature defoliation in nursery stock (Cummins and Fiorino, 1969; Larsen, 1971, 1973), snap beans (Palevitch, 1970; Davis *et al.,* 1972), and cotton (de Wilde, 1971). The activity of Ethrel as a defoliant has been increased by combining it with various substances. For example, Bromodine increased Ethrel activity as a result of greater leaf damage (Larsen, 1971). Endothall [7-oxabicyclo(2,2,1)heptane-2,3-dicarboxylic acid) also increased the defoliating effect of Ethrel, with the synergism resulting from inhibition of Ethrel translocation from the leaves (Sterrett *et al.,* 1973, 1974a). One to three applications of Ethrel at 200 to 400 ppm plus 1 to 2% D-WK surfactant were very effective in inducing defoliation of fruit-tree nursery stock, whereas Ethrel alone at the same concentrations was not (Larsen, 1973). Ethrel-endothall combinations also induced rapid foliar abscission in several species of deciduous forest trees, but were ineffective on coniferous species (Sterrett *et al.,* 1974b).

REFERENCES

Abeles, F. B. (1973). "Ethylene in Plant Biology." Academic Press, New York.
Abeles, F. B., Craker, L. E., and Leather, G. R. (1971). Abscission: The phytogerontological effects of ethylene. *Plant Physiol.* **47,** 7–9.

Addicott, F. T. (1970). Plant hormones in the control of abscission. *Biol. Rev. Cambridge Phil. Soc.* **45**, 485–524.

Addicott, F. T., and Lynch, R. S. (1955). Physiology of abscission. *Annu. Rev. Plant Physiol.* **6**, 211–238.

Addicott, F. T., and Lynch, R. S. (1957). Defoliation and desiccation: Harvest-aid practices. *Advan. Agron.* **9**, 67–93.

Addicott, F. T., and Lyon, J. L. (1973). The physiological ecology of abscission. *In* "Shedding of Plant Parts" (T. T. Kozlowski, ed.), pp. 475–524. Academic Press, New York.

Ahlgren, C. E., and Hansen, H. L. (1957). Some effects of temporary flooding on coniferous trees. *J. Forest.* **55**, 647–650.

Alben, A. O. (1958). Waterlogging of subsoil associated with scorching and defoliation of Stuart pecan trees. *Proc. Amer. Soc. Hort. Sci.* **72**, 219–223.

Albertson, F. W., and Weaver, J. E. (1945). Injury and death or recovery of trees in prairie climate. *Ecol. Monogr.* **15**, 393–433.

Armstrong, W. (1964). Oxygen diffusion from the roots of some British bog plants. *Nature (London)* **204**, 801–802.

Armstrong, W. (1968). Oxygen diffusion from the roots of woody species. *Physiol. Plant.* **21**, 539–543.

Arvier, A. C. (1962). Diquat—a new herbicide and desiccant of possible value to sugar cane. *Proc. Conf. Queensland Soc. Sugar Cane Technol., 29th, 1962* pp. 63–68.

Ashton, F. M., and Crafts, A. S. (1973). "Mode of Action of Herbicides." Wiley, New York.

Atzmon, G., and Volcani, R. (1968). Hay desiccaton—by means of Reglone (diquat). *Proc. Isr. Weed Contr. Conf., 3rd, 1968* p. 7

Austin, R. B., and Longden, P. C. (1968). The yield and quality of red beet seed as affected by desiccant sprays and harvest date. *Weed Res.* **8**, 336–345.

Barber, D. A., Ebert, M., and Evans, N. T. S. (1962). The movement of ^{15}O through barley and rice plants. *J. Exp. Bot.* **13**, 397–403.

Beard, J. S. (1946). The natural vegetation of Trinidad. *Oxford Forest. Mem.* **20**.

Beckman, C. H., Brun, W. A., and Buddenhagen, I. W. (1962). Water relations in banana plants infected with *Pseudomonas solanacearum*. *Phytopathology* **52**, 1144–1148.

Beckman, C. H., Kuntz, J. E., Riker, A. J., and Berbee, J. G. (1953). Host responses associated with the development of oak wilt. *Phytopathology* **43**, 448–454.

Beckman. C. H., Mace, M. E., Halmos, S., and McGahan, M. W. (1961). Physical barriers associated with resistance in *Fusarium* wilt of bananas. *Phytopathology* **51**, 507–515.

Blackhurst, F. M., and Wood, R. K. S. (1963). Resistance of tomato plants to *Verticillium albo-atrum*. *Trans. Brit. Mycol. Soc.* **46**, 385–392.

Bloodworth, M. E., Page, J. B., and Cowley, W. R. (1956). Some applications of the thermoelectric method for measuring flow rates in plants. *Agron. J.* **48**, 222–228.

Boeuf, F., and Genet, P. (1906). Les dessechement des orangers à la suite des sirrocas d'automne. *Rev. Hort. Tunis.* **5**, 20–27 and 57–58.

Böhning, R. H., and Lusanandana, B. (1952). A comparative study of gradual and abrupt changes in root temperature on water absorption. *Plant Physiol.* **27**, 475–488.

Boyce, J. S. (1961). "Forest Pathology." McGraw-Hill, New York.

Bruce, R. R., and Romkens, M. J. M. (1965). Fruiting and growth characteristics of cotton in relation to soil moisture tension. *Agron. J.* **57**, 135–140.

Burrows, W. J., and Carr, D. J. (1969). The effects of flooding the root system of sunflower plants on the cytokinin content in the xylem sap. *Physiol. Plant.* **22**, 1105–1112.

Cegna, P. (1965). Chemical desiccation. *Lotta Antiparass.* **17**, 12–13.

Chalot, C., and Deslandes, R. (1914). "Culture du Citronnier à la Dominique." Challamel, Paris.

Chang, H. T., and Loomis, W. E. (1945). Effect of carbon dioxide on absorption of water and nutrients by roots. *Plant Physiol.* **20**, 221–232.

Chen, J. C. P., and Liu, P. P. D. (1965). Evaluation of cane leaf desiccant "Gramoxone." *Sugar J.* **27**, 22–23 and 32.

Childers, N. F., and White, D. G. (1942). Influence of submersion of the roots on transpiration, apparent photosynthesis, and respiration of young apple trees. *Plant Physiol.* **17**, 603–618.

Cochrane, P. D., and Procter, G. C. (1970). All-weather weed control in sugar cane. *Sugar y Azucar* **65** (9), 28–29 and 50.

Cooper, W. C., and Henry, W. H. (1973). Chemical control of fruit abscission. *In* "Shedding of Plant Parts" (T. T. Kozlowski, ed.), pp. 475–524. Academic Press, New York.

Coster, C. (1923). Lauberneuerung und andere periodische Lebensprozesse in dem trockenen Monsun-gebiet Ost-Javas, *Ann. Jard. Bot. Buitenz.* **33**, 117–189.

Crawford, R. M. M. (1967). Alcohol dehydrogenase activity in relation to flooding tolerance in roots. *J. Exp. Bot.* **18**, 458–464.

Crawford, R. M. M., and McManmon, M. (1968). Inductive responses of alcohol and malic dehydrogenases in relation to flooding tolerance in roots. *J. Exp. Bot.* **19**, 435–441.

Crawford, R. M. M., and Tyler, P. D. (1969). Organic acid metabolism in relation to flooding tolerance in roots. *J. Ecol.* **57**, 235–244.

Cummins, J. N., and Fiorino, P. (1969). Preharvest defoliation of apple nursery stock using Ethrel. *HortScience* **4**, 339–344.

Cunningham, G. L., and Strain, B. R. (1969). Ecological significance of seasonal leaf variability in a desert shrub. *Ecology* **50**, 400–408.

Curry, J. R., and Church, T. W. (1952). Observations on winter drying of conifers in the Adirondacks. *J. Forest.* **59**, 114–116.

Darter, I. E., and Wright, N. (1963). Paraquat: A new herbicide and desiccant. *Pestic Abstr. C.* **9**, 203–206.

Davis. J. T., Sterrett, J. P., and Leather, G. R. (1972). Ethephon-endothall as a chemical abscissor of bean leaves. *HortScience* **7**, 478–479.

Denny, F. E., and Miller, L. P. (1935). Production of ethylene by plant tissue as indicated by the epinastic response of leaves. *Contrib. Boyce Thompson Inst.* **7**, 97–102.

De Roo, H. C. (1969). Sap stress and water uptake in detached shoots of wilt-diseased and normal rhododendrons. *HortScience* **4**, 51–52.

de Wilde, R. C. (1971). Practical applications of (2-chloroethyl) phosphonic acid in agricultural production. *HortScience* **6**, 364–370.

Dimond, A. E. (1955). Pathogenesis in the wilt diseases. *Annu. Rev. Plant Physiol.* **6**, 329–350.

Dimond, A. E. (1967). Physiology of wilt disease. *In* "The Dynamic Role of Molecular Constituents in Plant-parasite Interaction" (C. J. Mirocha and I. Uritani, eds.), pp. 100–120. Amer. Phytopathol. Soc., St. Paul, Minnesota.

Dimond, A. E. (1970). Biophysics and biochemistry of the vascular wilt syndrome. *Annu. Rev. Phytopathol.* **8**, 301–322.

Dimond, A. E., and Waggoner, P. E. (1953). The water economy of *Fusarium*-wilted tomato plants. *Phytopathology* **43**, 619–623.

Duniway, J. M. (1971). Water relations of Fusarium wilt in tomato. *Physiol. Plant Pathol.* **1**, 537–546.

Duniway, J. M. (1973). Pathogen-induced changes in host water relations. *Phytopathology* **63**, 458–466.

Eiten, G. (1972). The cerrado vegetation of Brazil. *Bot. Rev.* **38**, 201–341.

Evenari, M., Shanan, L., and Tadmor, N. H. (1971). "The Negev, the Challenge of a Desert." Harvard Univ. Press, Cambridge, Massachusetts.

Felt, E. P. (1943). Delayed winter injury. *Proc. Nat. Shade Tree Conf., 15th, 1943* pp. 19–22.

Ferri, M. G. (1961). Problems of water relations of some Brazilian vegetation types with special consideration of the concepts of xeromorphy and xerophytism. *In* "Plant-water Relationships in Arid and Semi-arid Conditions," pp. 191–197. UNESCO, Paris.

Freyman, S. (1970). Chemical curing of pinegrass with atrazine and paraquat. *Can. J. Plant Sci.* **50** 195–196.

Gagnon, C. (1967). Polyphenols and discoloration in the elm disease investigated by histochemical techniques. *Can. J. Bot.* **45**, 2119–2124.

Geary, T. F., and Kuntz, J. E. (1962). The effect of growth regulators on oak wilt development. Phytopathology, **52**, 733.

Georgii, W. (1920). Sirocco observations in the southwestern part of Palestine. *Mon. Weather Rev.* **48**, 40.

Gill, C. J. (1970). The flooding tolerance of woody species—a review. *Forest. Abstr.* **31**, 671–688.

Hall, T. F., and Smith, G. E. (1955). Effects of flooding on woody plants, West Sandy dewatering project, Kentucky Reservoir. *J. Forest.* **53**, 281–285.

Hallaway, H. M., and Osborne, D. J. (1969). Ethylene: A factor in defoliation by auxins. *Science* **163**, 1067–1068.

Handley, W. R. C. (1939). The effect of prolonged chilling on water movements and radial growth in trees. *Ann. Bot. (London)* **3**, 803–813.

Harris, D. G., and van Bavel, C. H. M. (1957). Growth, yield and water absorption of tobacco plants as affected by the composition of the root atmosphere. *Agron. J.* **49**, 11–14.

Harrison, A. T., Small, E., and Mooney, H. A. (1971). Drought relationships and distribution of two mediterranean-climate California plant communities. *Ecology* **52**, 869–875.

Harrison, J. A. C. (1970). Water deficit in potato plants infected with *Verticillium albo-atrum*. *Ann. Appl. Biol.* **66**, 225–231.

Harrison, J. A. C. (1971). Transpiration in potato plants infected with *Verticillium* spp. *Ann. Appl. Biol.* **68**, 159–168.

Hartley, C., and Merrill, T. (1915). Storm and drought injury to foliage of ornamental trees. *Phytopathology* **5**, 20–29.

Havis, J. R. (1971). Water movement in stems during freezing. *Cryobiology* **8**, 581–585.

Hedgecock, G. G. (1912). Winter-killing and smelter-injury in the forests of Montana. *Torreya* **12**, 25–30.

Heinicke, A. J. (1932). The effect of submerging the roots of apple trees at different seasons of the year. *Proc. Amer. Soc. Hort. Sci.* **29**, 205–207.

Helms, J. A., Cobb, F. W., Jr., and Whitney, H. S. (1971). Effect of infection by *Verticicladiella wagenerii* on the physiology of *Pinus ponderosa*. *Phytopathology* **61,** 920–925.

Henson, W. R. (1952). Chinook winds and redbelt injury to lodgepole pine in the Rocky Mountain parks area of Canada. *Forest. Chron.* **28,** 62–64.

Hermann, R. K., and Lavender, D. P. (1967). Tolerance of coniferous seedlings to silting. *J. Forest.* **65,** 824–825.

Hodgson, R., Peterson, W. H., and Riker, A. J. (1949). The toxicity of polysaccharides and other large molecules to tomato cuttings. *Phytopathology* **39,** 47–62.

Homer, R. F., Meer, G. C., and Tomlinson, T. E. (1960). Mode of action of dipiridyl quaternary salts as herbicides. *J. Sci. Food Agr.* **11,** 309–315.

Hook, D. D., Brown, C. L., and Kormanik, P. P. (1970). Lenticels and water root development of swamp tupelo under various flooding conditions. *Bot. Gaz. (Chicago)* **131,** 217–221.

Hook, D. D., Brown, C. L., and Kormanik, P. P. (1971). Inductive flood tolerance in swamp tupelo (*Nyssa sylvatica* var. *biflora* (Walt.) Sarg.). *J. Exp. Bot.* **22,** 178–189.

Hook, D. D., Brown, C. L., and Wetmore, R. H. (1972). Aeration in trees. *Bot. Gaz. (Chicago)* **133,** 443–454.

Hosner, J. F. (1958). The effects of complete inundation upon seedlings of six bottomland tree species. *Ecology* **39,** 371–373.

Hosner, J. F. (1959). Survival, root and shoot growth of six bottomland tree species following flooding. *J. Forest.* **57,** 927–928.

Hosner, J. F. (1960). Relative tolerance to complete inundation of fourteen bottomland tree species. *Forest Sci.* **6,** 246–251.

Hosner, J. F., and Boyce, S. G. (1962). Relative tolerance to water saturated soil of various bottomland hardwoods. *Forest Sci.* **8,** 180–186.

Hubert, E. E. (1930). How weather causes tree diseases. *Northwest Sci.* **6,** 10.

Huikari, O. (1954). Experiments on the effect of anaerobic media upon birch, pine, and spruce seedlings. *Commun. Inst. Forest. Fenn.* **42,** 1–13.

Hunt, F. M. (1951). Effects of flooded soil on growth of pine seedlings. *Plant Physiol.* **26,** 363–368.

Isaac, I., and Harrison, J. A. C. (1968). The symptoms and causal agents of early-dying disease (*Verticillium* wilt) of potatoes. *Ann. Appl. Biol.* **61,** 231–241.

Jackson, M. B., and Osborne, D. J. (1970). Ethylene, the natural regulator of leaf abscission. *Nature (London)* **225,** 1019–1022.

Jackson, W. T. (1955). The role of adventitious roots in recovery of shoots, following flooding of the original root systems. *Amer. J. Bot.* **42,** 816–819.

Jackson, W. T. (1956). Flooding injury studied by approach-graft and split root system techniques. *Amer. J. Bot.* **43,** 496–502.

Jacorossi, F. (1965). Weed control in lucerne. 2. *Lotta Antiparass.* **17,** 3–7.

Johnston, R. D. (1959). Control of water movement by stem chilling. *Aust. J. Bot.* **7,** 97–108.

Jordan, W. R., Morgan, P. W., and Davenport, T. L. (1972). Water stress enhances ethylene-mediated leaf abscission in cotton. *Plant Physiol.* **50,** 756–758.

Kawase, M. (1972). Effect of flooding on ethylene concentration in horticultural plants. *J. Amer. Soc. Hort. Sci.* **97,** 584–588.

Kawase, M. (1974). Role of ethylene in induction of flooding damage in sunflower. *Physiol. Plant.* **31,** 29–38.

Kincer, J. B. (1934). Data on the drought. *Science* **80,** 179.

Kovachich, W. G. (1948). A preliminary anatomical note on vascular wilt disease of the oil palm (*Elaeis guineensis*). *Ann. Bot. (London)* [N.S.] 12, 327–329.

Kozlowski, T. T. (1943). Transpiration rates of some forest tree species during the dormant season. *Plant Physiol.* 18, 252–260.

Kozlowski, T. T. (1955). Tree growth, action and interaction of soil and other factors. *J. Forest.* 53, 508–512.

Kozlowski, T. T. (1958). Water relations and growth of trees. *J. Forest.* 56, 498–502.

Kozlowski, T. T. (1964). "Water Metabolism in Plants." Harper, New York.

Kozlowski, T. T. (1967). Water relations of trees. *Midwest Chapt. Int. Shade Tree Conf. Proc., 22nd Ann. Conf.* pp. 34–42.

Kozlowski, T. T. (1968a). Introduction. *In* "Water Deficits and Plant Growth" (T. T. Kozlowski, ed.), Vol. 1, pp. 1–21. Academic Press, New York.

Kozlowski, T. T. (1968b). Water balance in shade trees. *Proc. Int. Shade Tree Conf., 44th, 1968* pp. 29–41.

Kozlowski, T. T. (1970). Role of environment in plant propagation: Water relations. *Proc. Int. Plant Propagators Soc. Annu. Meet.* pp. 123–139.

Kozlowski, T. T. (1971). "Growth and Development of Trees," Vol. I. Academic Press, New York.

Kozlowski, T. T. (1972). Physiology of water stress. *In* "Wildland Shrubs—Their Biology and Utilization" (C. M. McKell, J. R. Blaisdell, and J. R. Goodin, eds.), Gen. Tech. Rep. INT-1. pp. 229–244. U. S. Dept. Agr. Forest Serv., Ogden Utah.

Kozlowski, T. T. (1973). Extent and significance of shedding of plant parts. *In* "Shedding of Plant Parts (T. T. Kozlowski, ed.), pp. 1–21. Academic Press, New York.

Kozlowski, T. T., Kuntz, J. E., and Winget, C. H. (1962). Effect of oak wilt on cambial activity. *J. Forest.* 60, 558–561.

Kramer, P. J. (1940). Root resistance as a cause of decreased water absorption by plants at low temperatures. *Plant Physiol.* 15, 63–79.

Kramer, P. J. (1942). Species differences with respect to water absorption at low soil temperatures. *Amer. J. Bot.* 29, 828–832.

Kramer, P. J. (1951). Causes of injury to plants resulting from flooding of the soil. *Plant Physiol.* 26, 722–736.

Kramer, P. J. (1969). "Plant and Soil Water Relationships. A Modern Synthesis." McGraw-Hill, New York.

Kramer, P. J., and Jackson, W. T. (1954). Causes of injury to flooded tobacco plants. *Plant Physiol.* 29, 241–245.

Kramer, P. J., and Kozlowski, T. T. (1960). "Physiology of Trees." McGraw-Hill, New York.

Larcher, W. (1963). Zur spätwinterlichen Erschwerung der Wasserbilanz von Holzpflanzen an der Waldgrenze. *Ber. Naturwiss.-Med. Ver. Innsbruck* 53, 125–137.

Larcher, W. (1972). Der Wasserhaushalt immergrünen Pflazen im Winter. *Ber. Deut. Bot. Ges.* 85, 315–317.

Larsen, F. E. (1971). Prestorage promotion of leaf abscission of deciduous tree fruit with bromodine-ethephon mixtures. *HortScience* 6, 135–137.

Larsen, F. E. (1973). Stimulation of leaf abscission of tree fruit nursery stock with ethephon-surfactant mixtures. *J. Amer. Soc. Hort. Sci.* 98, 34–36.

Leyton, L., and Rousseau, L. F. (1958). Root growth of tree seedlings in relation to aeration. *In* "Physiology of Forest Trees" (K. V. Thimann, ed.), pp. 467–475. Ronald Press, New York.

Lindsay, J. H. (1967). The effect of environmental factors on the leaf water balance of conifers. M.S. Thesis, University of Wyoming, Laramie.

Lynch, J. M. (1972). Identification of substrates and isolation of microorganisms responsible for ethylene production in the soil. *Nature* (*London*) **240**, 45–46.

McAlpine, R. G. (1961). Yellow-poplar seedlings intolerant to flooding. *J. Forest.* **59**, 566–568.

McCleary, J. A. (1968). The biology of desert plants. *In* "Desert Biology" (G. W. Brown, ed.), Vol. 1, pp. 141–194. Academic Press, New York.

McClure, T. T. (1950). Anatomical aspects of the Fusarium wilt of sweet potatoes. *Phytopathology* **40**, 769–775.

McDermott, R. E. (1954). Effects of saturated soil on seedling growth of some bottomland hardwood species. *Ecology* **35**, 36–41.

MacHardy, W. E., and Beckman, C. H. (1973). Water relations in American elm infected with *Ceratocystis ulmi*. *Phytopathology* **63**, 98–103.

MacHattie, L. B. (1963). Winter injury to lodgepole pine foliage. *Weather* **18**, 301–307.

McMichael, B. L., Jordan, W. R., and Powell, R. D. (1972). An effect of water stress on ethylene production by intact cotton petioles. *Plant Physiol.* **49**, 658–660.

McMichael, B. L., Jordan, W. R., and Powell, R. D. (1973). Abscission processes in cotton: Induction by plant water deficit. *Agron. J.* **65**, 202–204.

Marsden, D. H. (1950). Dry weather and tree troubles in Massachusetts. *Plant Dis. Rep.* **34**, 400–401.

Mees, G. C. (1960). Experiments on the herbicidal action of 1,1,-ethylene-2,2-dipyridylium dibromide. *Ann. Appl. Biol.* **48**, 601–612.

Melrose, G. P. (1919). Red-belt injury in British Columbia. *Can. Forest J.* **15**, 164.

Merrill, E. D. 1945. "Plant Life of the Pacific World." Macmillan, New York.

Michael, G. (1963). Ein Beitrag zum Frosttrocknisproblem. *Naturwissenschaften* **50**, 382.

Michael, G. (1966). Untersuchungen über die winterliche Dürreresistenz eininger immergrüner Gehölze im Hinblick auf eine Frosttrocknisgefahr. *Flora* (*Jena*), *Abt B* **156**, 350–372.

Millborrow, B. V. (1974). The chemistry and physiology of abscisic acid. *Annu. Rev. Plant Physiol.* **25**, 259–307.

Mooney, H. A., and Dunn, E. L. (1970). Convergent evolution of Mediterranean-climate evergreen sclerophyll shrubs. *Evolution* **24**, 292–303.

Mooney, H. A., Dunn, E. L., Shropshire, F., and Song, L. (1970). Vegetation comparison between the Mediterranean climatic areas of California and Chile. *Flora* (*Jena*) **159**, 480–496.

Moreland, D. E. (1967). Mechanisms of action of herbicides. *Annu. Rev. Plant Physiol.* **18**, 365–386.

Murphy, H. J. (1968) Potato vine killing. *Amer. Potato J.* **45**, 472–478.

Nair, V. M. G., Kuntz, J. E., and Sachs, I. B. (1966). Tyloses induced by *Ceratocystis fagacearum* in oak wilt development. *Phytopathology* **57**, 823–824.

Nelson, D. C., and Nylund, R. E. (1969). Effect of chemicals on vine kill, yield and quality of potatoes in the Red River Valley. *Amer. Potato J.* **46**, 315–322.

Okonoue, M., and Sasaki, O. (1960). Depth of frozen ground on slope with no snow cover in winter. *J. Jap. Forest. Soc.* **42**, 339–342.

Oppenheimer, H. R. (1960). Adaptation to drought: Xerophytism. *In* "Plant-water

Relationships in Arid and Semi-arid Conditions," Rev. Res., pp. 105–138. UNESCO, Paris.

Orshan, G. (1972). Morphological and physiological plasticity in relation to drought. *In* "Wildland Shrubs—Their Biology and Utilization" (C. M. McKell, J. P. Blaisdell, and J. R. Goodin, eds.), Gen. Tech. Rep. INT-1, pp. 245–254. U. S. Dep. Agr. Forest Serv., Ogden, Utah.

Orshansky, G. (1954). Surface reduction and its significance as a hydroecological factor. *J. Ecol.* **42,** 442–444.

Osborne, D. J. (1973). Internal factors regulating abscission. *In* "Shedding of Plant Parts" (T. T. Kozlowski, ed.), pp. 125–147. Academic Press, New York.

Ouellette, G. B. (1962). Studies on the infection process of *Ceratocystis ulmi* (Buism.) C. Moreau in American elm trees. *Can. J. Bot.* **40,** 1567–1575.

Palevitch, D. (1970). Defoliation of snap beans with preharvest treatments of 2-chloroethylphosphonic acid. *HortScience* **5,** 224–226.

Parker, J. (1950). The effect of flooding on the transpiration and survival of some northeastern forest tree species. *Plant Physiol.* **25,** 453–460.

Parker, J. (1951). Moisture retention in leaves of conifers of the northern Rocky Mountains. *Bot. Gaz. (Chicago)* **113,** 210–216.

Parker, J. (1965). Physiological diseases of trees and shrubs. *Advan. Front. Plant Sci.* **12,** 97–248.

Phillips, I. D. J. (1964a). Root-shoot hormone relations. I. The importance of an aerated root system in the relation of growth hormone levels in the shoot of *Helianthus annuus*. *Ann. Bot. (London)* [N.S.] **28,** 17–35.

Phillips, I. D. J. (1964b). Root-shoot hormone relations. II. Changes in endogenous auxin concentration produced by flooding of the root system in *Helianthus annuus*. *Ann. Bot. (London)* [N.S.] **28,** 38–45.

Pierson. C. F., Gothoskar, S. S., Walker, J. C., and Stahmann, M. A. (1955). Histological studies on the role of pectic enzymes in the development of *Fusarium* wilt symptoms in tomato. *Phytopathology* **45,** 524–527.

Pirone, P. P. (1972). "Tree Maintenance." Oxford Univ. Press, London and New York.

Pisek, A., and Larcher, W. (1954). Zusammenhang zwischen Austrocknungs Resistenz und Frosthärte bei Immergrünen. *Protoplasma* **44,** 30–46.

Pisek, A., and Winkler, E. (1953). Schliessbewegung der Stomata bei ökologisch verschieden Pflazentypen in Abhängigkeit von Wassersättigungszustand der Blätter und vom Licht. *Planta* **42,** 253–278.

Pook, E. W., Costin, A. B., and Moore, C. W. E. (1966). Water stress in native vegetation during the drought of 1965. *Aust. J. Bot.* **14,** 257–267.

Pool, R. J. (1913). Some effects of drought on vegetation. *Science* **38,** 822–825.

Powers, H. K., Jr. (1954). The mechanism of wilting in tobacco plants affected by black shank. *Phytopathology* **44,** 513–521.

Reed, H. S., and Bartholomew, E. T. (1930). The effects of desiccating winds on citrus trees. *Calif., Agr. Exp. Sta., Bull.* **484,** 1–59.

Reid, D. M., and Crozier, A. (1971). Effects of waterlogging on the gibberellin content and growth of tomato plants. *J. Exp. Bot.* **22,** 39–48.

Reid, D. M., Crozier, A., and Harvey, B. M. R. (1969). The effects of flooding on the export of gibberellins from the root to shoot. *Planta* **89,** 376–379.

Richards, P. W. (1952). "The Tropical Rain Forest." Cambridge Univ. Press, London and New York.

Roberts, B. R. (1966). Transpiration of elm seedlings as influenced by inoculation with *Ceratocystis ulmi*. *Forest Sci.* **12**, 44–47.

Roberty, G. (1946). Les associations végétales de la vallée moyenne du Niger Veröff. *Geobot. Inst. Rubel.* **22.**

Rowe, R. N., and Beardsell, D. V. (1973). Waterlogging of fruit trees. *Hort. Abstr.* **43**, 534–548.

Sachs, I. B., Nair, V. M. G., and Kuntz, J. E. (1970). Penetration and degradation of cell walls in oaks infected with *Ceratocystis fagacearum*. *Phytopathology* **60**, 1399–1404.

Sakai, A. (1968). Mechanism of desiccation damage of forest trees in winter. *Contrib. Inst. Low Temp. Sci., Hokkaido Univ., Ser. B* **15**, 15–35.

Sakai, A. (1970). Mechanism of desiccation damage of conifers wintering in soil-frozen areas. *Ecology* **51**, 657–664.

Schramm, R. J., and Wolf, F. T. (1954). The transpiration of black shank-infected tobacco. *J. Elisha Mitchell Sci. Soc.* **70**, 255–261.

Scudder, W. T. (1966). Desiccation of potato vines. *Fla., Agr. Exp. Sta., Rep.* **223.**

Smith, K. A., and Restall, S. W. F. (1971). The occurrence of ethylene in anaerobic soil. *J. Soil Sci.* **22**, 430–443.

Smith, K. A., and Russell, R. S. (1969). Occurrence of ethylene and its significance in anaerobic soil. *Nature (London)* **222**, 769–771.

Sneva, F. A. (1967). Chemical curing of range grasses with paraquat. *J. Range Manage.* **20**, 389–394.

Sterrett, J. P., Leather, G. R., and Tozer, W. E. (1973). Defoliation response of woody seedlings to endothall/ethephon. *HortScience* **8**, 387–388.

Sterrett, J. P., Leather, G. R., and Tozer, W. E. (1974a). An explanation for the synergistic interaction of endothall and ethephon on foliar abscission. *J. Amer. Soc. Hort. Sci.* **99**, 395–397.

Sterrett, J. P., Leather, G. R., Tozer, W. E., Foster, W. D., and Webb, D. T. (1974b). Foliar abscission of woody plants with combinations of endothall and ethephon. *Weed Sci.* **22**, 608–614.

Stockton, J. R., Doneen, L. D., and Walhood, V. T. (1961). Boll shedding and growth of the cotton plant in relation to irrigation frequency. *Agron. J.* **53**, 272–275.

Struckmeyer, B. E., Beckman, C. H., Kuntz, J. E., and Riker, A. J. (1954). Plugging of vessels by tyloses and gums in wilting oaks. *Phytopathology* **44**, 148–153.

Subramanian, D., and Saraswathi-Devi, L. (1959). Water is deficient. *In* "Plant Pathology" (J. G. Horsfall and A. E. Dimond, eds.), Vol. 1, pp. 313–348. Academic Press, New York.

Sveshnikova, M. V., and Zalensky, O. V. (1956). "Water Regime of Plants of Arid Territories in Central Asia and Kazakstan," 18th Int. Georg. Congr. USSSR. Acad. Sci. (Moscow, 1956), pp. 227–237.

Talboys, P. W. (1958a). Some mechanisms contributing to *Verticillium* resistance in the hop root. *Trans. Brit. Mycol. Soc.* **41**, 227–241.

Talboys, P. W. (1958b). Association of tyloses and hyperplasia of the xylem with vascular invasion of the hop by *Verticillium albo-atrum*. *Trans. Brit. Mycol. Soc.* **41**, 249–260.

Talboys, P. W. (1968). Water deficit in vascular disease. *In* "Water Deficits and Plant Growth" (T. T. Kozlowski, ed.), Vol. 2, pp. 255–311. Academic Press, New York.

Te Beest, D., Durbin, R. D., and Kuntz, J. E. (1973). Anatomy of leaf abscission induced by oak wilt. *Phytopathology* **63**, 232–256.

Threlfall, R. J. (1959). Physiological studies on the *Verticillium* wilt disease of tomato. *Ann. Appl. Biol.* **47**, 57–77.

Tranquillini, W. (1963). Climate and water relations of plants in the subalpine region. *In* "The Water Relations of Plants" (A. J. Rutter and F. H. Whitehead, eds.), pp. 153–167. Blackwell, Oxford.

Tranquillini, W. (1964). The physiology of plants at high altitudes. *Annu. Rev. Plant Physiol.* **15**, 345–362.

Van der Meer, J. H. H. (1925). Verticillium wilt of herbaceous and woody plants. *Meded. Landbouwhogesch. Wageningen* **28**, 1–82.

Voigt, G. K. (1951). Causes of injury to conifers during the winter of 1947–48 in Wisconsin. *Trans. Wis. Acad. Sci., Arts Lett.* **40**, 241–243.

Warner, H. L., and Leopold, A. C. (1967). Plant growth regulation by stimulation of ethylene production. *BioScience* **17**, 722.

Warner, H. L., and Leopold, A. C. (1969). Ethylene evolution from 2-chloroethylphosphonic acid. *Plant Physiol.* **44**, 156–158.

Weaver, J. E., and Albertson, F. W. (1936). Effects of the great drought on the prairies of Iowa, Nebraska, and Kansas. *Ecology* **17**, 567–639.

Weaver, J. E., Stoddart, L. A., and Noll, W. (1935). Response of the prairie to the great drought of 1934. *Ecology* **16**, 612–629.

White, W. C., and Weiser, J. (1964). The relation of tissue desiccation, extreme cold, and rapid temperature fluctuation to winter injury of American arborvitae. *Proc. Amer. Soc. Hort. Sci.* **85**, 554–563.

Willey, C. R. (1970). Effects of short periods of anaerobic and near-anaerobic conditions on water uptake by tomato roots. *Agron. J.* **62**, 224–229.

Williamson, R. E., and Splinter, W. E. (1968). Effect of gaseous composition of root environment upon root development and growth of *Nicotiana tabacum* L. *Agron. J.* **60**, 365–368.

Winslow, A. C., and Havis. J. R. (1967). Water movement in stems of American holly at low temperatures. *HortScience* **2**, 24–25.

Wright, H. (1905). Foliar periodicity of endemic and indigenous trees in Ceylon. *Ann. Bot. Gard. Peradeniya* **2**, 415–517.

Yelenosky, G. (1963). Soil aeration and tree growth. *Proc. Int. Shade Tree Conf., 39th. 1963* pp. 16–25.

Zimmermann, M. H. (1964). Effect of low temperature on ascent of sap in trees. *Plant Physiol.* **39**, 568–572.

Zimmermann, P. W., Hitchcock, A. E., and Wilcoxon, F. (1939). Responses of plants to growth substances applied as solutions and as vapors. *Contrib. Boyce Thompson Inst.* **10**, 363–376.

CHAPTER 6

WATER DEFICITS AND
FLOW OF LATEX

B. R. Buttery

AGRICULTURE CANADA RESEARCH STATION, HARROW, ONTARIO, CANADA

and

S. G. Boatman

ICI PLANT PROTECTION DIVISION., JEALOTT'S HILL RESEARCH STATION,
NEAR BRACKNELL., BERKS., ENGLAND

233

I. INTRODUCTION

Latex flow is for the most part an artificial phenomenon in that it is induced by deliberate and controlled wounding known as tapping. Latex occurs in many different species, although the major rubber producing plants belong to one or other of only five families: Euphorbiaceae, Moraceae, Apocynaceae, Asclepiadaceae, and Compositeae. By far the greatest amount of attention has been given to the rubber tree of commerce, *Hevea brasiliensis,* Muell., Arg., with which this chapter is largely concerned.

The past decade has seen notable advances in understanding factors controlling flow of latex in *Hevea.* The great majority of this work has been carried out at a very few research centers in the Far East and West Africa. All these centers are mainly concerned with developing methods of profitably exploiting rubber trees; the more basic studies that have emerged have been undertaken to obtain a better understanding of factors limiting production and, hopefully, to find some way of circumventing these limitations. Many of the results have been published in specialized journals, such as the *Revue Generale des Caoutchoucs et Plastiques* and the *Journal of The Rubber Research Institute of Malaya,* and perhaps, have not reached a wide audience.

A great deal of the early work on rubber trees, including water relations and latex flow, was summarized by Dijkman (1951). More recent, and briefer reviews are those of Blackman (1965), Southorn (1969), Boatman (1970), and Abraham *et al.* (1971a).

II. OCCURRENCE AND COMPOSITION OF LATEX

A. TYPES OF LATICES AND LATEX VESSELS

Latex is highly variable in composition and is frequently, but not necessarily, milky in appearance. Latex is found in laticifers which may be (a) articulated, in which a series of cells is united through partial or complete dissolution of cross walls; these laticifers may be anastomosing or not; or (b) nonarticulated, consisting of single cells often very elongated and usually much branched (Esau, 1965).

Sheldrake (1969) found a striking correlation between cellulase activity in latex and the presence of articulated as opposed to nonarticulated laticifers. He suggested that the cellulase is instrumental in breaking down the cross walls during formation of the articulated type.

Latex may be present in ordinary parenchymatous cells, as in guayale (*Parthenium argentatum*) (Bonner and Galston, 1947), or it may be found in a branching (*Euphorbia*) or anastomosing (*Hevea*) system of tubes.

Laticifers also intergrade with certain idioblasts that contain tannins, mucilages, proteinaceous or other compounds. The situation is further complicated by the occurrence of schizogenous canals containing latex (Kisser, 1958). Thus, laticifers cannot be delimited precisely (Esau 1965).

Latex-containing plants are estimated to include some 12,500 species in 900 genera (van Die, 1955). These range from small herbaceous annuals such as the spurges (*Euphorbia*) to large trees like *Hevea* and *Arctocarpus* and tropical climbers like *Cryptostegia* and *Landolphia*.

Resins and rubber are characteristic components of latex in many plants. Terpenes occur in different amounts depending on the kind of plant, and rubber itself is sometimes entirely lacking. Latex may contain a large amount of protein (*Ficus callosa*), sugar (Compositae), or tannins (*Musa,* Aroideae). The latex of some Papaveraceae contains important alkaloids (e.g., *Papaver somniferum;* Fairbairn and Kapoor, 1960) and that of *Carica papaya* is outstanding for the occurrence of the proteolytic enzyme papain. Crystals of oxalates and malates may be abundant in latex. Certain plants contain starch grains in the laticifers, often together with the enzyme diastase. The starch grains of the genus *Euphorbia* may attain large size and assume various, sometimes peculiar, shapes (Mahlberg, 1973).

Gutta, a lower molecular weight trans isomer of rubber, is obtained mainly from *Palaquium gutta, Mimusops balata,* and *Achras sapota* (chicle), all members of the family Sapotaceae.

The latex of various plants may be clear (*Morus, Nerium*) or milky (*Asclepias, Euphorbia, Ficus*). It is yellow-brown in *Cannabis* and yellow or orange in the Papaveraceae. The turbidity or milkiness of latex does not depend directly on its composition, but results from the difference between the refractive indices of the particles and the dispersion medium (Esau, 1965).

In the nonarticulated laticifers of many plants, the nuclei are known to undergo divisions resulting in a multinucleate coenocytic condition (Mahlberg, 1959).

B. ANATOMY

Only a few laticiferous plants have been studied in detail.

The latex of *Taraxacum kok-saghyz* is found in anastomosing concentric cylinders. The root is the main source of rubber for commercial extraction (Krotkov, 1945). In *Euphorbia wulfenii,* latex tubes are confined to the phloem and run almost entirely in a longitudinal direction with only occasional bifurcations. The latex tubes are large, averaging 30 μm in diameter and capable of exuding copious quantities of latex (Spencer, 1939a).

Cryptostegia grandiflora and *C. madagascariensis* are tropical vines of semiarid areas. In the stems, rubber is found only in latex ducts; none has been observed in cortex chlorenchyma. The ducts are scattered vertically throughout the pith and bark (phloem, phloem rays, and cortex), although a small number extend laterally in the wood rays and make possible an interchange of latex between pith and bark. The ducts, occasionally branched and essentially without septa, range from 5–35 μm in diameter (average 25 μm).

Cryptostegia develops long semileafless stems known as "whips" in which the pith ducts are larger and of more uniform size than those of the bark (Whittenberger *et al.,* 1945). Some 3200 latex tubes may be found in a cross section of a whip of average vigor: 800 in the pith and 2400 in the outer phloem and cortex. The pith tubes average 33.6 μm in diameter while the bark tubes are only 20 μm. The total volume of the latex tubes in a normal internode is therefore 0.18–0.24 ml since the internodes in the central portion of a whip average 12–16 cm in length (Curtis and Blondeau, 1946). Rubber is found in nonseptate laticiferous ducts of the leaf petiole and blade. The ducts, present in approximately equal numbers in the tissue above and below the veins, parallel and branch with the veins.

Globules of rubber also occur in all green cell types in the leaf (e.g., palisade parenchyma, spongy parenchyma); these globules may comprise 80% of total rubber in the leaf (Whittenberger *et al.,* 1945).

In the leaves of *Hevea,* the latex vessels are principally confined to the phloem, but as Haberlandt (1914) has recorded for other laticiferous plants, numerous branches pass out from the main system and make their way up between and in close contact with the cells of the palisade layer and end blindly at the epidermis (Spencer, 1939b).

In the region of the abscission layer, the latex tubes were completely occluded in many places by a highly refractive unstained transparent substance (callose) in contact with which the adjacent latex often formed marked concave menisci. The deposition of callose occurred at the time of secondary thickening of the petioles and of the stem on which they were borne, and was completed by the time leaves attained final green color and posture. This lends no support to Haberlandt's theory of a translocatory function of laticifers.

The gynoecium of *H. brasiliensis* is of the typical euphorbeaceous type, consisting of three carpels united into a trilocular ovary with a single suspended anatropous ovule in each loculus. After fertilization, the ovary wall becomes differentiated into two parts: an inner portion composed of cells, which lengthen greatly in the radial direction and lignify to form the hard wall of the capsule, and an outer layer which retains its parenchymatous condition and in which a rich laticiferous system is developed. The laticifers are not so extensive in the septa and central column bearing the placentas. The funicle and raphe of the developing seed are almost devoid of laticiferous tubes. A feature of *Hevea* seeds is the laticiferous system developed just beneath the inner limiting layer of cells of the inner integument. The walls between these laticiferous cells break down and processes grow out of many of them: some of the processes are branched. This laticiferous system of the inner integument consists of a number of cells communicating one with another due to the partial dissolution of the intervening walls and of unsegmented processes emitted from these cells. It seems to be a combination of articulated and nonarticulated types of laticifer (Parkin, 1900).

In the vegetative parts of *H. brasiliensis,* the latex vessels are confined almost entirely to the secondary phloem region of the trunk, branches, and roots (Fig. 1). They are of articulated anastomosing type arising as longitudinally contiguous rows of cells from which most of the cross walls eventually disappear, partially or completely, during differentiation and maturation. The vessels run more or less longitudinally, parallel with the main axis of the tree. Since they are differentiated at fairly uniform time intervals by the cambium, they are arranged as rings (cylinders) concentric

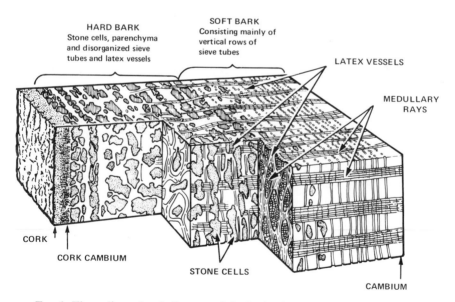

FIG. 1. Three-dimensional diagram of the bark of a mature tree of *Hevea brasiliensis.* The sieve tubes that make up the bulk of the soft bark, and parenchyma in both soft and hard bark, are not shown in the diagram. Redrawn from Riches and Gooding (1952).

with it and the outer bark: anastomoses are frequent where adjacent vessels within a ring touch and fuse, but they are rare between adjacent cylinders (Dickenson, 1969).

In *Hevea brasiliensis,* there is a close correlation of the alignment of the latex vessels in the bark with that of the peripheral wood fibers, which are usually arranged in a gentle spiral inclined from lower left to upper right if facing a standing tree. The angle of inclination is a clonal characteristic. In 3 of 27 clones examined, the angle inclined from lower right to upper left (Gomez and Chen, 1967).

The number of latex vessel rings has been used as an indicator of future productivity (Bobilioff, 1923), but it does not adequately represent the quantity of laticiferous tissue involved in exploitation of the tree. The type of planting material, whether budded or seedling, age of tissue, and height of sampling influence the distribution and frequency of latex vessel rings, and the density and diameter of latex vessels all need to be considered. Over the past thirty years, breeding and selection for higher yield have resulted in an increase in average number of rings of latex vessels from 13 to 27 (Wycherley, 1969; Gomez *et al.,* 1972; Narayanan *et al.,* 1973).

The latex vessels of branches do not have a direct influence on the production of a tree; Arisz (1918, quoted by Dïjkman, 1951) showed that in all the higher branches the latex vessels between stem and branches were broken, apparently mechanically as a result of the bending tension.

The phelloderm of virgin bark consists of only a few rows of cells, while that formed by wound phellogen after tapping becomes very thick. The number of latex vessels formed by the cambium after the bark has been tapped under normal conditions of growth is larger than originally. Almost all the newly formed latex vessels are functional, in contrast with those in the original bark. In the latter, the latex vessel cylinders are broken up by tangential pull with the formation of nonlaticiferous cell complexes and stone cells wedging into the latex vessel cylinders.

A large proportion of *Hevea* trees in commercial production are buddings: a clonal scion on a seedling rootstock. The number of latex vessel cylinders in the scion does not show a clear relationship to those in the stock. Often a number of latex vessel cylinders from the scion end blindly in the bast of the stock (Bobilioff, 1923).

Most evidence suggests that there is little or no movement of latex in an untapped tree. Schweizer (1949) examined a budded tree in which the scion produced white latex, but that in the stock was yellow. When punctures were made in the bark on each side of the graft union, even when the punctures were as close as 1–2 cm, the first drops of exuded latex were white and yellow, respectively. After a few drops the latex from the scion became tinged with yellow. Thus, although there appears to be continuity between the vessels, the synthetic processes appear to be disparate in the two tissues. Again, in the two species of *Taraxacum,* the rubber particles can be easily distinguished by their shape, and it has been established that when the shoot of one species is grafted on to the root of the other, both parts continue to produce their characteristic particles.

C. Cytology; Latex Composition

In the earliest stages of initiation and development, the latex vessel in *Hevea* is recognizably organized as a living cellular system in possession of nucleus, cytoplasm, and associated characteristic organelles and membrane systems. As maturation proceeds, these features become progressively confined to the peripheral regions of the vessel, eventually to become vestigial, until at full maturity latex as a cytoplasmic derivative occupies the major portion of the lumen (Archer *et al.,* 1963; Dickenson, 1969).

Mitochondria occur frequently and fairly evenly dispersed throughout the differentiating cytoplasm of very young latex vessels. As differentiation proceeds and the central portion progressively acquires a structure characteristic of latex, the mitochondria are confined mainly to the persistent peripheral cytoplasm. As the vessels age further, the peripheral cytoplasm degenerates and the number of mitochondria diminishes considerably. Initially, nuclei persist in their original positions and numbers as the latex vessels are formed by the dissolution of cross walls in columns of cells. The nuclei degenerate progressively during maturation of the vessels so that they become increasingly infrequent until they disappear altogether. Nuclei are very rarely found in latex extracted from mature trees.

As latex vessels age, there is a marked degeneration of cytological structure so that the oldest vessels in the outer bark are little more than repositories for rubber particles (Archer and Audley, 1967). This ultrastructural transition in the young vessel is accompanied also by a physical change from the gel structure of the peripheral cytoplasm to the more fluid sol state of the latex, or, as has been suggested, to a reversible gel/sol state (Archer *et al.,* 1963). The main constituents of tapped latex are given in Table I.

The rubber particles in latex obtained by tapping are usually entirely spherical, but in some clones the larger ones may be ovoid or eccentrically pear-shaped. They may be as large as 5–6 μm in diameter or probably somewhat less than 100 Å. They are strongly protected in suspension by a thin film of absorbed protein and phospholipid: a negatively charged envelope of hydrophilic colloids. Electron micrographs of small rubber particles appear to show their molecular structure (Fig. 2A).

TABLE I

ORGANIC NONRUBBER CONSTITUENTS OF TAPPED LATEX FROM *Hevea brasiliensis*[a]

Fresh Latex[b]

Serum 48	Rubber phase 37	Bottom fraction 15	Frey-Wyssling particles
Inositols 1.0–1.5	Protein 0.5	Proteins 0.2	Carotenoids
Carbohydrates	Phospholipids 0.6	Phospholipids	Plastochromanol
Proteins 0.5	Tocotrienols (free	Plastoquinone	Other lipids
Glutathione 0.01	and esterified) 0.09	Ubiquinol	
Free amino acids 0.08	Sterols and sterol	Sterols	
Ascorbic acid 0.02	esters	Trigonelline 0.07	
Other organic acids	Fats and waxes	Ergothioneine 0.05	
Nitrogenous bases 0.04			
Mononucleotides 0.02			
Nucleic acids 0.002			

[a] Archer *et al.* (1969).
[b] The numbers next to the components indicate their approximate concentrations in gm/100 gm latex.

There are two main nonrubber particles in latex. The *lutoid* (Homans *et al.,* 1948; Ruinen, 1950) is a fragile, osmotically sensitive vesicle, 2–10 μm in diameter, with a liquid content, and bounded by a unit membrane 80–100 Å thick, with a complex ultrastructure. Electron microscopy reveals ordered arrays of ringlike structures consistent with current micellar concepts of membrane architecture (Gomez and Southorn, 1969; Dickenson, 1969). The vesicles gradually acquire, as a transition phase, the protein microfibrillar inclusions characteristic of the young lutoids of *Hevea brasiliensis,* but not all species of the genus (Fig. 2B). The function of these fibrils is unknown (Archer *et al.,* 1963; Audley, 1965, 1966).

Dickenson (1965) has shown that similar microfibrils may be present in latices of plants belonging to other families; in *Ficus* for example, although they are distinguished by the same features of great length and helical structure, the helices are smaller in diameter and pitch than those of *Hevea brasiliensis.*

The lutoids possess features in common with vacuoles such as the capacity to imbibe and concentrate such substances as neutral red; Wiersum (1958) suggested they should be regarded as a polydisperse vacuome. Pujarniscle (1968, 1970) pointed out that they are in many ways analogous with lysosomes of animal cells, possessing a range of hydrolytic enzymes enclosed by a membrane with selective permeability. Accumulation within the lutoid serum of Mg^{2+}, Ca^{2+}, Cu^{2+}, citrate, and phosphate accounts

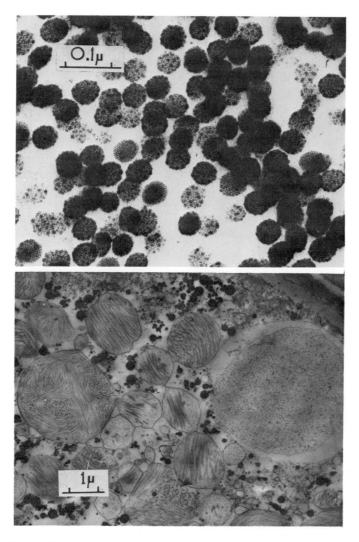

FIG. 2. (A) Upper electron micrograph shows small rubber particles from the Frey-Wyssling layer of centrifuged latex from *Hevea brasiliensis*. Granular appearance is consistent with a structure of rubber molecules (100,000 daltons) in a nonrubber matrix, possibly protein. (B) Lower electron micrograph shows section of a young latex vessel, with lutoids containing microfibrils, and scattered rubber particles (dense). The unit membrane surrounding the lutoids is contorted because of slight plasmolysis. In older vessels, the microfibrils disappear and the proportion of rubber increases. From Dickenson (1969).

for the increased acidity of the lutoid serum and, as suggested by Ribaillier *et al.* (1971), the lutoids may play a critical role in metabolic control (especially glycolysis) by the selective release or uptake of these compounds.

It is unclear how the lutoid population is maintained at high density despite continual losses during tapping. No evidence of regeneration of lutoids in the latex of young or mature vessels by division or budding has been shown. The alternative may be that, as has been postulated for the presumptive site of the biogenesis of rubber, the latex vessel primordia and the peripheral cytoplasm, constantly developing and differentiating as part of the incremental growth process, may comprise the continuously available sources from which, by autoreproduction, the lutoid population of the drainage area is maintained (Dickenson, 1965, 1969).

The second nonrubber particle (the "Frey-Wyssling particle") is mainly composed of lipid material, yellowish-brown or orange in color due to presence of carotenoids, of high refractive index, denser than rubber, and of the same size order as the larger rubber particles. It exists as inclusions in a larger elaborately constructed organelle rare in comparison to lutoids, termed the Frey-Wyssling complex.

The basic structure is of a relatively large spherical vesicle, 4–6 μm in diameter, fragile and deformable under moderate centrifugal forces, but mechanically more robust and less osmotically sensitive than the lutoid. It is bounded by a typical double membrane, and in the interior, in addition to the variable number of Frey-Wyssling globules, there are structures of great complexity, the elaborate nature of which suggests multiple activities. The constant association of these complex structures with liquid/carotenoid Frey-Wyssling globules and with membrane-bound rubber or lipid inclusions indicates their probable involvement in synthesis of one or more of these substances. Centrifuge-enriched fractions containing these organelles, when incubated with ^{14}C-mevalonate, achieve its incorporation into ^{14}C-β-carotene (Dickenson, 1969).

Fresh latex after tapping from the tree may be divided into three main fractions by centrifugation; these are a white upper layer, an aqueous serum and the "bottom fraction" (Homans *et al.,* 1948; Archer *et al.,* 1969). The top layer comprises rubber particles stabilized by an adsorbed layer of protein and phospholipids. The serum contains most of the soluble substances normally found in plant cells, including amino acids, proteins, carbohydrates, organic acids, inorganic salts, and nucleotidic materials. The bottom fraction consists largely of lutoids, but also includes varying amounts of rubber, lipid-containing Frey-Wyssling particles, mitochondria, and other particulate components having a density greater than that of the serum, e.g., ribosomes (McMullen, 1962).

Latex contained roughly 44% rubber, 46% latex serum, 9% lutoid serum, and 1% miscellaneous solids by volume (Southorn and Edwin, 1968). This may be an underestimate of lutoid (or bottom fraction) serum since some of the lutoids had burst during collection of the latex.

Latices from the Apocynaceae examined by van Die (1955) did not contain lutoids. Only about 1% of the centrifuged latex was bottom fraction and this consisted only of granular solids.

Kopaczewak (1951, quoted by van Die, 1955) stated that on centrifuging (at 7000 rpm), latex from *Euphorbia resinifera* and *E. helioscopia,* a yellowish bottom fraction (4-mm thick) was found together with an elastic top layer (1-mm thick)—a picture greatly resembling that of *Hevea*. Sterol-like compounds are found in both fractions of the *Euphorbia* and in the top fraction, a rubber-like substance as well. Matile *et al.* (1970) reported vacuoles in *Chelidonium* latex that are similar to *Hevea* lutoids, but contain characteristic alkaloids.

D. BIOSYNTHESIS OF RUBBER

The biosynthesis of rubber has been reviewed on a number of occasions (Archer and Audley, 1967; Lynen, 1969), but since continued synthesis is necessary for continued flow of latex, a brief account is in order here. The main pathway seems to be fairly clear, starting with acetyl-CoA, through mevalonate to isopentenyl pyrophosphate (IPP), which can be used either for chain extension in existing rubber particles or *de novo* synthesis of new particles. Polymerase activity at the interphase between serum and rubber particles appears to facilitate the production of rubber by a system in which the hydrophilic substrates, dimethylallyl pyrophosphate and IPP, are converted to a lipophilic end product. The soluble enzymes required for the steps up to IPP were all found in the latex serum. Thus, IPP formed in the serum could also be used for production of sterols and carotenoids. There seems to be a large variation in the capacity of different clones of *Hevea brasiliensis* to produce rubber. This may be due to differences in the efficiencies of enzymes controlling rate-limiting steps, or (in trees in tapping) to differences in rates of replacement of enzymes lost during latex flow.

The overall process of rubber synthesis requires three components: acetyl-CoA as building blocks, TPNH as a reducing agent, and ATP as an energy source. All three components are generated by the degradation of carbohydrates. This clearly suggests that *in situ* carbohydrate metabolism may become a limiting factor in rubber biosynthesis. D'Auzac (1964) found that endogenous sugars gradually disappeared in tapped latex. Glycolysis must be one of the main sources of ATP required for rubber synthesis (Jacob, 1970). All the enzymes of the Embden-Meyerhof pathway have been observed in latex. TPNH is generated by the hexose monophosphate "shunt" (Bealing, 1965). The low rates of incorporation of acetate and pyruvate into rubber *in vitro* may be the result of too few mitochondria in tapped latex compared to situations within the cell, which leads to a reduction in generation of acetyl-CoA.

Hydroxymethyl-glutaryl-CoA reductase activity had the lowest activity of the serum enzymes and could be the rate-limiting enzyme, as it is in the formation of cholesterol in mammals (Lynen, 1969).

Tupy (1969a) and Tupy and Resing (1969) showed that the rate of carbohydrate metabolism in latex was determined by the invertase-catalyzed hydrolysis of sucrose. Tupy (1969a) found that the rate-limiting step depends specifically on unfavorable conditions for invertase activity rather than a particular scarcity of the enzyme; latex pH is critical.

Utilization of sucrose in regeneration of lost latex constituents indicates that the capacity of the tree to supply sucrose to the laticiferous system is an important determinant of the latex-producing capacity of regularly tapped trees (Tupy, 1973a). Since the activity of invertase limits the overall metabolism of latex, it may influence production by controlling rubber formation. On the other hand, there is evidence suggesting a direct relationship between the invertase-dependent metabolic activity of the latex and latex flow (Tupy, 1973c). High-yielding clones were found to incorporate substrates into rubber more effectively than low-yielding varieties. Individual variations were, however, too pronounced to provide a reliable criterion for selection of potentially high-yielding trees (Woo and Edwin, 1970).

The bottom fraction is considered essential for conversion of acetate, but not mevalonate, into rubber. The particulate structures in latex may thus be important in enzymes catalyzing the metabolism of acetate to mevalonate: it was shown that mevalonate was incorporated more readily than acetate and also that removal of the bottom fraction seriously reduced incorporation of acetate but had, if anything, a beneficial effect on incorporation of mevalonate. With clone RRIM600, the whole latex incorporated 3.16% acetate and 63.5% mevalonate, while the top fraction (rubber plus serum) incorporated only 0.77% acetate but 80% mevalonate (Woo and Edwin, 1970). It has also been shown that conversion of β-hydroxyl β-methyl-glutaryl-CoA to rubber, *in vitro,* depends on the presence of bottom fraction particles (Archer and Audley, 1967).

Studies by d'Auzac (1965) of selected trees of one clone (PR107) showed a correlation between latex yield and rate of incorporation of acetate.

E. Function of Latex

A number of possible functions of latex were listed by Parkin (1900):
(1) transport and storage of food materials; (2) protection of the plant
by sealing wounds or checking the ravages of insects; and (3) reserve water
supply. Parkin favored the third alternative, noting the development of
a system of tubes running throughout the plant that could be filled with
water during the wet season and then gradually drawn upon during times
of drought. The genus *Euphorbia* chiefly inhabits dry regions and is one
of the richest in latex. Sen and Chawan (1972) regard the latex system
as a very important ecophysiological factor in the adaptation of *Euphorbia
caducifolia* to the extreme arid environment of the Indian desert. Fernando
and Tambiah (1970a) suggested that the correlation of yield of *Hevea*
with rainfall and temperature indicates that latex functions as a system
regulating water within the plant. The ability of the plant to survive and
grow satisfactorily under drought conditions appears to improve with increased capacity for latex production. They also pointed out that *Hevea*
is remarkably free from virus diseases, and this may be due to the protective effect of the latex system against insect vectors.

Haberlandt (1914) noted that the latex system (1) formed part of
the conducting strands and (2) had branches ramifying among the palisade
tissue of the leaves. Although admitting that direct experimental results
were inconclusive, he surmised that latex is most probably a nutritive fluid
because of the large quantities of plastic substances therein, but that latex
also contained materials of essentially excretory nature. Haberlandt (1914)
cited two ecologically important functions: (1) latex readily coagulates
on exposure to air, and is thus well fitted to serve as an occluding material
in the case of mechanical injury, and (2) protection against animal foes,
especially when the latex contains toxins, as it frequently does.

In rubber trees, at least, the latex vessels in petioles become ob-
structed by callose at a fairly early stage and thus could play no part in
transporting photosynthate from the leaves (Bobilioff, 1923; Spencer,
1939b). Bobilioff observed movement of particles in intact vessels of *Ficus
elastica,* but there appeared to be no set direction of flow, which was of
short duration and appeared to be due to causes outside the latex vessels,
chiefly changes in turgor pressure on the wall from adjacent cells. The
movements were thus localized phenomena.

Spence and McCallum (1935) showed that the rubber content of
Parthenium argentatum decreased under stress conditions and may therefore serve as a reserve food store. However, later observations demonstrated that conditions conducive to active growth of *Parthenium* promote
accumulation of rubber rather than its utilization (Traub, 1946; Benedict,

1949, 1950). A similar correlation of growth and rubber formation was found for *Hevea* seedlings by de Haan and van Aggelen-Bot (1948) and, again, there was no depletion of the rubber content even in etiolated plants after nearly 60% of the carbohydrate reserves had disappeared.

Bealing (1965) pointed out a relationship between rubber synthesis and pentose phosphate cycle metabolism. He suggested two rather distinct aspects of rubber formation: (1) as an auxiliary respiratory mechanism in plants subject to periodic oxygen starvation (especially in leaves of xerophytes and within latex vessels of trees) and (2) as a specialized device for sustaining the supply of pentose cycle metabolites which is prerequisite for active growth. In many plants, particularly in xerophytic species, rubber tends to be produced indiscriminately by all the cells of a tissue, whereas in others it occurs only in the well-defined components of a laticiferous system. That such localization is more frequent among the rubber-producing representatives of rapidly growing tropical vegetation led Bealing to the conclusion that a major function of rubber-containing latex, or even of latex in general, is to provide surrounding tissues with a continuous supply of pentose cycle derivatives.

F. Water Relations of Latex Vessels in Intact Plants

Observations on latex obtained by conventional tapping methods do not necessarily reflect composition of the latex *in situ*. Ferrand (1941) developed a method of obtaining small samples of latex more representative of that in the latex vessels. He considered that if a small puncture be made into the bark, the first few drops of latex that exude would be squeezed out by contraction of the severed latex vessels and will have undergone little or no dilution. Gooding (1952a) could detect little or no change in percentage total solids in the first six drops obtained by this technique.

However, using a sensitive microosmometer, Pakianathan (1967) showed a progressive decrease in osmotic concentration from drops 1 to 7 collected by Ferrand's technique. The first drop gave a value of 445 mOsm/liter and this decreased to 375 mOsm/liter for the seventh drop. Extrapolation back to zero time indicated an osmotic concentration of the latex *in situ* as 451 mOsm/liter. This suggests that dilution, that is, inflow of water from surrounding tissues, commences as soon as latex begins to flow. Thus measurements made on exuded latex need to be interpreted with caution.

Buttery and Boatman (1966) used a direct method of measuring turgor pressure in the bark of *Hevea* and other laticiferous trees. Simple capillary manometers were used, similar to those developed by Bourdeau and Schopmeyer (1958) for oleoresin. A small hole was bored in the bark down to the wood and a fine steel tube fitted to a sealed glass capillary was inserted into the hole. Latex flowed in rapidly and the pressure was estimated from the difference between the initial and final lengths of the air column in the capillary.

In *Hevea brasiliensis,* early morning turgor pressures were in the range 8–15 atm, decreasing during the day and increasing at night (Table II). These diurnal pressure changes were positively correlated with relative humidity of the atmosphere and negatively correlated with changes in temperature, evaporation from atmometers, leaf water deficit, and stomatal opening. They are not displayed by trees devoid of leaves (Fig. 3). Thus, the loss in turgor during the day probably is the result of withdrawal of water from the phloem tissues under transpirational stress.

Pressures at the base of the tree stem normally exceed those at the top (Fig. 4), the gradient approximating 1 atm/10 m at night, and increasing up to six times this value during the day. This increase probably reflects development of tension gradients in the xylem during transpiration. There was no evidence for seasonal changes in turgor under Malayan conditions (Buttery and Boatman, 1964, 1966).

Measurements on other laticiferous trees generally gave slightly lower values than those on *Hevea.* Some of these low values may have been brought about by leakage of latex into the bark tissue surrounding the hole bored for the manometer (Table II).

A hydrostatic pressure gradient of approximately 1 atm/10 m may be expected in the xylem even under conditions of full turgor, and the

TABLE II

TURGOR PRESSURES OBSERVED IN THE LATICIFEROUS SYSTEM
OF VARIOUS SPECIES[a,b]

Species	No. of trees recorded	Pressure range observed
Hevea brasiliensis, Muell-Arg.	>100	7.9–15.0
Ficus elastica, Roxb.	3	8.0–10.4
Alstonia scholaris, R. Br.	4	4.2–8.5[c]
Plumeria acutifolia, Poir	2	7.4–8.2
Cerbera manghas, L.	2	6.6–8.1[c]
Euphorbia pulcherrima, Willd.	2	6.9–7.5

[a] Buttery and Boatman, 1966.

[b] Measurements made between 0600 and 0800 hr at heights not exceeding 1.5 m from the ground.

[c] Extensive latex bleeding probably resulting in underestimates.

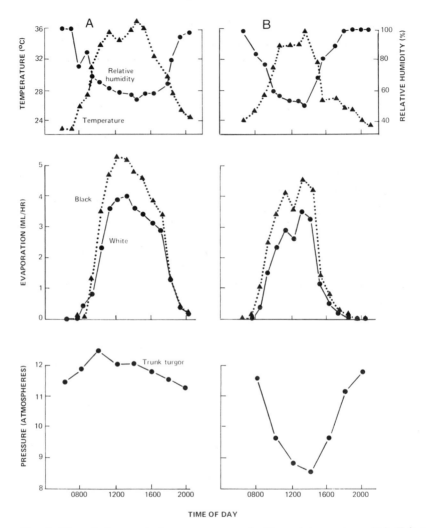

Fɪɢ. 3. Diurnal changes in turgor presssures in *Hevea* latex system. (A) February 21, 1964. Trees "wintering" and devoid of leaves. (B) March 24, 1964. Trees refoliated. Air temperature, relative humidity, and evaporation from black and white atmometers are included to facilitate comparisons between days. Buttery and Boatman (1966).

existence of such a xylem gradient would be sufficient to stabilize a similar one in the laticiferous phloem tissue.

It was suggested that the fall in turgor pressure of the latex during the day is brought about by the transfer of water from phloem tissue to xylem under conditions of rapid transpiration. This implies that (1) the

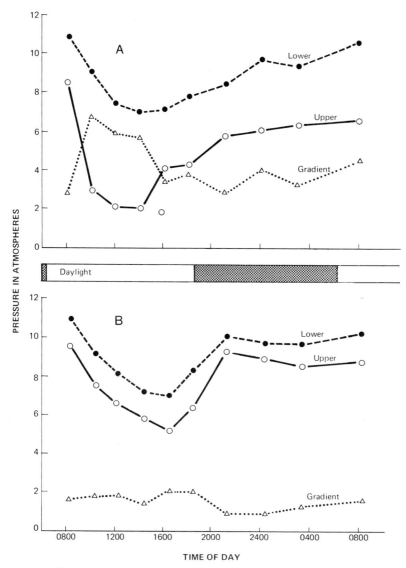

PRESSURE IN ATMOSPHERES

TIME OF DAY

Fɪɢ. 4. Diurnal changes in turgor pressure at two positions (upper and lower) 900 cm apart on the trunk of *Hevea brasiliensis* trees. Measurements over the same 24-hr period on two trees (A) and (B). The gradient in turgor pressure between the upper and lower positions is given in atm/10 m. Buttery and Boatman (1966).

chemical potential of the water in the xylem falls during transpiration, i.e., is under tension; (2) water can move freely between xylem and phloem tissues; and that (3) the latex vessel system behaves as an osmometer.

The first proposition is well supported (e.g., Slatyer, 1967) and the second seems a reasonable assumption. The third point is supported by the observed sensitivity of turgor pressures to meteorological conditions that would be expected to affect transpiration rates, by the relationship to leaf-water deficit and stomatal aperture, and by the marked differences in behavior on leafy and leafless trees.

Since the gradient in observed turgor pressure in the laticiferous system at night corresponds roughly to that expected in the xylem under conditions of full turgor, and since the magnitude of diurnal changes in observed turgor appears to be dependent on rate of transpiration, it seems probable that, if the conditions discussed earlier are satisfied, the increased gradients in turgor pressure measured during the day are a reflection of the development of xylem tensions of similar magnitude.

Following an afternoon shower, hydrostatic pressures increased in stem bark without any apparent decrease in leaf water deficit (LWD). The entry of water into leaves from the outer surface occurs only slowly in most species investigated (Slatyer, 1967), so we may assume that the major immediate effect of the rain was to reduce transpiration. It would appear that this, in turn, permitted a relaxation of xylem tension and a partial restoration of turgor in the trunk before any significant reduction in LWD occurred. This must imply that either the relaxation in xylem tension takes place relatively slowly from the base to the crown of the tree, or that equilibration between xylem and the other tissues concerned occurs more rapidly in the stem than in the leaves.

The interrelation between turgor pressure in the latex system and water potential of the xylem has been demonstrated directly, albeit under artificial conditions. B. R. Buttery (unpublished), using an isolated length of stem 1.2 m long, showed that a 0.27 M sucrose solution poured into the xylem decreased turgor pressure in the latex system by half, while a doubling of the sugar osmolarity reduced turgor to very low values. Replacement of the sugar solution by water brought about a considerable, but not complete, recovery in turgor of the latex over a period of 2 to 3 hr.

In a few cases, turgor pressures have been measured at the same time as osmotic pressures of latex obtained by microtapping of *Hevea*. The difference between the two gives the diffusion pressure deficit (DPD, or suction pressure). Even with measurements taken in early morning, these values for DPD were always above zero, ranging from 0.9–2.9 atm (water potentials of approximately −0.9 to −2.9 bars). The "within tree" variation was small, so that values for one tree ranged from 0.9–1.3 atm, while in another they varied from 2.2–2.9 atm (S. W. Pakianathian and G. F. Milford, unpublished).

In another test, measurements at 0400 hr, 0830 hr, and 1230 hr revealed turgor pressures of 12.3, 11.4, and 9.9 atm and osmotic pressures of 13.6, 13.2, and 13.2 atm, respectively. Thus the DPD increased from a value of 1.3 atm at 04.00 to 1.8 at 0830 hr and to 3.5 atm by midday.

Data on the variation of DPD with height do not seem to be available. But there is a problem that needs resolving here. If there is a positive hydrostatic pressure in the xylem at dawn of 1 to 2 atm (depending on tree height), this would be expected to lead to a negative DPD, i.e., a latex turgor pressure in excess of the osmotic pressure. The data in fact indicate that even at dawn the tree is not fully turgid and that there may still be a tension in the xylem. This is rather surprising especially in view of the rapid adjustments that occur in turgor pressure and in stem diameter in response to the fluctuation in weather. Scholander *et al.* (1965) observed that the size of the negative pressures in the xylem sap of vascular plants depended very much on the osmotic pressure of the root medium, as well as on the height above ground and relative humidity of the atmosphere. It seems likely that the xylem of *Hevea* is under negative pressure, even in the absence of transpiration, in order to balance the osmotic pressure of the soil water.

Pyke (1941) showed that *Hevea* stems underwent diurnal variations in diameter similar to those found by McDougal (1924) for many trees of the temperate zone. A rapid decrease in diameter started soon after sunrise and continued at a diminishing rate throughout the morning and afternoon, tailing off to a minimum between 2 and 3 PM followed by a rise which continued during the evening and very slowly throughout the night, reaching a maximum shortly before dawn. Pyke reported that this response closely followed changes in atmospheric humidity. The daily variation was greatest on hot bright days, when the reduction at 1.6 m above the ground of a trunk 20 cm in diameter might be 0.25 mm between 8 AM and 2 PM. On rainy or overcast days, the total reduction might be only half as much. The trees responded very quickly to changes in insolation; thus, in one observation during overclouding of the sun at noon for 22 min, a stem increased in diameter by 0.007 mm, although previously it had been steadily decreasing. This is in line with the rapid adjustment in turgor during a shower, observed by Buttery and Boatman (1966).

III. TAPPING

A. Methods of Inducing Latex Flow

Latex can be induced to flow in many plants by cutting open, or tapping, the appropriate tissue. Only in a few has flow been sufficiently prolonged and yield of rubber high enough to occasion study.

Bangham (1934) observed that piercing the fruit stalk of *Cryptostegia grandiflora* produced a jet of latex over a meter long, and maintained it for several seconds, indicating a high internal hydrostatic pressure. The semileafless thin stems or "whips" of *Cryptostegia* contain latex vessels which ramify horizontally and vertically in the pith and bark like a fungal mycelium. When the apex is cut off, there is at first a high rate of flow that very rapidly falls off and stops after 2–12 min. By localized freezing of the whips at different distances from the tip, it has been shown that latex is withdrawn from a distance of at least 70 cm. Additional latex appears if a small portion (2–6 cm) is cut off the following day (but several days rest is more usual). This process may be repeated until the base of the whip is reached (Curtis and Blondeau, 1946; Blackman, 1965). In mature latex vessels, such as those in the pith, regeneration of rubber is seemingly either negligible or very slow, and it appears that formation is restricted to the new vessels that are laid down in the bark. Normal yield of rubber is about 130 kg/ha/yr.

In *Castilloa,* the latex system is confined to the bark and since the vessels consist of elongated single cells with no interconnections, all vessels must be cut for maximum yield. This, in practice, entails cutting through the cambium to the wood. During the process of regeneration of the bark, a hard and extensive callus is formed which is difficult to cut. As a result, subsequent tapping cuts are either made transversely or if parallel, at a considerable distance from the previous cut. The flow of latex is very rapid and may continue as long as 2 hr. This rapid and prolonged flow may be associated with low viscosity but high stability of the latex and, therefore, little or no plugging of the vessels. A large volume of latex is collected at each tap, but owing to the very low rate at which the contents of the vessels are replenished, trees can only be tapped one to four times per year. Under optimal conditions a yield of 900–1100 kg of rubber per hectare per year may be obtained (Blackman, 1965).

Hevea has the great advantage that it can be tapped frequently since latex regenerates rapidly. Tapping does not touch the cambium and, hence, bark can regenerate smoothly and eventually be tapped again. Some of the better clones can produce over 3500 kg/ha/yr.

Tapping of *Hevea* reduces growth. Over the first year of tapping, the average reduction in the rate of girth increment of six high-yielding clones was 29%, but the variation between clones was considerable, the extremes being 10 and 63% (Rubber Research Institute of Malaya, 1962). This suggests that regeneration of latex consumes food supplies that would otherwise be used in growth. The rubber harvested during the first two years of tapping accounted for 3–11% of the dry weight accumulation in different clones, but this proportion probably increases to more than 20% in later years (Templeton, 1969).

The most common method of exploitation is to cut a spiral groove in the bark around half the circumference of the trunk, sloping down from left to right at an angle of 25°–30° from the horizontal. The cut is generally made to within 1–1.5 mm of the cambium. Latex flows down the cut to a metal spout and then into an earthenware or glass cup. Trees are normally tapped in the early morning and flow usually continues for a number of hours. The cup is emptied in the late morning and the latex taken for processing. The coagulated material on the cut and any latex that dripped into the cup after it had been emptied and coagulated there provide an additional lower-grade product which is collected at the next tapping, normally 2 days later. The cut is then reopened by removal of 1.5 mm of tissue from its lower surface and the whole process is repeated.

The tapping cut slowly progresses down the trunk until the ground or bud-graft union is reached, a process that normally takes 5 years. A new cut is then opened on the opposite side of the tree in the untapped bark. By the time this cut also reaches the base of the tree, cambial activity will have renewed the bark of the original tapping panel sufficiently for it to be tapped again.

In some regions, notably Indo-China and parts of West Africa, a "full spiral" cut is favored. This extends right round the tree and is tapped every fourth or fifth day.

B. Sequence of Events during Latex Flow

1. Water Relations

Cutting a latex vessel results in a rapid elastic expulsion of the fluid cytoplasm, a loss of turgor, and flow through the vessel along the pressure gradient. Simultaneously, water enters the vessel from surrounding tissues, owing to the gradient in water potential, and causes the dilution of latex referred to earlier (Frey-Wyssling, 1952; Riches and Gooding, 1952).

According to the law of Poiseuille, the flow in capillaries is given by volumetric flow rate

$$dQ/dt = \pi PR^4/8LN$$

i.e., the volumetric flow rate dQ/dt is directly proportional to the pressure gradient P/L and to the fourth power of the radius R and inversely proportional to the viscosity N of the streaming liquid.

Elastic tubes do not only allow capillary flow, but, at the same time, they are reservoirs of the streaming liquid. Thus the loss of latex causes a decrease of P and R so that a change of tube profile results. Further, the quantities Q, P, and R change locally within the capillary as they increase with the distance from the tapping cut; they also depend on the time elapsed from opening the vessels.

In the beginning of the flow, the elastic contraction of the latex tubes predominates (exponential decrease of the discharge), whereas after a while the flow is regulated by the increase of the capillary length (hyperbolic decrease of the flow rate), until it finally ceases as the latex coagulates.

The cell walls of the latex tubes are not only extensible, but permeable. The decrease in turgor pressure in the latex tubes caused by tapping is followed by infiltration of water from surrounding tissues. Frey-Wyssling showed that the water-soluble and insoluble components of latex were diluted in the same proportion, which suggests that the dilution was caused by pure water. In a bark "island," water intrusions of up to 40% were observed; while tapping of intact bark caused dilutions of only 10%.

The viscosity of fresh latex drops with decreasing concentration of dry matter in the latex (Frey-Wyssling, 1952), so that with concentrations of 45, 40, and 35% dry matter, relative viscosities of 13.0, 8.7, and 6.0 were found.

None of the quantities controlling the Poiseuille flow is constant during the discharge of latex, as the radius R, the pressure potential P/L, and the viscosity N change considerably when the latex tubes are tapped. Notwithstanding this handicap for a mathematical analysis of the latex flow, Frey-Wyssling (1952) showed that, at the beginning, the latex discharge is governed by the elastic contraction of the tubes (elastic discharge), while later, the increasing capillary length regulates flow.

The most important quantity is the radius R of the capillary as it appears in the fourth power. During the flow, R diminishes by one-quarter of its original value. The capillary length L increases parabolically with time; in isolated bark areas its maximum value is given by the length of the area, but for normal tapping, L cannot be determined as the streamlines diverge below the tapping cut. The turgor pressure (P) is the only quantity that holds its value, if places sufficiently distant from the tapping cut are considered. The pressure potential (P/L) decreases hyperbolically with time. The latex may be diluted by up to 40% during the flow from a bark island: this causes a drop of the viscosity to about one-half of its original value.

While the decrease of the radius (R) and the pressure potential (P/L) diminish the flow intensity, the decrease in viscosity works antagonistically since it facilitates flow.

Boatman (1966) reported that latex flow may be greatly prolonged by removal of additional slices of tissue (1–2 mm thick) from the surface of the cut at short time intervals (5–15 min) after initial tapping. This resistance to flow built up very close to the cut was termed plugging. So that, in addition to elastic contraction of the vessel and gradual increase of the latex vessel area contributing to flow, plugging of the ends of the cut vessels, a third factor, regulates flow. As Paardekooper and Samosorn (1969) pointed out, it is unlikely that a simple mathematical equation can represent accurately the flow rate from beginning to end in such a complex situation. Despite this, they found that several relatively simple models of the latex flow can almost correspond with empirically determined flow curves.

Perhaps the most successful equation for flow has been the "die-away" expression

$$y = be^{-at}$$

in which the flow rate (y) at a given time (t) is a function of the initial flow rate (b) and the time-flow constant (a). Application of this model to data from two clones resulted in reasonably straight lines, although the flow immediately after tapping deviated from the model. Ninane (1967a) and Ninane and David (1971) derived a similar equation independently.

2. Metabolic and Biochemical Changes during Flow

When a tree is tapped for the first time, the latex obtained is very viscous with a high particle frequency and flow ceases rapidly. Successive tappings at regular intervals (usually alternate days) result in increasing yields of more dilute latex until a more-or-less steady equilibrium position is reached (Fig. 5). The total weight of rubber obtained actually increases

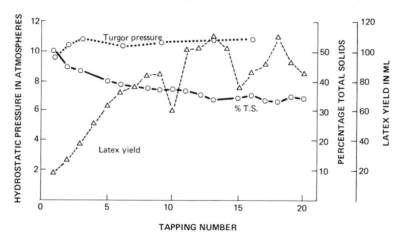

Fig. 5. Effect of alternate daily tapping on turgor pressure under the tapping cut measured immediately before tapping, volume of latex yielded, and percentage total solids content of latex. Buttery and Boatman (1967).

under regular tapping, since the increase in volume more than compensates for dilution.

The results of Gooding (1952b) support the suggestion of Arisz (1928) and Schweizer (1949) that the rate of formation of rubber in the bark bears an inverse relationship to the concentration of rubber in latex. During a period of tapping there was a more or less constant level of total solids in the latex. Withdrawal of latex by tapping did not appear to reduce the capacity of the tree to yield similar quantities (both in volume and contents) at subsequent tappings; latex was being regenerated as fast as it was removed.

As soon as the tree was rested, the concentration of the latex increased, rapidly at first, then more slowly, and finally reaching a maximum.

In general, tapping systems of high intensity (i.e., frequent taps and/or longer cuts) produce tapped latex of lower concentration than do systems of lower intensity. Gooding compared trees tapped on a half-spiral, alternate daily system throughout, with a set that was moved to a third-spiral, alternate daily system, after a calibration period on half-spiral. Those on the third-spiral tapping showed (a) higher percentage of total solids in the latex *in situ* (Ferrand method), (b) higher percentage of total solids in tapped latex, (c) no significant difference in dilution index, (d) lower volume of latex per tapping.

Since the dilution index is the same for the two tapping systems, it cannot be argued that the more dilute latex obtained by tapping at the higher intensity is due to a more intense dilution reaction. The similar dilution indices imply that the actual influx of water to a given volume of the original latex in the vessels is approximately the same in each case. The lower concentration of the tapped latex from the half-spiral cuts must result directly from the lower concentration of the latex in the vessels.

The changes in total solids during tapping reflect the variations in chemical composition as well as colloidal and biochemical properties that take place during the period of flow (Pujarniscle *et al.*, 1970). The rate of flow falls very rapidly at first, then more slowly; the rubber content falls smoothly to about 75% of the initial level; the magnesium level also drops continually from beginning to end of flow. The biosynthetic activity, low in the first three to four fractions, increases very considerably toward the end. However, a similar experiment in Vietnam by d'Auzac (1964), using much older trees, showed exactly the opposite trend. The stability of lutoids increases with time during flow; the volume of bottom fraction increases rapidly over the first three fractions and then remains fairly constant.

At the first tapping, most characters studied, such as rate of biosynthesis and hydrolytic activity, were at their lowest level, but the rubber content, plugging index, and index of bursting of lutoids, were at a maximum.

Tapping disturbs the equilibrium within the latex vessels, not only by loss of latex vessel contents, but perhaps also by a shock reaction provoked by wounding. These events lead to (a) an increase in protein synthesis, of rubber synthesis, and probably of lutoid synthesis, judging by the rapid increase in hydrolases and especially phosphatase and phosphodiesterases; these attain their maximum from 11 to 15 tappings (full-spiral, twice weekly), and (b) a strong destabilization of the lutoids at the time of first tapping.

Having passed a maximum, most of the properties measured declined slowly to a new equilibrium value but still above that of the first tapping. But for rubber content, bursting index, and plugging index, a steady decline to a new level was observed (Pujarniscle and Ribaillier, 1970).

Thus wounding (tapping), by disturbing the equilibrium, starts to activate metabolic processes and protein synthesis as well as destabilization of the lutoids. Eventually, the tree appears to become accustomed to these repeated trauma. The lutoids recover their stability, following perhaps a modification of their structure or the composition of their membrane; the metabolic processes stabilize themselves at a new higher level in line with new requirements for rubber synthesis.

Tupy (1973a) examined effects of tapping on sucrose concentration in the latex. A tree was brought into tapping and at each tap the first 10-ml fraction of latex was sampled for analysis. The small amount of latex that flowed at initial tappings resulted in a strong decrease of sucrose level in the latex vessels; the lowest point was at the fourth tap. Later, the sucrose level and total production of organic matter per tapping increased progressively.

Thus, in spite of the formation of a concentration sink, the sucrose import to the area of latex outflow does not seem to be greatly enhanced during the period between the first and fourth initial tappings, but subsequently, in spite of latex depletion of this sink, the import of sucrose increases greatly as indicated not only by the progressive rise of its level, but also by increase in the quantity of latex components regenerated between tappings.

In addition, the regular damage to the bark in the process of tapping induces a metabolic sink involved in bark regeneration. The existence of this additional mobilizing agent would explain a small, but generally observed, elevation of sucrose concentration in the latex close to the incision. The speed of decarboxylation of ^{14}C-glutamic acid is also elevated near the cut—some ten times higher than at 50 cm above or below the cut (Lustinec et al., 1966).

Possibly the sucrose in tapped latex could be from the sieve tubes,

258 B. R. Buttery and S. G. Boatman

since these are also tapped. But the similarity of the results found by analyzing successive samples of tapped latex with those obtained from samples
taken by microtapping at different stem heights, seems to exclude this possibility. It is difficult to imagine that sucrose-containing solutes would enter
in a similar proportion, both the first small volume taken by puncturing
the bark and the latex obtained by a long tapping cut.

Tupy's results show that sucrose concentration in the latex drainage
area (= sink) is depressed with respect to its level in more distant sites
and that this decrease is caused mainly by sucrose utilization in metabolic
processes involved in latex regeneration.

When a tree is newly brought into regular tapping, the general metabolic activity of laticiferous tissue participating in latex regeneration increases gradually, as indicated by a progressive rise in RNA level and enzyme activity of the latex (Tupy, 1969b). RNA concentration and synthesis adjust to the level of latex regeneration required by the tapping system.
The first three taps on a full circumference spiral cut bring about a drastic
decline in sucrose content in the latex vessels, whereas, with a further six
taps, the sucrose level gradually increases. The yield per tap increases almost linearly over the first nine taps. Thus the sucrose level seems to follow the gradual adjustment of the laticiferous tissue to the regular wounding and draining, with the movement of sucrose into the regenerating tissue
being determined mainly by a metabolic sink per se and not simply a concentration sink (gradient).

By the third day after latex outflow, the mean level becomes more
than twice the level of the first day; on the fifth day, a 3.6 times rise is
found. This implies that the fall in sucrose content after tapping is several
times greater than that of the latex as a whole. It follows that aqueous
dilution during flow can participate only slightly in depletion of sucrose
after tapping. Similarly, it was shown that the first few initial taps drasti-
cally reduced the sucrose level, whereas the total solids content was only
slightly reduced. Thus, the observed depletion of sucrose due to tapping
should be considered principally as a consequence of its utilization in renewal of the lost latex constituents.

During tapping, the activity of invertase decreases greatly and is very
low at cessation of flow. Does this indicate that invertase concentration
is highest near the tapping cut and decreases the further one goes from
it? The area of high invertase activity is extended by the use of chemicals
which prolong latex flow. Rutherford and Bard (1971) found a close relationship between invertase activity and increase in water uptake by chicory
(*Cichorium intylus* L.) root cells. Thus there is a possible implication of
invertase-dependent metabolism with permeability to water flow. In laticiferous tissue, such a relation is suggested by the finding that the treatments

enhancing invertase activity increase (under certain conditions) both the hydrostatic pressure in bark (Buttery and Boatman, 1967) and the dilution of latex during its flow after tapping (Lustinec *et al.,* 1967).

In conditions favorable to high metabolic activity of the latex, several punctures made into the bark with a thin pin were sufficient to obtain a higher latex yield than that given by normal tapping on a full spiral cut when invertase activity was limited by low sucrose levels (Tupy, 1973c).

C. PLUGGING INDEX

The time-flow constant of the "die-away" equation $y = be^{-at}$ may be regarded as an index of latex vessel plugging: the greater the extent of plugging, the more rapid the decline in flow rate with time and the larger the time-flow constant. This constant a can be estimated from two simple yield measurements (Paardekooper and Samosorn, 1969).

$$a = \frac{\text{Mean initial flow rate (ml/min)}}{\text{Total Yield Volume (ml)}}$$

For convenience, the plugging index is taken as $100a$. Milford *et al.* (1969) showed that the plugging index was relatively constant within a clone, independent of tree age, and relatively little affected by site factors. Paardekooper and Samosorn (1969), however, showed that in South Thailand there was a marked seasonal effect, with the index in December being half as great as in May; therefore, the total yield during this period doubled. The usually observed increase in latex yields in the second half of the year, after wintering, appears to be entirely the result of a longer duration of flow.

The initial flow rate is positively correlated with total yield, while the plugging index is negatively correlated with total yield. Yield differences between trees of the same clone are mainly caused by differences in initial flow rate, while differences between clones are mainly associated with differences in flow time.

There is a high positive correlation between plugging index and latex rubber content on a clonal basis. This suggests that a higher rubber content (and therefore a more viscous latex) causes an earlier plugging of the vessels. Milford *et al.* (1969) suggest, however, that other characteristics such as lutoid destabilization control plugging, which in turn could influence the latex rubber content. Thus, the changes in rubber content may be a consequence of plugging rather than the reverse. The absence of a correlation between plugging index and rubber content for trees of a single clone lends support to this second explanation. Also, the observed seasonal variation in plugging index does not seem to be associated with corresponding changes in rubber content (Paardekooper and Samosorn, 1969).

Trees of clone Tjir 1, which have a high plugging index, show both a rapid rise in turgor (near the cut) within 15 min of tapping, shorter duration of flow, and increases in flow rate following repeated tapping.

Clone RRIM 501, on the other hand, has a low plugging index, slow increase in turgor, long flow, and little response to repeated tapping. It is well established that the volume of the laticiferous system, as defined by anatomical factors such as stem girth, number of latex vessel rows, etc., is important in relation to yield (Gomez et al., 1972; Narayanan et al., 1973). Latex vessel plugging is also important in yield since it determines the extent to which the laticiferous system is drained on tapping.

The relative importance of anatomical factors and plugging in yield was examined by Milford et al. (1969). The partial correlation coefficients indicated that both factors contributed independently to yield; number of latex vessel rows accounted for 14% of the yield variation, plugging for 23%, and the two together for 29%.

Full-spiral tapping tends to prolong flow, with clones with a high plugging index (e.g., Tjir 1) responding favorably to the system, whereas those with low plugging indices (e.g., RRIM 501) yield better on half-spiral cuts.

Ninane and David (1971) showed that duration of flow and the flow constant c (equivalent to the plugging index) were both affected by the season.

D. Estimation of Drainage Area

The laticiferous system is both the storage region from which latex is released on tapping and the site of the final stages in rubber synthesis. Its structure and extent are thus of direct relevance to productivity.

A number of methods have been used to estimate the area of bark from which latex flows on tapping, with measurements being made at various distances from the cut and various times from opening:

1. Changes in content of rubber, total solid or other components (sampling by Ferrand method)
2. Changes in stem diameter in relation to tapping
3. Displacement of latex labeled with radioisotopes
4. Changes in turgor pressure
5. Yield of microtappings (small punctures in the bark)

Three different, but related, factors need to be considered: (a) the region from which the latex flows during an individual tapping, (b) the area in which measurable disturbances occur (e.g., in turgor pressure) during the course of one tapping, and (c) the area affected by repetition of tapping in the normal sequence.

1. Gooding (1952a) took latex samples using Ferrand's technique of microtapping, both before tapping and 2 hr later, when flow had ceased. Latex concentration was considerably lowered at points 2, 30, and 60 cm below the cut on the panel side of the tree, at the level of the cut, and 60 cm below on the reverse side, 2 hr after tapping.

In other experiments with higher yielding trees, total solids above the cut were also depressed. Total solids recovered more quickly near the cut than further away (in line with recovery in turgor and trunk diameter observed elsewhere). Over a period of several months of regular tapping, a depression in total solids became apparent at considerable distances from the cut. By the twenty-fifth tapping, total solids content of the latex had fallen from near 60% to rather less than 30%, uniformly up the stem to a height of 9m above the ground. Thus it would appear that, in the long run, tapping draws on the resources of virtually the whole latex system of the tree. Bloomfield (1951) detected changes attributable to tapping in latex of branches 10 m distant from the cut. Thus, the lack of continuity between the laticifers of stem and branches described by Arisz (in Dijkman, 1951) is not complete.

It was not possible to detect any change in the concentration of latex at points distant from the cut resulting from day-to-day tapping after the general level of latex concentration in the tree had fallen to its lower equilibrium level. Gooding postulated that dilution of latex at any point must mean that there has been a reduction of turgor pressure at that point to allow influx of water. This presumably implies a movement of latex toward the tapping cut as turgor falls. Whenever dilution appears in the latex vessels as a result of tapping, it implies that there has been movement of latex toward the cut, and, in the trees under consideration, tapping has been followed by a movement of latex toward the cut from even the most remote parts of the tree.

2. It has been shown that stem contraction occurs below the cut (Pyke, 1941) and that the maximum contraction observed is less and time of occurrence later as distance from the cut increases (Gooding, 1952a). Although very small, stem contraction was detectable at a distance of 90 cm below the cut (Luštinec *et al.*, 1969). The shrinkage of the bark is less above the cut and its poorly defined maximum is some 30–40 cm away. Luštinec *et al.* (1969) attributed this to flow of latex from above the cut.

Following a mathematical analysis of their results, they deduced that the lateral flow of latex became relatively more important as yield increased.

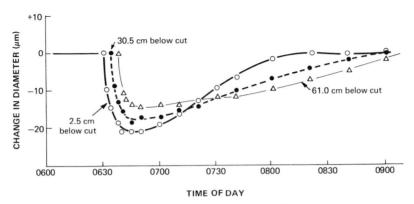

Fig. 6. Contraction of trunk of *Hevea brasiliensis* following tapping: tree leafless (wintering) to avoid effects of normal diurnal shrinkage through transpiration loss. Contraction at three distances below tapping cut, 2.5, 30.5, and 61.0 cm. Gooding (1952a).

It is apparent from one of Gooding's (1952a) figures that the bark near the cut begins to expand again quite rapidly—within about 20 min of opening (Fig. 6).

3. In another attempt to study the drainage area, labeled phosphate (^{32}P) or acetate (^{14}C) was injected into the bark at 10, 25, and 45 cm below the cut, and when tapped 2 days later, three peaks of radioactivity were detected in outflowing latex, which corresponded to recovery of radioactivity of 0.9, 0.5, and 0.01%. The low level of recovery at 45 cm presumably indicates that it is near the limit of the drained area (during one tapping), on a reduced spiral system (Fig. 7) (Luštinec and Resing, 1965).

Similar peaks of radioactivity were obtained from injections of radioisotopes made *above* the cut in renewing or virgin bark, though these were much lower recoveries (Luštinec and Resing, 1965). Movement of soluble salts does not necessarily mean that there was mass flow of latex. A more precise method would be to separate the rubber particles and measure incorporated ^{14}C from labeled acetate.

4. Turgor pressures near the cut are greatly reduced very soon after the cut is opened, while smaller pressure drops are observed at a distance from the cut, with some time lag (Buttery and Boatman, 1967).

Displacement area has been estimated from the fall in turgor pressures observed at various distances from the cut, before tapping and at 10–12 min afterward (S. W. Pakianathan, unpublished). In general, the fall in pressure extended above and below the cut and on the opposite side of the tree. The area of bark showing 10% or more fall in turgor pressure was shown to be highly correlated ($r = +0.83$; $P > 0.01$) with latex yield.

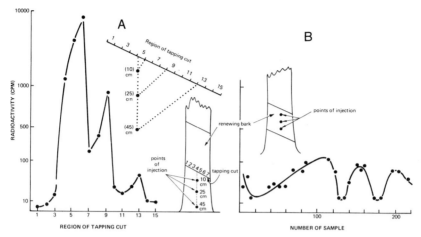

FIG. 7. Recovery of radioactivity in tapped latex. (A) One day after injection of NaH$_2$^{32}PO$_4$ at three sites below the cut. Latex was collected separately from different regions of the cut. Note that the quantity of ^{32}P recovered depends on the distance from the cut and that the region of recovery moves progressively to the right for the lower sites of injection. (B) different tree, two days after injection of NaH$_2$^{32}PO$_4$ at three sites in renewing bark above the cut. Latex was collected from the whole cut in successive fractions. Redrawn from Luštinec and Resing (1965).

The linear regression line did not pass through the origin, but indicated a displacement area of 2345 cm^2 at zero yield. However, if the displacement area was restricted to the area below the cut, the regression line against yield fitted nearly as well (+0.77) and could now be fitted through the origin. Thus, Pakinathan suggests that the displacement area above the cut does not have a direct influence on yield.

The fall in turgor above the cut may be mostly due to the loss of water from the vessels (i.e., uncut innermost cylinders) and other tissues. The latex vessel rings above the cut are thought not to be directly interconnected with those tapped, though presumably the older rings of latex vessels in the renewing bark are connected, laterally, with concurrent rings of vessels below the cut. It seems likely that water is lost osmotically to the drainage area under the cut; water must move from higher up the tree to replace this.

When the tapping cut was near the stock-scion union, irregular pressure gradients were observed extending both into the tap root and laterals.

5. "Microtapping" was used by Luštinec and Resing (1965) to investigate effects of regular tapping on latex properties at various distances from the tapping cut. In this method, small holes are made simultaneously into the wood at the required locations. Small collecting vessels are held under each hole and the latex collected, dried, and weighed. The yield

from microtaps (before normal tapping) at 10-cm intervals along a line passing through the main tapping cut showed low levels at 35–55 cm below the cut and 40–100 cm above the cut. Yields from microtapping were high near the cut itself and at greater distances from the cut. This was on bud-grafted trees that had been in tapping 1 mo. With older trees of seedling origin (tapped 4 yr), this elevated yield around the cut was not detectable; highest yields from microtappings were found near ground level, below the cut.

The vertical extent of the displacement area is more or less equal from one end of the cut to the other. For young trees yielding 15 gm dry matter on a half-spiral cut, the length was about 55 cm, while for a full-spiral cut yielding 25 gm, the displacement area was 75 cm in length.

With high yields from a half-spiral cut, the displacement area may extend right round the trunk (Luštinec et al., 1966).

E. BARK CHARACTERISTICS AFFECTING YIELD

The studies of Frey-Wyssling (1952) and Riches and Gooding (1952) relate the influence of the diameter of the latex vessels on the rate of flow of latex to Poiseuille's equation for viscous flow in a capillary, where it is shown that the volume of flow is proportional to the fourth power of the radius of the capillary. The mean diameter of vessels in eight clones ranged from 21.6–29.7 μm corresponding to a potential difference in flow of more than three times between the smallest and largest vessels (Gomez et al., 1972).

An assessment of numbers of rings cut was made by Gomez et al. (1972), who showed that about 50% of the latex vessel rings were not cut in normal tapping of young trees, but the proportion became less in older trees. Fernando and Tambiah (1970b) have also shown that high-yielding trees tend to have sieve tubes with larger diameters. Possibly, a cambium that cuts off wide sieve tubes also cuts off wide latex vessels, which favors higher yields.

Schweizer (1949) showed that, among other things, the pH varied depending on depth of tapping. The outer latex cylinders gave latex with pH 6.8, while the inner rings yielded pH 5.9. At the same time, the rubber content fell from 55.7 to 31.6%.

It has not been possible to estimate the contribution to the exuded latex from the sieve tubes. Schweizer (1949) suggested it could be considerable.

De Jonge (1969) demonstrated a pronounced beneficial effect of depth of tapping on yield, over at least the first 4 yr of tapping. Thus, with clone GT1, the 3-yr yield of rubber was 2232 lb with shallow tapping and 2880 with deep tapping. This was accompanied by an increase in late dripping (after normal collection) from 9.7% to 16.3% and a fall in dry rubber content from 34.8 to 30.9%. Yearly changes of tapping panel from one side of the tree to the other (using a half-spiral cut) caused a yield reduction if tapping was shallow, but a yield increase with deep tapping.

F. Diurnal and Seasonal Changes in Latex Flow

The flow of latex varies both in volume and in concentration according to the time of day and season of the year. The diurnal changes seem to be mainly due to changes in transpiration rate; seasonal effects can be partly attributed to variation in rainfall, partly to the physiological demands of "wintering" (leaf fall) and refoliation.

Dïjkman (1951) quotes yields of 100, 96, and 85% for tappings made at 0700, 0900, and 1100 hr. He stated that the actual depression in yield varies from day to day and from place to place—presumably depending on the balance between transpiration rate and water supply.

Gooding (1952b) compared tapping at 0800 and 1100 hr for trees tapped on alternate days on a half-spiral; the latex yield was 15% lower at 1100 while for trees tapped on a third spiral the yield was 25% lower at 1100 than at 0800 hr. In both cases, rubber content at 1100 hr was 1.5 points higher than at 0800 hr.

Ninane (1967a), using trees of clone PR107, showed a decrease in yield from 100% at 0600 hr to 86% at 1400 hr, with a slight recovery by 1600 hr. In another experiment, yield was about 81% of the early morning value, between noon and 1430, with no clear indication of recovery between 1430 and 1730 hr. The rubber content was about 2 points higher in the afternoon compared with 0600 hr. Paardekooper and Sookmark (1969) tapped trees at intervals during a 24-hr period. No differences in yield were found between 2000 and 0700. As the day progressed, the yield gradually declined to a minimum of about 70% of the night yield around 1300 hr, after which the yield increased again (Fig. 8). The rubber content followed an opposite trend and at noon was up to 4 points higher than during the night.

Ninane (1967b) showed that transpiration of young rubber plants, being influenced both by temperature and the relative humidity of the surrounding air, was highly correlated with saturation deficit of the air.

Diurnal variation in yield per tapping follows the same pattern as the daily course of vapor pressure deficit (VPD) (Fig. 8). It appears that the recovery of yield in late afternoon anticipates the reduction in VPD. The relative yield reaches the "night" level by 2000 hr, at which hour the VPD is still 3 mm. A possible explanation is that the total yield depends

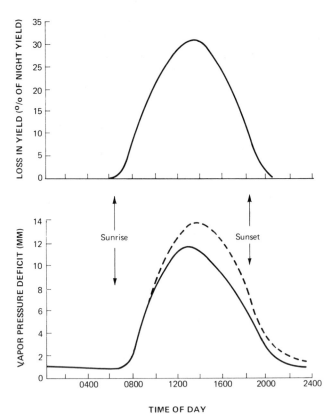

FIG. 8. Diurnal variation in yield expressed as percentage loss in yield compared with the diurnal variation in vapor pressure deficit of the air at Haadyai, Thailand. Solid line (in lower figure) represents average of all days from May to October; broken line represents average of days without rain. Paardekooper and Sookmark (1969).

not only on turgor pressure in the vessels at the time of tapping, but also on atmospheric conditions during the course of flow.

Ninane (1967a) suggested that the yield decreases sharply whenever the VPD reaches 8 mm. Ninane also found that restoration of yield in the afternoon lags behind the decrease in VPD. Paardekooper and Sook- mark (1969) did not find this to be so.

Given the close relation between VPD and relative yield, one may expect smaller differences between "night" and "day" yields on days with low VPD (cool or humid days). Examination of detailed experimental records shows that this is indeed the case (Paardekooper and Sookmark, 1969). Possibly the VPD is generally higher in Thailand than in Malaya or Indonesia, and this may account for the different levels of yield decrease at midday noted above.

Yields of latex and rubber from whips, fruits, and trunks of *Cryptostegia grandiflora* reach a minimum value around midday. There is a strong negative correlation between vapor pressure deficit and latex yield, especially from the whips. Total latex production from a 400-acre plantation was always low during periods when a local "sirocco" induced conditions of extreme aridity. Curtis and Blondeau (1946) considered that the rate of evaporation from the leaves was greater than the uptake of water from the roots and that this deficit causes withdrawal of water from the latex system, and hence a much reduced flow of latex after tapping. However, the maximum latex yield from the trunk occurred between 0900 and 1000 hr—somewhat after time of maximum turgor. The explanation for this is obscure, but may be connected with rising temperatures reducing latex viscosity and thus reducing resistance to flow.

The sapodilla tree (*Achras zapota*) also shows a diurnal rhythm in flow rates, with maximum flow occurring around dawn, which coincides with maximum turgidity and maximum stem diameter (Karling, 1934).

Dïjkman (1951), after studying the data of Arisz and Schweizer, concluded that the lowest rubber-forming intensity was always correlated with the period in which new leaves were developed. Defoliation invariably caused considerable decrease in yield. Schweizer (1949) found that the typical effects of wintering occurred in a stock budded higher than usual (although the latices of the stock and scion retained their peculiarities of composition). After 2 yr of observation, a ring of bark was removed, which effectively separated the bark of stock and scion. From then on, the typical effects of wintering and refoliation could be detected only in the scion; the latex of the stock remained unchanged. In other words, the ring barking prevented the scion from imposing its influence on the stock.

In the Ivory Coast, latex production reaches a peak in August just after the main rainy season in May-June. A secondary peak of production occurs in November to early December after the minor rainy season (Ribaillier, 1971). It appears that in periods of lowered yields (dry spells), the total solids content of the latex, lutoid bursting index, and plugging index are all high; they are all indicative of unfavorable factors for flow.

In the wet districts of Ceylon, the plugging index reaches a maximum after winter defoliation and declines gradually until the onset of the next wintering period (Waidyanatha and Pathiratne, 1971).

In most years, quantity of rain and its seasonal distribution in Malaya are probably sufficient for maximum yields. However, rainfall has an effect on harvesting of latex and hence on the quantity and quality that may be collected (Wycherley, 1963). Thus near Kuala Lumpur, the periods of highest rainfall, April and October, were in fact associated with decreases in yields of top quality latex, but increases in "scrap."

IV. MECHANISMS INVOLVED IN THE CESSATION OF LATEX FLOW

A. Plugging

With rubber trees in regular tapping, the flow of latex ceases some one to several hours after opening the cut. Boatman's (1966) results showed that a major factor was the rapid buildup of a resistance to flow very near the cut surface—this has become known as "plugging." Milford et al. (1969) and Paardekooper and Samosorn (1969) confirmed these results and demonstrated that the effectiveness of plugging varies considerably from clone to clone. A considerable amount of recent experimentation has been directed at identification of the plugging mechanism.

One possibility is that the release of turgor results in a localized collapse of the vessels. Southorn (1967) showed that bark contraction very near the cut was 70–120 μm (some 5 min after tapping), which corresponded with a contraction per vessel of 3.5–6.0 μm. These are maximum values: no doubt some of the contraction should be attributed to cells in the bark, other than the latex vessels. The values are of the same order as those of Boatman (1966) for positions much further away from the cut. Recovery began within 10–20 min.

A contraction of 25% of vessel diameter would have the effect of reducing flow rate by about two-thirds; if Poiseuille's law is applicable, flow is proportional to the fourth power of the vessel radius. Some compensation for this would be brought about by the reduction in viscosity caused by inflow of water. Also, the effective bore of the vessels may be less than that measured microscopically because of the lining of viscous cytoplasm. However, Frey-Wyssling pointed out that according to Poiseuille, flow would in any case be zero adjacent to the capillary walls.

Latex vessels contract to the same extent very close to the cut as they do further away. There is no evidence for a localized blockage due to a localized collapse of vessels. Thus, the plugging is almost certainly a property of the latex itself, not the vessels.

Microscopic examination of the tapping region shows that the plugs within the latex vessels are formed largely of rubber and need not be physically part of the "cap" that eventually forms over the entire tapping cut. Electron micrographs taken a few minutes after tapping show some vessels completely free and others with well-developed internal plugs of coagulum. This suggests that plug formation occurs rather suddenly in individual vessels, the flow from the cut falling off as an increasing number of pathways become blocked. In some sections, it appeared that material was still being forced out of partially blocked vessels since stream lines were visible. The flowing material in these cases was severely distorted into elongated shapes along the flow lines. There was distorted rubber particles and appreciable amounts of long threadlike processes that appear in some cases to be tubular. Whether these are fragments of endoplasmic reticulum or artifacts of shear arising in the vessels after tapping is uncertain (Southorn, 1968a).

Both in the vessel plugs and the coagulum cap, damaged lutoids are regularly found. The serum from these lutoids will have been released. Wiersum (1958) was probably the first to suggest that the dilution reaction which occurs in the tree on tapping may cause swelling of the osmotically sensitive lutoids remaining in the latex vessel and that this may be one of the factors leading to cessation of latex flow. Homans et al. (1948) had shown that addition of water to fresh latex or to the separated lutoid fraction caused an increase in viscosity and coagulation. In the absence of the lutoids, the top fraction from the centifuge (i.e., the rubber particles) remained stable for days.

Southorn and Edwin (1968) showed that the B-serum extracted from lutoids has an extremely fast and complete flocculating action on aqueous suspensions of rubber particles. In whole latex, the situation is complicated by the fact that the B-serum and C-serum (the main latex serum) react with each other. The effect of B-serum released into whole latex therefore depends on a balance of activities between the two sera. If all the lutoids rupture and release their contents, there will not usually be complete destabilization owing to the protective action of the latex serum. However, whenever a lutoid ruptures, there will be a localized high concentration of B-serum that could produce a microfloc of destabilized rubber particles. Ultrasonic treatment of whole latex which ruptures the lutoids produces large numbers of microflocs. If lutoids are previously removed, such treatments do not produce any microflocs (Southorn and Edwin, 1968).

The destabilization process of lutoids seems to occur in two steps. First these organelles remain intact, but the membrane permeability toward substrates increases. Second, lutoids burst and release their proteins, although some of them remain attached to the membrane by ionic bonds. The immediate consequence of a release of hydrolases would be a decrease in biosynthesis of rubber (Pujarniscle, 1969).

B-serum has a large content of cationically active material, both as divalent inorganic ions (Ca^{2+} and Mg^{2+}) and as proteins of high isoelectric point. C-serum, on the other hand, has a preponderance of proteins of relatively low isoelectric point and a low concentration of divalent metal cations. The difference in charge distribution of the proteins of the two sera shows clearly in gel electrophoresis patterns (Moir and Tata, 1960). The particles of rubber are coated with colloidally protective envelopes, probably phospholipid protein (Cockbain and Philpott, 1963). The particles are all negatively charged and thus repel one another.

Release of B-serum into latex reduces this electrostatic stability factor; if enough B-serum is added, the charge at the shear interface of the rubber particle (Zeta potential) can be brought from its normal —45 mV to near zero. B-serum and C-serum proteins also interact rather similarly and are thrown out of solution together (Southorn and Yip, 1968).

If free metallic cations are removed from B-serum by dialysis, the serum still retains most of its fast clotting activity, so that the proteins must play a major role. Tata and Yip (1968) have shown that the protein activity pertains to a particular protein band of the B-serum starch-gel electrophoresis pattern. They were able to separate enough of this protein to demonstrate that it retained its clotting activity, while the dialyzed B-serum from which it had been removed showed no significant activity.

In practice, the colloid may lose stability with a partial reduction in charge, rather than complete elimination. The size and configuration of the proteins may be of some importance: the smallest rubber particles may, in fact, be smaller than some of the serum proteins and there is some electron microscopical evidence that protein bridges can be formed between particles, which then leads to flocculation. As noted by Ries and Meyers (1968) for other colloid systems, charge neutralization and bridging may occur simultaneously.

Experiments have been carried out in which the formation of a coagulum cap has been prevented by washing the cut, with no detectable effects on flow (Southorn, 1964). The plugging that usually occurs during the period of high-pressure gradients cannot be ascribed to the coagulum cap, though the cap may play a part when flow has slowed down from other causes. Boatman (1966) concluded that plugging occurs during rapid flow, being detectable within 1.5 min after tapping. Plugging probably ceases abruptly in an individual vessel as á microfloc is formed. Flow through the remaining vessels would be faster than indicated by total flow divided by total number of severed vessels in cross section.

Pakianathan et al. (1966) observed that damage to latex bottom-fraction particles (mainly lutoids) is much greater in the latex that first emerges after tapping than it is in latex collected during later stages of flow, but Pakianathan and Milford (1973) showed that if the fractions were collected into hypertonic solutions the reverse situation was found.

Influx of water into latex vessels during flow produces a decline in rubber content and of osmolarity of successive samples collected, though there is usually a recovery in concentration toward the end of flow (Gooding, 1952a; Boatman, 1966). The overall lowering of osmotic concentration is not such as to cause general rupturing of lutoids (Pujarniscle *et al.*, 1970), but Pakianathan *et al.* (1966) suggested that the rapidity of entry of water could cause localized "osmotic shock," thus damaging the lutoids.

Osmotic changes in latex after collection do not necessarily correspond to changes occurring in the vessel. Early samples of latex contain more damaged lutoids than later ones, but much of the bottom-fraction damage observed must have occurred after the latex had left the vessel. Damage to bottom-fraction particles was reduced or prevented by collecting directly into buffered sucrose (Pakianathan *et al.*, 1966). It seems that osmotic changes during flow will help to damage lutoids; whether this is the only or main cause of lutoid damage is uncertain.

Pakianathan and Milford (1973) showed that within the vessels (bark punctures) both the osmotic pressure and bottom-fraction particles declined by 13–14% after tapping. But in latex from the tapping cut, the bottom-fraction particles decreased more or less continuously during flow until some 50% were lost. The osmotic pressure of tapped latex was two atmospheres lower than that within the vessels. Judged by the osmotic pressure measurements, dilution within the latex vessel system is more or less constant from 5 to 60 cm below the cut and does not fall below 400 mOsm/liter; this may be sufficiently low to sensitize the lutoids, but not to cause bursting. However the much lower values observed in the latex from the cut suggest there is a rapid influx of water (or loss of osmotic material) in the 5 cm below the cut; i.e., this is the region of highest stress in terms of osmotic changes as well as in shear forces.

Optical and electron micrographs of latex vessels sometimes show distortion of particles along flow lines, which suggests that shear stresses could be important (Southern, 1968a). The flow properties of fresh latex are greatly modified by its bottom-fraction particles (van Gils, 1951; Verhaar, 1956).

Using a microviscometer, Southern (1968b) showed that the viscosity of fresh latex varied according to the time allowed for "rest." A reading taken after 10 min rest gave values of 60–70 centipoise, but repeated measurements with the rolling ball viscometer caused viscosity to fall to around 20 centipoise. This could be shown repeatedly on the same latex sample. It seems that during the rest period some structure builds up in the latex to which this type of instrument is extremely sensitive. The structure seems to be broken down by repeated passages of the ball (shear effects). Thus it seems a clear case of thixotropy.

Yip and Southorn (1973) further demonstrated that latex samples collected during the later stages of flow were capable of greater responses to shear than those obtained immediately after tapping. This was attributed to a higher proportion of relatively undamaged lutoid particles in the later samples.

Separation of the latex components by centrifugation and study of their viscosities established that the thixotropy was associated with the lutoid fraction.

Thus, lutoids could influence flow in two ways. Their direct influence on rheology as intact particles in latex vessels of 20–50 μm bore could affect flow rate under a given pressure gradient. Their rupture would release the destabilizing B-serum into the latex.

Yip and Southorn (1968) investigated flow of fresh *Hevea* latex at high pressures through glass capillaries with internal diameters of 22 to 80 μm. In 15 out of 25 trials, blockage of the capillaries occurred with pressure gradients ranging from 0.2 to 1.0 atm/mm. With latex from which the lutoid fraction had been removed by centrifugation, no blockage was observed in any of the 21 experiments. In the absence of lutoids, a straight-line relationship was found between flow rate and pressure gradient up to the maximum used, 12.9 atm/mm (Fig. 9).

With whole latex under high pressure gradients, examination of latex outflow just before blockage occurred showed that there were very few lutoids but large quantities of microflocs and reduced zeta potential, with

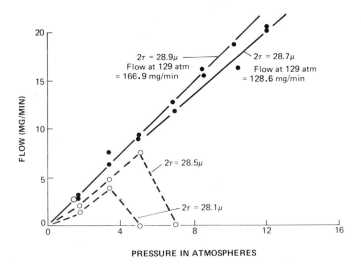

FIG. 9. Flow of whole fresh latex (broken lines) and of latex minus bottom fraction (solid lines) through glass capillaries of 1-cm length under varying pressure. Latex minus bottom fraction maintained a more-or-less linear relationship between flow rate and pressure up to the maximum attainable, 129 atm. Yip and Southorn (1968).

the particles showing less Brownian motion than with latex outflow under low pressure.

These experiments suggest that lutoids are ruptured when the pressure gradient reaches a critical level with consequent release of B-serum, leading to plugging of the capillary, and that such rupture of lutoids by shear within the latex vessels during the early stages of rapid flow is a possible mechanism for latex vessel plugging.

The ratio between the activity of free and total acid phosphatase activity has been taken as a measure of lutoid integrity; this is because acid phosphatase occurs mainly within the lutoids (Ribaillier, 1968).

$$\text{bursting index} = \frac{\text{activity of free acid phosphatase}}{\text{activity of total acid phosphatase}} \times 100$$

Ribaillier established a highly significant correlation between the bursting index and level of Mg^{2+} in latex; also between the activity of free acid phosphatase and Mg^{2+} level and an inverse correlation between bursting index and latex yield. He concluded that Ca and Mg concentrations play a part in stability of the lutoids.

The proportion of lutoids and their instability increases in the latex nearer the cambium. This was shown by repeated microtappings on the same site (Ribaillier, 1970). Despite this, tapping "to the wood" resulted in a considerable increase in flow. Ribaillier suggested that this is because the less stable latex from above the panel is in this case excluded from the flow area. However, it seems likely that most latex reaching the cut from above will do so by flowing around the sides, not "under" the cut: thus tapping-to-the-wood would not alter the situation in that respect. An alternative explanation is that tapping to the wood increases the total number of vessels severed leading to a greater reduction in turgor, and less plugging, as shown by Southorn and Gomez (1970) for increased length of cut.

If shear is an important factor in vessel plugging, then the maintenance of high pressure gradients near the cut would be expected to have an appreciable effect. The faster the initial high pressure gradient disappears, the less effective this plugging mechanism would be.

Southorn and Gomez (1970) studied the effect of length of tapping cut on plugging index: that is the increase in flow obtained by reopening the cut after 20 min. The mean percentage recovery of flow (and by implication of turgor pressure) is higher for shorter cuts than for longer ones. In successive cuts on the same tree, the trend is toward successively lowered flow recovery for each reopening, which suggests that the cumulative fluid losses are affecting turgor recovery. This trend is more noticeable for long cuts, presumably because there the total fluid loss is greater (Fig. 10).

The yield per latex vessel from a microcut was more than ten times that from a conventional half-spiral cut, which shows that expulsion forces were very much higher during flow from the shorter cut. Very short cuts were associated with intense plugging, well-maintained turgor pressures, and high flow rates per vessel (and, hence, high pressure gradients for an

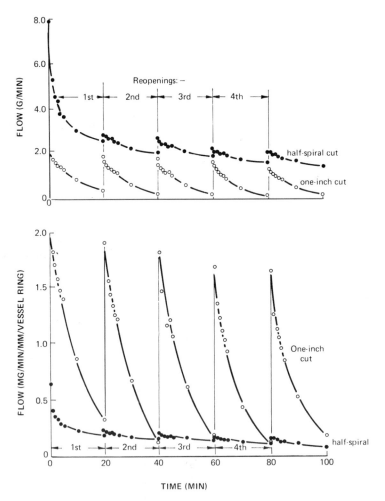

FIG. 10. Flow curves for repeated tapping of a tree of clone PR 107 for half-spiral and one-in. cuts. Upper figure shows the flow from each of the two cuts; lower figure shows the flow per millimeter of cut per vessel ring for each of the two cuts. Southorn and Gomez (1970).

appreciable period within the vessels). This is in accord with a hypothesis stressing the importance of shear effects on lutoid damage leading to plugging.

The numerical value of plugging index as determined by the method of Milford *et al.* (1969) will depend for a given tree on length of cut employed.

Although plugging may be the main factor terminating latex flow in half-spiral tapping of most clones, this mechanism becomes less effective with longer cuts because of the rapid and generalized loss of turgor leading to reduced "shear." In this case, general loss of turgor would eventually retard flow to a level where coagulation along the bark and back diffusion of coagulants could seal the vessels.

Pakianathan *et al.* (1966) found that osmolarity of the first 5 ml of flow was 0.348 *M*, whereas that of latex from end of flow was 0.491 *M*. Damage to bottom fraction was high in the first fraction, but very low in the last. Addition of serum from the first fraction to latex from the last fraction resulted in considerable bottom-fraction damage. Something other than shear must be operating here, possibly osmotic or perhaps chemical.

A possible trigger for the plugging might be a depolarization of particle membranes brought about by the electrical responses to tapping wounds. When *Hevea* is wounded, electrical responses are produced that exhibit similarity in polarity, magnitude, and duration to those observed in *Mimosa pudica*. These responses are transmitted over small distances, but are much more localized than in *Mimosa* (Lim *et al.*, 1969). Electrical responses are likely to be associated with transfer of materials across membranes leading to subtle changes in cell physiology. Thus, the rubber tree has a fast "awareness-to-wounding" mechanism available. But its involvement in latex flow and plugging is unclear.

Southorn (1969) suggests that shear may play an important part in lutoid damage and that internal plugging occurs mainly during the fast initial flow, whereas coagulation on the surface of the cut is effective in the final closure of the cut; when the rate of flow is low. It is important to distinguish, however, between rates of movement through individual latex vessels and the overall rate of flow from the cut. It seems likely that the latex obtained during the closing stages of flow comes from relatively few unsealed vessels and that flow rates through these may, in fact, be high; especially as turgor pressures often recover substantially at this stage (Buttery and Boatman, 1967). Thus, there seems no substantial reason for supposing that the two types of sealing are really different. However, since damage to bottom fraction changes with time, there must be at least quantitative changes in the sealing mechanism during flow.

It is interesting that the cessation of flow from sieve tubes bears, at least, a superficial resemblance to that from latex vessels. Tammes *et al.* (1969, 1971) report two phases in the outflow of sieve tube sap from *Yucca*.

    1. an expulsion for a few seconds of sap every time a new slice is cut from the wound surface, which is also observed at $0°C$.
    2. a continuous flow of exudate at a lower rate which stops gradually over approximately 24 hr at $0°C$.

The closing mechanism of the sieve tubes of *Yucca* is sited very near the wound, at a distance of 1 mm or less (probably the length of 1 sieve element). Thus, although the structure and functions of these two phloem systems differ considerably, the effective response to wounding is similar.

B. OTHER FACTORS

In addition to plugging, which in many cases stops flow during an individual tapping, and to rate of biosynthesis, which sets an upper limit to the quantity of latex that can be extracted in the long run, latex yield may also be limited by "brown bast."

The malady of tapped trees known as brown bast, which renders the affected trees partially or completely nonyielding, has been recognized for over 50 yr (Rands, 1921). In this condition, a part or whole of the tapping cut ceases to yield latex and the inner phloem assumes a characteristic brown-gray coloration. At a later stage, some of the afflicted trees may show extensive meristematic activity which, in severe cases, results in extreme deformation of the entire stem, which renders further tapping impracticable.

The incidence of dryness becomes greater when tapping intensity is increased. In experiments employing intensive tapping systems, the frequency of tapping played a more important role than the length of cut in inducing dryness, but the combination of both increased frequency of tapping and length of cut resulted in a considerable increase in the number of dry trees. Heavy exploitation resulted in a decrease in the volume, rubber content, and total solids content of the latex (Chua, 1966, 1967).

Carbohydrate levels in the bark did not seem to be reduced by heavy tapping, but protein nitrogen and total nitrogen were lowered appreciably. This seemed to be associated with excessive loss of "serum solids" which consist largely of soluble proteins and some RNA.

There is an internal breakdown in the tissues of dry trees where there is a general disintegration of cellular organization of the latex vessels and sieve tubes. The cork cambium becomes more active and the damaged tissues are abscised.

Chua (1967) suggested that trees become dry because the latex vessels are depleted of enzymes to the extent that they can no longer maintain vital processes. Boatman (1970) was skeptical that the proposed mechanism would result in the very sudden onset of the condition which is so characteristic. In a later study, Bealing and Chua (1972) reported that although the content of many compounds was much reduced in the latex as a result of very heavy tapping, the overall composition of the bark was only slightly affected. They suggested that tapping results in diminished permeability of the latex vessels, the effect being directly proportional to tapping intensity. The onset of brown-bast was attributed to *in situ* coagulation of the latex as the result of a critical reduction in permeability of vessel membranes.

V. METHODS OF INCREASING LATEX FLOW

A number of chemicals have been shown to prolong latex flow after tapping in *Hevea*. These include copper sulfate (Compagnon and Tixier, 1950), 2,4-dichlorophenoxyacetic acid (2,4-D) (Chapman, 1951), 2,4,5-trichlorophenoxyacetic acid (2,4,5-T) (Baptiste and de Jonge, 1955), ethylene oxide (Taysum, 1961a), various bactericidal compounds (Taysum, 1961b), a range of growth regulators, herbicides, organomercurials, acetylene, ethylene, halogenparaffins and 2-chloroethanephosphonic acid (ethephon) (Abraham *et al.*, 1968a,b; Banchi and Polinière, 1969).

Chacko *et al.* (1972) showed that ethephon application could more than double the yield of tapped latex from papaya fruit. Amchem Products Incorporated (1969) reported latex yield increases in ethephon treated trees of *Ficus elastica* as well as *Hevea*.

Effects on latex yield can be detected only over a narrow range of concentrations. The difference between the yield increase induced by a very active and slightly active compound is small (less than tenfold) compared with a range of about a 1000-fold in most bioassays of extension growth (Fawcett *et al.*, 1955; Pybus *et al.*, 1959).

Both 2,4-D and 2,4,5-T have been used extensively in Malaysia for a number of years. The active material is usually dissolved in a palm-oil carrier at a concentration of 1–1.5% acid equivalent (Baptiste and de Jonge, 1955; Puddy and Warrior, 1961) and painted on to a lightly scraped area of bark immediately below the tapping cut. The width of the strip to be treated is chosen so that it is completely excised by the normal process of tapping within 4 mo, which avoids any possible damage due to the induction of meristematic activity by the growth regulator. Application to the newly tapped bark above the cut is also effective and requires no prior scraping to assist penetration, but it carries some risk of uneven growth of the renewing bark (Abraham and Boatman, 1964).

Application of ethephon increases yield and does so more consistently than auxin stimulants (Abraham *et al.,* 1971b,c), and there is an absence of undesirable side effects. Thus, Ribailler and d'Auzac (1970) report that ethephon does not induce the disorganized proliferation of tissue caused by 2,4,5-T and 2,4-D.

Figure 11 shows a typical response to 2,4,5-T treatment. Two groups, each of 10 trees of clone Tjir 1, were tapped every second day and the volume of latex obtained and its rubber content recorded. After the third recorded tapping, the trees were lightly scraped below the cut and the experimental trees were painted with 1% 2,4,5-T, while the controls received only the palm-oil carrier. This latter treatment often causes a small transient yield increase (Baptiste, 1955), but this was negligible compared with the response to the growth regulator. The volumes of latex obtained increased by two to three times and duration of flow was correspondingly extended, while there was a comparatively small decline in rubber content. The duration of this response is rather variable but, in general, the yield declines to the control level within 4 mo. In one experiment, applications were continued at 6-mo intervals for 9 yr, which gives an overall yield increase of 37%, without any discernible harmful effects. In general, the best responses to 2,4-D and 2,4,5-T are obtained from trees tapped at a

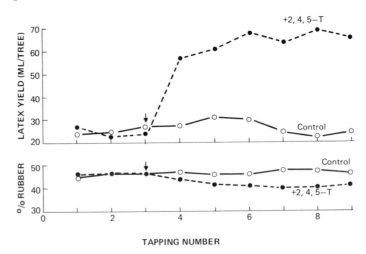

Fig. 11. Effect of 2,4,5-T on the volume and rubber content of latex obtained by alternate daily tapping. Treatments were applied after the third recorded tapping as indicated by the arrows. The controls received the carrier oil only, while the experimental trees were treated with 1% 2,4,5-T. Boatman (1970).

moderate intensity, and small, or even negative responses have been reported from heavily tapped fields (de Jonge, 1961). Thus, the use of chemical "yield stimulants" is best regarded as a technique for reducing labor costs and bark consumption entailed by high-intensity tapping.

Treatment with chemical "stimulants" intensifies the reduction in stem growth caused by tapping and is, therefore, usually avoided during the early years of exploitation, when growth rates are still high. Attempts are in progress, however, to combine low-intensity systems with the use of ethephon on young trees, with the object of combining satisfactory growth and yield with a reduction in labor costs (Rubber Research Institute of Malaya, 1972).

Boatman (1966) and Buttery and Boatman (1967) produced evidence that 2,4,5-T delays the plugging of cut latex vessels, and Milford *et al.* (1969) reported a strong positive correlation between plugging index and response. Pakianathan *et al.* (1966) reported that the proportion of damaged lutoids in latex samples collected during flow increased rather than decreased following 2,4,5-T treatment. However, Pakianathan and Milford (1973) have since shown that in trees treated with ethephon or 2,4,5-T the loss of lutoids did not commence until some 45 min after tapping, but thereafter increased until by the end of flow nearly 50% of the lutoids were lost (as with unstimulated trees). Thus, in some way, yield stimulants appear to delay the retention of lutoids by the cut vessel ends (or the cut itself) and, hence, delay plugging and prolong flow. The percentage of lutoids bursting during latex flow is reported to decrease after ethephon treatment from about 13 to 6% (Ribaillier, 1970). Ribaillier (1972) has shown that yield stimulants increase the permeability of lutoid membranes and also have a stabilizing effect on lutoids during latex flow.

Application of 2,4,5-T to clone Tjir 1 brought about an increase in yield and a decrease in plugging index (Boatman, 1966). The initial flow rate was not affected, so the yield difference can be attributed entirely to increased time of flow. In some clones, however, there is a suggestion of increased initial flow rates (RRIM501 and PB5/63). Boatman (1966) suggested that two separate mechanisms may be involved.

Treatment of the bark of *Hevea brasiliensis* with 2,4-dichlorophenoxy-acetic acid (2,4-D) or 1-naphthylacetic acid (NAA) greatly increases sucrose level, invertase activity, and sucrose utilization in the latex; the efficacy of 2,4-D is considerably greater than that of NAA. The greater sucrose utilization is the consequence of increased invertase activity. The changes occur as soon as the first tapping following bark treatment. It was suggested that the rise in both sucrose level and utilization in the latex vessels mediated the effect of auxins on latex production. This was thought to be related to increases in the osmotic and turgor pressures in the laticiferous tissue which cause greater flow on tapping, as well as to enhanced regeneration of latex (Tupy, 1969).

Evidence was obtained that the first step, necessary for the mobilizing action of auxins, was the activation of the overall metabolism of the treated tissue. It was further concluded that, owing to this activated metabolism, sucrose was imported into the induced sink, actively, even against a concentration and turgor gradient (Tupy, 1973b).

In previously untapped trees, both the introduction of tapping and application of 2,4-D brought about an increase in the level of invertase. In regularly tapped trees, the amount of latex invertase is several times higher than in untapped trees and evidence was obtained that its activity is regulated by the variation of latex pH. There was no increase in the content of latex invertase when trees adapted to regular tapping were treated with 2,4-D. The effect of auxin on actual invertase activity was essentially mediated through related increase of latex pH. Ethephon and bark scraping were also shown to increase latex pH and sucrose level in tapped trees (Tupy, 1973e), though the sucrose level fell below the check after several tappings (Tupy, 1973d).

In clones responding well to the chemical stimulants, the activity of invertase is limited predominantly by low latex pH. On the other hand, the more dependent the activity of latex invertase on sucrose concentration, the smaller is the increase of latex production after stimulation. A positive correlation was found between the initial response to treatment with 2-chloroethylphosphonic acid and sucrose level in the latex (Tupy, 1973c).

Tupy supposes that for naturally high-yielding clones, known for easy flow of their latex, the maximum production will be determined principally by the capacity of trees to supply sucrose to laticifers. Naturally, this capacity depends on availability of reserve carbohydrates, on photosynthetic capacity, on transport of sucrose and demands of other tissues in the tree for the same material. He suggests that 2,4-D has a less favorable effect than ethephon on sucrose for the formation of latex.

The yield of microtappings was only affected by 2,4-D immediately below the cut—the site of application—where the yield of microtappings increased two- to threefold. A similar (microtap) yield increase was observed at the site of application when this was 65 cm below the cut (Luštinec et al., 1967). This seems likely to be related to localized reduction in ability to form plugs.

An increase in displacement area was observed when the yield increase due to 2,4-D application was large. This seemed to be more apparent if the hormone was applied at some distance from the cut (Luštinec et al., 1967). The levels of N, P, K and Mg of the latex obtained by microtapping and its decarboxylation activity (glutamate) were elevated in the trees treated with 2,4-D. Mineral composition was affected by 2,4-D even if the yield response was weak, but decarboxylation (and hence, presumably, rubber biosynthesis) was only significantly increased when the response to stimulation was strong (Luštinec *et al.*, 1967). It is possible that the increased biosynthetic activity could bring about an alteration in particle-size distribution or the proportion of lutoids to other particles (Gomez, 1966).

Since the displacement area is little changed for weak to moderate yield increases (as response to stimulation), Luštinec *et al.* (1967) suggested that either the latex vessels nearer the cambium must be brought into action or that the same vessels as usual are drained more thoroughly.

Taysum (1961b) explained the activities of bactericidal compounds such as neomycin by their effects on the bacterial population that inhabits the surface of the tapping cut and penetrates the cut ends of latex vessels. However, a direct effect on metabolism of trees is also possible especially in view of the heterogeneous range of active compounds. Abraham *et al.* (1968a) suggested that the common factor may be selective and/or limited toxicity, and it now seems likely (Abraham *et al.*, 1968b) that all active compounds have the capacity to produce ethylene or induce its formation in plant tissues. The report by Baur and Morgan (1969) that picloram stimulates production of ethylene by plants is also consistent with the theory.

Pratt and Goeschl (1969) note that endogenous ethylene production may occur normally during the development of a tissue or in response to external stimuli including treatment with auxin analogs, application of toxic substances, or simple physical injury. It is of interest to recall that physical injury to *Hevea* is well known to stimulate latex flow. Examples are the first opening of a tree and its subsequent regular tapping (Buttery and Boatman, 1967) and scraping of the bark below the cut *without* any chemical treatment (Kamerun, 1912, as cited by Baptiste, 1955). Finally, experimental evidence is now available that *Hevea* tissues, like those of other plants, produce ethylene after application of 2,4,5-T or 2,4-D (Archer *et al.*, 1969). Thus some, and possibly all, yield stimulants act via ethylene. Acetylene probably owes its activity to its close structural analogy to ethylene (Abraham *et al.*, 1971a).

Osborne and Sargent (1974) have proposed a model for the mechanism of promotion of latex flow by ethylene. They suggest that ethylene treatment results in wider latex vessels with thicker, more rigid, walls. This, in turn, leads to less constriction of the cut vessels after tapping, reduced shearing forces at the orifice, a lower proportion of damaged lutoid particles, and less enzymatic coagulation of rubber particles. The delayed occlusion of the vessel orifice results in prolonged flow.

The main objection to this model is that only the innermost two or three latex vessel cylinders are likely to be in a stage of growth susceptible to ethylene, and these cylinders are not cut during normal tapping. The mature tree contains 20–30 cylinders of latex vessels and only the outermost 50–60% of these are opened at each tapping (Gomez *et al.*, 1972).

This proposed mechanism should be susceptible to experimental proof and will doubtless act as a spur to further investigations.

REFERENCES

Abraham, P. D., and Boatman, S. G. (1964). The influence of formulation on yield response and bark damage following the application of yield stimulants above the tapping cut. *J. Rubber Res. Inst. Malaya* **18**, 211–230.

Abraham, P. D., Boatman, S. G., Blackman, G. E., and Powell, R. G. (1968a). Effects of plant growth regulators and other compounds on the flow of latex in *Hevea brasiliensis*. *Ann. Appl. Biol.* **62**, 159–173.

Abraham, P. D., Wycherley, P. R., and Pakianathan, S. W. (1968b). Stimulation of latex flow in *Hevea brasiliensis* by 4-amino-3,5,6-trichloropicolinic acid and 2-chloroethanephosphonic acid. *J. Rubber Res. Inst. Malaya* **20**, 291–305.

Abraham, P. D., Blencowe, J. W., Chua, S. E., Gomez, J. B., Moir, G. F. J., Pakianathan, S. W., Sekhar, B. C., Southorn, W. A., and Wycherley, P. R. (1971a). Novel stimulants and procedures in the exploitation of *Hevea*. I. Introductory review. *J. Rubber Res. Inst. Malaya* **23**, 85–89.

Abraham, P. D., Blencowe, J. W., Chua, S. E., Gomez, J. B., Moir, G. F. J., Pakianathan, S. W., Sekhar, B. C., Southorn, W. A., and Wycherley, P. R. (1971b). Novel stimulants and procedures in the exploitation of *Hevea*. II. Pilot trial using 2-chloroethyl-phosphonic acid (ethephon) and acetylene with various tapping systems. *J. Rubber Res. Inst. Malaya* **23**, 90–113.

Abraham, P. D., Blencowe, J. W., Chua, S. E., Gomez, J. B., Moir, G. F. J., Pakianathan, S. W., Sekhar, B. C., Southorn, W. A., and Wycherley, P. R. (1971c). Novel stimulants and procedures in the exploitation of *Hevea*. III. Comparison of alternative methods of applying stimulants. *J. Rubber Res. Inst. Malaya* **23**, 114–137.

Amchem Products Incorporated. (1969). "Ethrel," Tech. Serv. Data Sheet, E-172. Amchem Products Inc., Ambler, Pennsylvania.

Archer, B. L., and Audley, B. G. (1967). The biosynthesis of rubber. *Advan. Enzymol.* **29**, 221–257.

Archer, B. L., Barnard, D., Cockbain, E. G., Dickenson, P. B., and McMullen, A. I. (1963). Structure, composition and biochemistry of *Hevea* latex. *In* "The Chemistry and Physics of Rubber-like Substances" (L. Bateman, ed.), pp. 41–72. Maclaren, London.

Archer, B. L., Audley, B. G., McSweeney, G. P., and Tan Chee Hong (1969). Studies on composition of latex serum and "bottom fraction" particles. *J. Rubber Res. Inst. Malaya* **21**, 560–569.

Arisz, W. H. (1928). Physiology van het tappen. *Archf. Rubbercult. Ned.-Indië* **12**, 220–241.

Audley, B. G. (1965). Studies of an organelle in *Hevea* latex containing helical protein microfibrils. *In* "Proceeding of the Natural Rubber Producers Research

Association Jubilee Conference, Cambridge, 1964" (L. Mullins ed.), pp. 43–51. Maclaren, London.

Audley, B. G. (1966). The isolation and composition of helical protein microfibrils from *Hevea brasiliensis* latex. *Biochem. J.* **98**, 335–341.

Banchi, Y., and Polinière, J. P. (1969). Effects of minerals introduced directly into the wood and of acetylene applied to the bark of *Hevea*. *J. Rubber Res. Inst. Malaya* **21**, 192–206.

Bangham, W. N. (1934). Internal pressure in latex systems. *Science* **80**, 290.

Baptiste, E. D. C. (1955). Stimulation of yield in *Hevea brasiliensis*. I. Pre-war experiments with vegetable oils. *J. Rubber Res. Inst. Malaya* **14**, 355–361.

Baptiste, E. D. C., and de Jonge, P. (1955). Stimulation of yield in *Hevea brasiliensis*. II. Effect of synthetic growth substances on yield and on bark renewal. *J. Rubber Res. Inst. Malaya* **14**, 362–382.

Baur, J. R., and Morgan, P. H. (1969). Effects of picloram and ethylene on leaf movement in huisache and mesquite seedlings. *Plant Physiol.* **44**, 831–838.

Bealing, F. J. (1965). Role of rubber and other terpenoids in plant metabolism. *In* "Proceedings of the Natural Rubber Producers Research Association Jubilee Conference, Cambridge, 1964" (L. Mullins, ed.), pp. 113–122. Maclaren, London.

Bealing, F. J., and Chua, S. E. (1972). Output, composition and metabolic activity of *Hevea* latex in relation to tapping intensity and the onset of brown bast. *J. Rubber Res. Inst. Malaya* **23**, 204–231.

Benedict, H. M. (1949). A further study of the non-utilization of rubber as a food reserve by guayule. *Bot. Gaz.* (*Chicago*) **111**, 36–43.

Benedict, H. M. (1950). Factors affecting the accumulation of rubber in seedling guayule plants. *Bot. Gaz.* (*Chicago*) **112**, 86–95.

Blackman, G. E. (1965). Factors affecting the production of latex. *In* "Proceedings of the Natural Rubber Producers Research Association Jubilee Conference, Cambridge, 1964" (L. Mullins, ed.), pp. 43–51. Maclaren, London.

Bloomfield, G. F. (1951). Studies in *Hevea* rubber. VI. Characteristics of rubber in latex of untapped trees and branches of trees in regular tapping. *J. Rubber Res. Inst. Malaya* **13**, 10–34.

Boatman, S. G. (1966). Preliminary physiological studies on the promotion of latex flow by plant growth substances. *J. Rubber Res. Inst. Malaya* **19**, 243–258.

Boatman, S. G. (1970). Physiological aspects of the exploitation of rubber trees. *In* "Physiology of Tree Crops" (L. C. Luckwill and C. V. Cutting, eds.), pp. 323–333. Academic Press, New York.

Bobilioff, W. (1923). "Anatomy of *Hevea brasiliensis*" (C. Yampolsky, transl.) Orell Fussli, Zurich.

Bonner, J., and Galston, A. W. (1947). The physiology and biochemistry of rubber formation in plants. *Bot. Rev.* **13**, 543–596.

Bordeau, P. F., and Schopmeyer, C. S. (1958). Oleoresin exudation pressure in slash pine; its measurement, heritability and relation to oleoresin yield. *In* "The Physiology of Forest Trees" (K. V. Thimann, ed.), pp. 313–319. Ronald Press, New York.

Buttery, B. R., and Boatman, S. G. (1964). Turgor pressures in phloem: Measurements on *Hevea* latex. *Science* **145**, 285–286.

Buttery, B. R., and Boatman, S. G. (1966). Manometric measurement of turgor pressures in laticiferous phloem tissues. *J. Exp. Bot.* **17**, 283–296.

Buttery, B. R., and Boatman, S. G. (1967). Effects of tapping, wounding and growth regulators on turgor pressure in *Hevea brasiliensis* Muell., Arg. *J. Exp. Bot.* **18,** 644–659.

Chacko, E. K., Randhawa, G. S., Menon, M. A., and Negi, S. P. (1972). Stimulation of latex production in Papaya (*Carica papaya* L.) by 2-chloroethane-phosphonic acid. *Curr. Sci.* **41,** 465.

Chapman, G. W. (1951). Plant hormones and yield in *Hevea brasiliensis*. *J. Rubber Res. Inst. Malaya* **13,** 167–176.

Chua, S. E. (1966). Physiological changes in *Hevea brasiliensis* tapping panels during the induction of dryness by interruption of phloem transport. I. Changes in latex. *J. Rubber Res. Inst. Malaya* **19,** 277–281.

Chua, S. E. (1967). Physiological changes in *Hevea* trees under intensive tapping. *J. Rubber Res. Inst. Malaya* **20,** 100–105.

Cockbain, E. G., and Philpott. M. W. (1963). Colloidal properties of latex. *In* "The Chemistry and Physics of Rubber-like Substances" (L. Bateman, ed.), pp. 73–95. Maclaren, London.

Compagnon, P., and Tixier, P. (1950). Sur une possibilité d'améliorer la production d'*Hevea brasiliensis* par l'apport d'oligo-éléments. *Rev. Gen. Caout.* **27,** 525–526, 591–594, and 663–665.

Curtis, J. T., and Blondeau, R. (1946). Influence of time of day on latex flow from *Cryptostegia grandiflora*. *Amer. J. Bot.* **33,** 264–270

d'Auzac, J. (1964). Variations de l'activité enzymatique et de quelques constituants du latex durant la saignée de l'*Hevea brasiliensis*. *Rev. Gen. Caout. Plast.* **41,** 840–842.

d'Auzac, J. (1965). Relations entre la composition biochimique du latex, l'intensité de quelques réactions metaboliques et la productivité de l'*Hevea brasiliensis*. *Rev. Gen. Caout. Plast.* **42,** 1027–1036.

de Haan, I., and van Aggelen-Bot, G. M. (1948). De vorming van rubber bij *Hevea brasiliensis*. *Archf. Rubbercult.* **26,** 121–180.

de Jonge, P. (1961). Intensive tapping of mature rubber. *Proc. Natur. Rubber Res. Conf., 1960* pp. 211–223.

de Jonge, P. (1969). Influence of depth of tapping cut on growth and yield. *J. Rubber Res. Inst. Malaya* **21,** 348–352.

Dickenson, P. B. (1965). The ultrastructure of the latex vessels of *Hevea brasiliensis*. *In* "The Proceedings of the Natural Rubber Producers Research Association Jubilee Conference, Cambridge, 1964" (L. Mullins, ed.), pp. 52–66. Maclaren, London.

Dickenson, P. B. (1969). Electron microscopical studies of latex vessels of *Hevea brasiliensis*. *J. Rubber Res. Inst. Malaya* **21,** 543–558.

Dïjkman, M. L. (1951). "*Hevea*: Thirty Years of Research in the Far East." Univ. of Miami Press, Coral Cables, Florida.

Esau, K. (1965). "Plant Anatomy." Wiley, New York.

Fairbain, J. W., and Kapoor, L. D. (1960). The laticiferous vessels of *Papaver somniferum* L. *Planta Med.* **8,** 49–61.

Fawcett, C. H., Wain, R. L., and Wightman, F. (1955). Studies on plant growth-regulating substances. VIII. The growth-promoting activity of certain aryloxy- and arylthio-alkane-carboxylic acids. *Ann. Appl. Biol.* **43,** 342–354.

Fernando, D. M., and Tambiah, M. S. (1970a). A study of the significance of latex in *Hevea* species. *Rubber Res. Inst. Ceylon, Quart. J.* **46,** 69–77.

Fernando, D. M., and Tambiah, M. S. (1970b). Sieve tube diameters and yields in *Hevea* spp. A preliminary study. *Rubber Res. Inst. Ceylon, Quart, J.* **46,** 89–92.

Ferrand, M. (1941). Nouvelle méthode permettant la détermination de la concentration du latex *in situ,* chez les plantes à laticifères et en particulier chez *Hevea brasiliensis. Acta Biol. Belg.* **1**, 193–197.

Frey-Wyssling, A. (1952). Latex flow. *In* "Deformation and Flow in Biological Systems" (A. Frey-Wyssling, ed.), pp. 322–349. North-Holland Publ., Amsterdam.

Gomez, J. B. (1966). Electron microscopic studies on the development of latex vessels in *Hevea brasiliensis* Muell-Arg. Ph.D. Thesis, University of Leeds.

Gomez, J. B., and Chen, K. T. (1967). Alignment of anatomical elements. *J. Rubber Res. Inst. Malaya* **20**, 91–99.

Gomez, J. B., and Southorn, W. A. (1969). Studies in lutoid membrane ultrastructure. *J. Rubber Res. Inst. Malaya* **21**, 513–522.

Gomez, J. B., Narayanan, R., and Chen, K. T. (1972). Some structural factors affecting the productivity of *Hevea brasiliensis*. I. Quantitative determination of the laticiferous tissue. *J. Rubber Res. Inst. Malaya* **23**, 193–203.

Gooding, E. G. B. (1952a). Studies on the physiology of latex. II. Latex flow on tapping *Hevea brasiliensis:* Associated changes in trunk diameter and latex concentration. *New Phytol.* **51**, 11–29.

Gooding, E. G. B. (1952b). Studies on the physiology of latex. III. Effects of various factors on the concentration of latex of *Hevea brasiliensis. New Phytol.* **51**, 139–153.

Haberlandt, G. (1914). "Physiological Plant Anatomy" (M. Drummond, transl.), 3rd ed. Macmillan, New York.

Homans, L. N. S., van Dalfsen, J. W., and van Gils, G. E. (1948). Complexity of fresh *Hevea* latex. *Nature (London)* **161**, 177–178.

Jacob, J. L. (1970). Particularités de la glycolyse et de sa régulation au sein du latex d'*Hévéa brasiliensis. Physiol. Veg.* **8**, 395–411.

Karling, J. S. (1934). Dendrograph studies of *Achras zapota* in relation to the optimum conditions for tapping. *Amer. J. Bot.* **21**, 161–193.

Kisser, J. (1958). Die ausscheidung von ätherischen Ölen und Harzen. *In* "Handbuch der Pflanzenphysiologie" (W. Ruhland, ed.), Vol. 10, pp. 91–131. Springer-Verlag, Berlin and New York.

Krotkov, G. (1945). A review of the literature on *Taraxacum kok-sagkyz. Bot. Rev.* **11**, 417–461.

Lim, C. M., Southorn, W. A., Gomez, J. B., and Yip, E. (1969). Electrophysiological phenomena in *Hevea brasiliensis. J. Rubber Res. Inst. Malaya* **21**, 524–542.

Luštinec, J., and Resing, W. L. (1965). Méthodes pour la délimitation de l'aire drainée a l'aide des microsaignées et des radioisotopes. *Rev. Gen. Caout. Plast.* **42**, 1161–1165.

Luštinec, J., Chai, K. C., and Resing, W. L. (1966). L'aire drainée chez les jeunes arbres de l'*Hevea brasiliensis. Rev. Gen. Caout. Plast.* **43**, 1343–1354.

Luštinec, J., Langlois, S., Resing, W. L., and Chai, K. C. (1967). La stimulation de l'*Hévéa* par les acides chlorophénoxyacétiques et son influence sur l'aire drainée. *Rev. Gen. Caout. Plast.* **44**, 635–641.

Luštinec, J., Simmer, J., and Resing, W. L. (1969). Trunk contraction of *Hevea. Biol. Plant.* **11**, 236–244.

Lynen, F. (1969). Biochemical problems of rubber synthesis. *J. Rubber Res. Inst. Malaya* **21**, 389–406.

McDougal, D. T. (1924). Growth in trees and massive organs of plants. *Carnegie Inst. Wash., Publ.* **350.**

McMullen, A. I. (1962). Particulate ribonucleo-protein components of *Hevea brasiliensis* latex. *Biochem. J.* **85**, 491–495.

Mahlberg, P. G. (1959). Karyokinesis in the non-articulated laticifers of *Nerium oleander* L. *Phytomorphology* **9**, 110–118.

Mahlberg, P. G. (1973). Scanning electron microscopy of starch grains from latex of *Euphorbia terracina* and *E. tirucalli*. *Planta* 110, 77–80.

Matile, P., Jans, B., and Rickenbacker, R. (1970). Vacuoles of *Chelidonium* latex: Lysosomal property and accumulation of alkaloids. *Biochem. Physiol. Pflanzen* **161**, 447–458.

Milford, G. F. J., Paardekooper, E. C., and Ho, C. Y. (1969). Latex vessel plugging; its importance to yield and clonal behaviour. *J. Rubber Res. Inst. Malaya* **21**, 274–282.

Moir, G. F. J., and Tata, S. J. (1960). The proteins of *Hevea brasiliensis* latex. III. The soluble proteins of 'bottom fraction.' *J. Rubber Res. Inst. Malaya* **16**, 155–165.

Narayanan, R., Gomez, J. B., and Chen, K. T. (1973). Some structural factors affecting the productivity of *Hevea brasiliensis*. II. Correlation studies between structural factors and yield. *J. Rubber Res. Inst. Malaya* **23**, 285–297.

Ninane, F. (1967a). Relations entre les facteurs écologiques et les variations journalières dans la physiologie et les rendements de l'*Hevea brasiliensis*. Possibilities d'applications pratiques. *Opusc. Techq.* No. 12/67.

Ninane, F. (1967b). Evapotranspiration réelle et croissance de jeunes hévéas soumis à différentes humidités du sol. *Rev. Gen. Caout. Plast.* **44**, 207–212.

Ninane, F., and David, R. (1971). Problèmes relatifs aux fonctions mathématiques de l'écoulement du latex chez l'*Hevea brasiliensis Rev. Gen. Caout. Plast.* **48**, 285–289.

Osborne, D. J., and Sargent, J. A. (1974). A model for the mechanism of stimulation of latex flow in *Hevea brasiliensis* by ethylene. *Ann. Appl. Biol.* **78**, 83–88.

Paardekooper, E. C., and Samosorn, S. (1969). Clonal variation in latex flow patterns. *J. Rubber Res. Inst. Malaya* **21**, 264–273.

Paardekooper, E. C., and Sookmark, S. (1969). Diurnal variation in latex yield. *J. Rubber Res. Inst. Malaya* **21**, 341–347.

Pakianathan, S. W. (1967). Determination of osmolarity in small latex samples by vapour pressure osmometer. *J. Rubber Res. Inst. Malaya* **20**, 23–26.

Pakianathan, S. W., and Milford, G. F. J. (1973). Changes in the bottom fraction contents of latex during flow in *Hevea brasiliensis*. *J. Rubber Res. Inst. Malaya* **23**, 391–400.

Pakianathan, S. W., Boatman, S. G., and Taysum, D. H. (1966). Particle aggregation following dilution of *Hevea* latex: A possible mechanism for the closure of latex vessels after tapping. *J. Rubber Res. Inst. Malaya* **19**, 259–271.

Parkin, J. (1900). Observations on latex and its functions. *Ann. Bot. (London)* **14**, 193–214.

Pratt, H. K., and Goeschl, J. D. (1969). Physiological roles of ethylene in plants. *Annu. Rev. Plant Physiol.* **20**, 541–584.

Puddy, C. A., and Warriar, S. M. (1961). Yield stimulation of *Hevea brasiliensis* by 2,4-dichlorophenoxyacetic acid. *Proc. Natur. Rubber Res. Conf.*, 1960 pp. 194–210.

Pujarniscle, S. (1968). Caractère lysosomal des lutoïdes du latex d'*Hevea brasiliensis* Muell.-Arg. *Physiol. Veg.* **6**, 27–46.

Pujarniscle, S. (1969). Etude de quelques facteurs intervenant sur la perméabilité et la stabilité de la membrane des lutoïdes du latex d'*Hevea brasilensis*. *Physiol. Veg.* 7, 391–403.

Pujarniscle, S. (1970). Etude biochemique des lutoïdes du latex d'*Hevea brasiliensis* différence et analogie avec les lysosomes. *Rev. Gen. Caout. Plast.* **47,** 175–178.

Pujarniscle, S., and Ribaillier, D. (1970). Du rôle des lutoïdes dans l'écoulement du latex chez l'*Hevea brasiliensis. I.* Etude de l'evolution des hydrolases lutoidiques et de quelques propriétés du latex lors de la mise en saignée d'arbres vierges. *Rev. Gen. Caout. Plast.* **47,** 1001–1003.

Pujarniscle, S., Ribaillier, D., and d'Auzac, J. (1970). Du rôle des lutoïdes dans l'écoulement du latex chez l'*Hevea brasiliensis.* II. Evolution des hydrolases lutoïdiques et de quelques propriétés du latex au cours de la saignée. *Rev. Gen. Caout. Plast.* **47,** 1317–1321.

Pybus, M. B., Smith, M. S., Wain, R. L., and Wightman, F. (1959). Studies on plant growth-regulating substances. XIII. Chloro- and methylphenoxyacetic and benzoic acids. *Ann. Appl. Biol.* **47,** 173–181.

Pyke, E. E. (1941). Trunk diameter of trees of *Hevea brasiliensis:* Experiments with a new dendrometer. *Nature (London)* **148,** 51–52.

Rands, R. D. (1921). Brown Bast disease of plantation rubber, its cause and prevention. *Archf. Rubbercult. Ned.-Indië.* **5,** 233–271.

Ribaillier, D. (1968). Action *in vitro* de certains ions minéraux et composés organiques sur la stabilité des lutoïdes du latex d'*Hévéa. Rev. Gen. Caout. Plast.* **45,** 1395–1398.

Ribaillier, D. (1970). Importance des lutoïdes dans l'écoulement du latex: Action de la stimulation. *Rev. Gen. Caout. Plast.* **47,** 305–310.

Ribaillier, D. (1971). Etude de la variation saisonnière de quelques propriétés du latex d'*Hevea brasiliensis. Rev. Gen. Caout. Plast.* **48,** 1091–1093.

Ribaillier, D. (1972). Quelques aspects du rôle des lutoïdes dans la physiologie et l'écoulement du latex d'*Hevea brasiliensis* (Kanth) Muell., Arg. Action de produits liberant de l'ethylène. Ph.D. Thesis, Abidjan University, Ivory Coast; *Hort. Abstr.* **43,** 9216.

Ribaillier, D., and d'Auzac, J. (1970). Nouvelles perspectives de stimulation hormonale de la production chez l'*Hevea brasiliensis. Rev. Gen. Caout. Plast.* **47,** 433–439.

Ribaillier, D., Jacob, J. L., and d'Auzac, J. (1971). Sur certains charactères vacuolaires des lutoïdes du latex d'*Hevea brasiliensis* Muell. Arg. *Physiol. Veg.* **9,** 423–437.

Riches, J. P., and Gooding, E. G. B. (1952). Studies in the physiology of latex. I. Latex flow on tapping-theoretical considerations. *New Phytol.* **51,** 1–10.

Ries. H. E., and Meyers, B. L. (1968). Flocculation mechanism: Charge neutralization and bridging. *Science* **160,** 1449–1450.

Rubber Research Institute of Malaya. (1962). Annual Report, Effect of Tapping on Growth. pp. 60–61. Rubber Res. Inst., Malaya.

Rubber Research Institute of Malaya. (1972). Annual Report, Field Investigations on Latex Flow and Stimulation. pp. 80–81. Rubber Res. Inst., Malaya.

Ruinen, J. (1950). Microscopy of the lutoids in *Hevea* latex. *Ann. Bogor.* **1,** 27–45.

Rutherford, P. P., and Bard, D. R. (1971). Water uptake and invertase and hydrolase activities induced in chicory root disks by treatment with various plant growth regulators. *Phytochemistry* **10,** 1635–1638.

Scholander, P. F., Hammel, H. T., Bradstreet, E. D., and Hemmingsen, E. A. (1965). Sap pressure in vascular plants. *Science* **148,** 339–346.

Schweizer, J. (1949). *Hevea* latex as a biological substance. *Archf. Rubbercult.* **26,** 345–397.

Sen, D. N., and Chawan, D. D. (1972). Leafless *Euphorbias* on Rajasthan rocks. IV. Water relations of seedlings and adult plants. *Vegetatio, Haag* **24**, 193–214.

Sheldrake, A. R. (1969). Cellulase in latex and its possible significance in cell differentiation. *Planta* **89**, 82–84.

Slatyer, R. O. (1967). "Plant-Water Relationships." Academic Press, New York.

Southorn, W. A. (1964). A complex sub-cellular component of widespread occurrence in plant latices. *J. Exp. Bot.* **15**, 616–621.

Southorn, W. A. (1967). Local changes in bark dimensions of *Hevea brasiliensis* very close to the tapping cut. *J. Rubber Res. Inst. Malaya* **20**, 36–43.

Southorn, W. A. (1968a). Latex flow studies. I. Electron microscopy of *Hevea brasiliensis* in the region of the tapping cut. *J. Rubber Res. Inst. Malaya* **20**, 176–186.

Southorn, W. A. (1968b). Latex flow studies. IV. Thixotropy due to lutoids in fresh latex demonstrated by a microviscometer of new design. *J. Rubber Res. Inst. Malaya* **20**, 226–235.

Southorn, W. A. (1969). Physiology of *Hevea* (latex flow). *J. Rubber Res. Inst. Malaya* **21**, 494–512.

Southorn, W. A., and Edwin, E. E. (1968). Latex flow studies. II. Influence of lutoids on the stability and flow of *Hevea* latex. *J. Rubber Res. Inst. Malaya* **20**, 187–200.

Southorn, W. A., and Gomez, J. B. (1970). Latex flow studies. VII. Influence of length of tapping cut on latex flow pattern. *J. Rubber Res. Inst. Malaya* **23**, 15–22.

Southorn, W. A., and Yip, E. (1968). Latex flow studies. III. Electrostatic considerations in the colloidal stability of fresh *Hevea* latex. *J. Rubber Res. Inst. Malaya* **20**, 201–215.

Spence, D., and McCallum, W. J. (1935). The function of rubber hydrocarbons in the living plant. *Trans., Inst. Rubber Ind.* **11**, 119–134.

Spencer, H. J. (1939a). The effect of puncturing individual latex tubes of *Euphorbia wulfenii*. *Ann. Bot. (London)* [N.S.] **3**, 227–229.

Spencer, H. J. (1939b). On the nature of the blocking of the laticiferous system at the leaf base of *Hevea brasiliensis*. *Ann. Bot. (London)* [N.S.] **3**, 231–235.

Tammes, P. M. L., Vonk, C. R., and van Die, J. (1969). Studies on phloem exudation from *Yucca flaccida* Haw. VII. The effect of cooling on exudation. *Acta Bot. Neer.* **18**, 224–229.

Tammes, P. M. L., van Die, J., and Ie, T. S. (1971). Studies on phloem exudation from *Yucca flaccida* Haw. VIII. Fluid mechanics and exudation. *Acta Bot. Neer.* **20**, 245–252.

Tata, S. J., and Yip, E. (1968). A protein fraction from B-serum with strong destabilizing activity on latex. *Res. Arch. Rubber Res. Inst. Malaya, Doc.* **59**.

Taysum, D. H. (1961a). Effect of ethylene oxide on the tapping of *Hevea brasiliensis*. *Nature (London)* **191**, 1319–1320.

Taysum, D. H. (1961b). Yield increases by the treatment of *Hevea brasiliensis* with antibiotics. *Proc. Natur. Rubber Res. Conf., 1960* pp. 224–240.

Templeton, J. K. (1969). Partition of assimilates. *J. Rubber Res. Inst. Malaya* **21**, 259–263.

Traub, H. P. (1946). Concerning the function of rubber hydrocarbon (caoutchouc) in the guayule plant, *Parthenium argentatum* A. Gray. *Plant Physiol.* **21**, 425–444.

Tupy, J. (1969a). Stimulatory effects of 2,4-dichlorophenoxyacetic acid and 1-NAA on sucrose level, invertase activity and sucrose utilization in the latex of *Hevea brasiliensis*. *Planta* **88**, 144–153.

Tupy, J. (1969b). Nucleic acids in latex and the production of rubber in *Hevea brasiliensis J. Rubber Res. Inst. Malaya* **21**, 468–476.

Tupy, J. (1973a). The level and distribution pattern of latex sucrose along the trunk of *Hevea brasiliensis. Physiol. Veg.* **11**, 1–11.

Tupy, J. (1973b). The sucrose mobilizing effect of auxins in *Hevea brasiliensis. Physiol. Veg.* **11**, 13–23.

Tupy, J. (1973c). The activity of latex invertase and latex production in *Hevea brasiliensis. Physiol. Veg.* **11**, 633–641.

Tupy, J. (1973d). Influence de la stimulation hormonale de la production sur la teneur en saccharose du latex d'*Hevea brasiliensis. Rev. Gen. Caout. Plast.* **50**, 311–314.

Tupy, J. (1973e). The regulation of invertase activity in the latex of *Hevea brasiliensis* Muell. Arg. *J. Exp. Bot.* **24**, 516–524.

Tupy, J., and Resing, W. L. (1969). Substrate and metabolism of carbon dioxide formation in *Hevea* latex *in vitro. J. Rubber Res. Inst. Malaya* **21**, 456–467.

van Die, J. (1955). A comparative study of the particle fractions from Apocynaceae latices. *Ann. Bogor.* **2**, 1–124.

van Gils, G. E. (1951). The contribution of the yellow fraction to the viscosity of latex. *J. Rubber Res. Inst. Malaya* **13**, 131–132.

Verhaar, G. (1956). The structure of *Hevea* latex and its viscosity. *Rubber Chem. Technol.* **29**, 1484–1487.

Waidyanatha, U. P de S., and Pathiratne, L. S. S. (1971). Studies on latex flow patterns and plugging indices of clones. *Rubber Res. Inst. Ceylon, Quart. J.* **48**, 47–55.

Whittenberger, R. T., Brice, B. A., and Copley, M. J. (1945). Distribution of rubber in *Cryptostegia* as a factor in its recovery. *India Rubber World* **112**, 319–323.

Wiersum, L. K. (1958). Quelques aspects physiologiques de l'étude du latex. *Rev. Gen. Caout.* **35**, 276–280.

Woo, C. H., and Edwin, E. E. (1970). Relationship between latex yield of *Hevea* and rubber biosynthesis *in vitro. J. Rubber Res. Inst. Malaya* **23**, 68–73.

Wycherley, P. R. (1963). Variation in the performance of *Hevea* in Malaya. *J. Trop. Geogr.* **17**, 143–171.

Wycherley, P. R. (1969). Breeding of *Hevea. J. Rubber Res. Inst. Malaya* **21**, 38–55.

Yip, E., and Southorn, W. A. (1968). Latex flow studies. VI. Effects of high pressure gradients on flow of fresh *Hevea* latex in narrow bore capillaries. *J. Rubber Res. Inst. Malaya* **20**, 248–256.

Yip, E., and Southorn, W. A. (1973). Rheology of fresh latex from *Hevea* collected over successive intervals from tapping. *J. Rubber Res. Inst. Malaya* **23**, 277–284.

CHAPTER 7

WATER DEFICITS AND NITROGEN-FIXING ROOT NODULES

Janet I. Sprent

DEPARTMENT OF BIOLOGICAL SCIENCES, UNIVERSITY OF DUNDEE, SCOTLAND

I. INTRODUCTION

The ability to fix nitrogen is a property confined to certain prokaryotic organisms, which, in some cases, live symbiotically with other plant species. In this chapter we shall be concerned solely with these symbiotic systems. They can be divided into two groups: those such as lichens (e.g., *Peltigera* sp.), some *Bryophyta* (*Anthoceros, Blasia*) and a few angiosperms (e.g.,

Gunnera) in which the microsymbiont is a species of blue-green alga, frequently *Nostoc,* and those (always higher plants) where the microsymbiont
is bacterial. Legumes represent economically the most important type and,
in them, the endosymbiont is a species of *Rhizobium.* Nonleguminous
angiosperms have a "higher" bacterial component in their root nodules,
probably an *actinomycete.* They frequently occur as primarily colonizers
(e.g., *Caeonothus, Alnus*). Several detailed reviews of these associations
have been published recently (e.g., Quispel, 1974; Nutman, 1975). The
distinction between the legume and nonlegume type of nodule has become
blurred with the recent observation of Trinick (1974) that *Rhizobium*
can form a nitrogen-fixing association with a species of the nonlegume
genus *Trema,* described at first as *T. aspera,* but now known to be
T. cannabina.

The success of these various associations suggests that the metabolic
pathways of the two symbionts are closely integrated and it is not surprising
to find that their nitrogen-fixing activity is sensitive to environmental fluctuations. What is surprising is that, in the past, most research has concentrated on effects of temperature and to a lesser extent, of light on nitrogen
fixation. Isolated reports (Wilson, 1931, 1942; Doku, 1970) have looked
at aspects of the water relations of nitrogen fixation, but, until recently,
no detailed work has been carried out. To date, most of our knowledge
is of legumes (Kuo and Boersma, 1971; Sprent, 1971a,b, 1972a,b,c; Sprent
and Sylvester, 1973; Minchin and Pate, 1974; Pankhurst and Sprent,
1975a,b) although some information is available for nonlegumes (Paul
et al., 1971) and lichens (Hitch and Stewart, 1973; Kershaw, 1974). This
chapter will be concerned mainly with legumes.

Prior to 1966, estimates of nitrogen fixation were made either by total
nitrogen analyses or by measuring incorporation of $^{15}N_2$ (Burris and Wilson, 1957). These methods are both time-consuming and expensive, although still useful and indeed essential for some types of study. The
discovery that the nitrogenase enzyme can reduce acetylene to ethylene
(Dilworth, 1966; Schöllhorn and Burris, 1966) provided the basis for a
very rapid and inexpensive assay using gas chromatography. This has greatly
assisted work in the nitrogen-fixation field. In theory, since two electrons
are involved in reduction of C_2H_2 to C_2H_4 and six in reduction of N_2 to
$2NH_3$, dividing the acetylene reducing activity by three gives the nitro-
gen-reducing activity and by summation over time, total nitrogen fixed.
Some investigators found this type of numerical relationship (Hardy *et
al.,* 1971), but others found wide variation (Bergersen, 1970). It is probably safest to regard the acetylene reduction assay as a measure of nitrogen-
fixing potential and this practice will be adopted here. Specific activity
refers to activity per unit weight of nodule (see Hardy *et al.,* 1973).

II. EFFECTS ON NODULE FORMATION, MORPHOLOGY, AND STRUCTURE

A. Infection of Roots and Nodulation

Before infection and nodulation can occur, sufficient bacteria of the correct rhizobial species and strain for the relevant host cultivar must be available. Thus the effects of moisture on survival of rhizobia in soil are important. Often this has not been studied from the infection point of view, but rather in investigating survival of rhizobia in commercially prepared inocula, both before and after placement in soil. For example, Roughley and Vincent (1967) worked with peat cultures and found *Rhizobium meliloti* to be more drought resistant then *R. trifolii*. In sterilized soil, Foulds (1971) found the reverse to be the case. These differences reinforce Vincent's (1965) argument that survival in monoculture may be quite different from survival in natural soils where biotic and other factors are included. Chatel and Parker (1973) found that heat as well as dryness affected survival of rhizobia in Western Australian soils.

Apart from mere survival, water supply is thought to affect both multiplication and movement of rhizobia in soil. Working with the peanut (*Arachis hypogea* L.), Shimshi *et al.* (1967) found that shallow placement of inocula in soils gave best nodulation even when irrigation regimes were such that soil moisture content in the upper layers varied from field capacity to extreme drought. They concluded that rhizobia multiplied rapidly following irrigation and thereafter survived in the soil, probably in water films surrounding soil particles. Presumably following multiplication, rhizobia migrate in soil while sufficient moisture is available, since Hamdi (1970) found that moderate soil-water tensions (−8 bars) slowed movement of *Rhizobium trifolii*. When water-filled pores in this soil became discontinuous, migration ceased. It was suggested that this lack of migration could account for the poor nodulation often obtained in light soils. Brockwell and Whalley (1970) also found nodulation failure of *Medicago trunculata* and *M. littoralis* in dry soils, even when the rhizobia (*R. meliloti*) in seed pellets remained viable. They suggested that the marked wetting and drying cycles in areas of infrequent precipitation could lead to enhanced nitrification and, hence, produce soil nitrate contents sufficiently high to inhibit nodulation.

Several studies show links between soil moisture and legume nodulation (see, e.g., Masefield, 1952, 1961; McKee, 1961; Doku, 1970). For cowpea (*Vigna unguiculata*), the effects of soil moisture interact with both day length and host cultivar (Doku, 1970), which emphasizes the importance of field trials in the regions where the relevant crops are to be grown.

These reported effects on nodulation could be a consequence of effects on rhizobial survival, migration, and multiplication as described above, or they could indicate that soil moisture affects the infection and subsequent nodulation processes. Detailed studies on these latter aspects are urgently needed.

B. STRUCTURE OF NORMAL NODULES

Before the detailed effects of water stress on nodule structure are considered, a brief account of structure of unstressed nodules will be given. In most grain legumes (*Pisum, Vicia, Glycine,* etc.) a tap root develops quite extensively before a lateral root system is initiated. Nodules first occur on the tap root close to the transition zone between root and hypocotyl. Such nodules are often near the soil surface and may be particularly sensitive to water stress (see Section IV,A). Forage legumes (*Trifolium, Medicago,* etc.) tend to have root systems that branch earlier and nodules are soon formed on all branches.

Two general types of nodule structure are recognized; they may be described morphologically as (a) basically spherical and (b) basically elongated. The spherical type is found in species such as soybean (*Glycine max*) and various forms of *Phaseolus vulgaris.* In the early stages of growth, there is a shell of meristematic cells which maintains the spherical shape and which has only a limited period of activity. When the meristem ceases to divide, cell enlargement continues, so that the nodules continue to grow in size. Throughout, they maintain a fairly constant proportion of their volume as intercellular air space (Bergersen and Goodchild, 1973a). All the central, infected cells are at an approximately uniform state of differentiation. They are active for 3–5 wk and then become senescent: thus, the nodules have a limited active life and nitrogen fixation must be continued by newly formed nodules. Gas exchange is thought to occur at the nodule surface (Sprent, 1971b; Bergersen and Goodchild, 1973a), rather than by transport from the shoot system to root system as occurs in some legumes (Evans and Ebert, 1960; Greenwood, 1967). Pankhurst and Sprent (1975a) consider that gaseous exchange occurs principally via intercellular spaces between cells of the lenticel-like ridges that radiate from the region of attachment to the root.

The elongated type of nodule, found in clovers (*Trifolium* sp.), peas (*Pisum* sp.), *Vicia* sp. and many other genera, has an apical meristem that may dichotomize, which leads to prolonged growth, with the nodules sometimes being perennial. New nitrogen-fixing tissue is continually being formed and the older regions become senescent. Active regions in all nodules are detectable by their pink color, due to leghemoglobin; senescent regions appear greenish-brown due to bile-type pigments. Vascular traces in the elongated type of nodule contain transfer cells, whereas those of the spherical types examined so far do not (Pate *et al.,* 1969). The main features of these two types of nodule are illustrated in Fig. 1. Further details can be obtained from Bieberdorf (1938) and Fraser (1942). The two types differ markedly in some of their responses to water stress.

C. Nodule Development and Morphology under Stress

In *Phaseolus vulgaris,* Sprent (1975) found that nodule number and size were reduced by water stress, which implies an effect on infection-nodulation as well as nodule development. In a more detailed study on *Glycine max* (J. I. Sprent, unpublished), nodules formed under water stress had the same morphology and general anatomy as control nodules, although measurements of cell size were not made. In particular, the development of the lenticel-like ridges, which are the likely aeration pathway

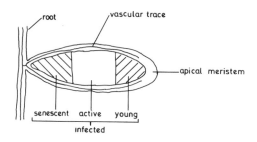

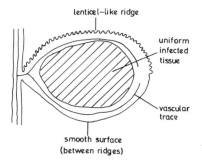

Fig. 1. Diagram of longitudinal sections through legume nodules of the elongate (A) and spherical (B) type.

for spherical nodules still occurred under stress. Nodule production was related to shoot development, as if the plant produced only those nodules that the shoot system could support (Fig. 2). The potential specific activity of nodules was not significantly affected by stress. Thus, although soybean is not considered to be a drought-tolerant legume (when compared, for example, with alfalfa, *Medicago sativa* (Cameron and Mullaly, 1972), it possesses considerable ability to adapt to its imposed moisture regime, as Furuhati and Monsi (1973) also found. White clover (*Trifolium repens*) produces fewer and smaller nodules when grown under stress (Engin and Sprent, 1973) and their specific activity is also depressed. Thus, legumes, not surprisingly, differ in how efficiently they nodulate under water stress.

Although nodules of soybean developed under continuous stress may appear normal, those stressed after being grown on an adequate supply of water show both morphological and anatomical (see Section II,E) effects. At the surface, cells of the lenticel-like ridges collapse and may partially occlude the air spaces, which restricts gaseous diffusion (Sprent, 1975; Pankhurst and Sprent, 1975a). No information has been found on the surface detail of elongate nodules, such as those of clover, under normal or stressed conditions.

D. Nodule Size and Shape in Soybean and Clover under Stress

As nodules lose water from their surfaces, they must inevitably shrink. Pankhurst and Sprent (1975b) estimated this shrinkage in detached soy-

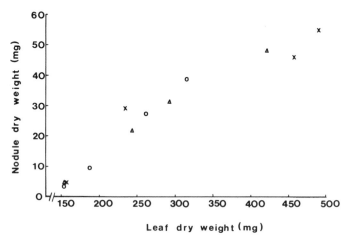

Fig. 2. Relationship between nodule and shoot dry weights for soybean (*Glycine max*) plants grown at three levels of water stress: X, control; O, 75%; and △, 50% of full water supply. Each point is the mean of 12 plants. Readings were taken weekly for 4 wks. Unpublished data of J. I. Sprent.

bean nodules and by simultaneous measurements of oxygen uptake concluded that nodule porosity (Table I) was also reduced; thus, once again,
gaseous diffusion might be restricted. Because of their small size, clover
nodules are more difficult to measure. Engin (1974) overcame this problem by using a modified projected photographic image method. This gave
estimates of nodule surface area as well as volume, which enabled both
these factors to be related to activity in stressed and control nodules. Measurements were made on 3-wk-old seedlings, over a period of a further
3 wk during which time plants were exposed to various levels of water
stress (Fig. 3). A 25% reduction in water supply resulted in much reduced

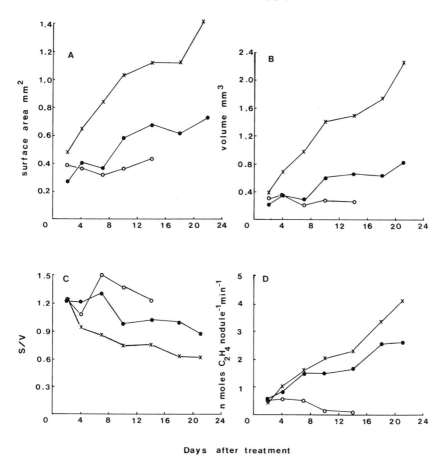

Fig. 3. Effect of water stress on surface area (A), volume (B), S/V (C), and
acetylene-reducing activity (D) of *Trifolium repens* nodules. Each point is the mean
of 9 readings: X, control, full water supply; ●, 75% water; ○. 0% water, i.e., no
water given after commencement of treatment. Data of Engin (1974).

TABLE I

THE EFFECT OF WATER STRESS ON THE VOLUME, SURFACE AREA, POROSITY, AND DIFFUSION OF OXYGEN INTO DETACHED SOYBEAN ROOT NODULES[a]

Nodule fresh weight (mg/nodule)	Nodule water content (FW/DW)	Nodule water potential (bars)[b]	Nodule volume (mm³)[c]	Nodule surface area (mm²)	Nodule porosity (%)[d]	Rate of oxygen diffusion into nodules at different pO_2 (nmoles O_2 mm^{-2} min^{-1})		
						0.1 atm	0.2 atm	0.4 atm
17.25	5.25	−1.5	18.10	33.34	3.12	0.31	0.38	0.58
15.36	4.62	−4.0	15.65	30.26	2.96	0.17	0.30	0.42
13.44	4.13	−9.0	13.95	28.00	—	0.07	0.15	0.26
10.69	3.27	−16.5	11.23	24.23	2.81	0.02	0.04	0.14

[a] All values given are the mean of four replicate samples.
[b] Determined by the Shardakov dye method (Knipling, 1967).
[c] Obtained from the displacement of water by the nodules.
[d] Determined by the pycnometer method of Jensen et al. (1969).

volume growth and nitrogen-fixing activity per nodule although the activity per mm³ of nodule was slightly increased (Fig. 3D). The surface to volume ratio fell more slowly with time in stressed nodules.

E. Nodule Structure in Soybean and White Clover under Stress

Work on this aspect of stress has been hampered both by the complexity of nodule structure and by the paucity of information on the effects of stress on the nearest structural equivalent, i.e., the root. However, from studies on white clover, soybean, and, to a lesser extent, French bean (*Phaseolus vulgaris*), a general picture is beginning to emerge.

In soybean, the physical connection between the nodule and its potential water supply, the root, is relatively small. Water loss is thought to occur principally from the nodule surface (Sprent, 1971b) and, although a sphere presents the minimum surface:volume and tends to conserve water, it is apparent that in dry soils water loss exceeds supply, particularly in the upper regions of the root system. In elongated and branched nodules, the situation may be even more acute.

At the point where stress is sufficient to depress nitrogen fixation completely in soybean, a number of major changes can be seen in nodules (Sprent, 1971b). In particular, plasmodesmata between infected and uninfected cells of the central region are broken and cells of the cortex show gross disruption. In a more detailed study (Sprent, 1972a), a gradation of effects was obtained, with variation occurring with different amounts of stress and different types of cells. Vacuolate cells showed much greater disruption than nonvacuolate cells for the same amount of stress. Infected cells, which in soybean are nonvacuolate, showed no obvious effects. Progressive cytoplasmic disruption was observed in other cells, with organelles such as mitochondria showing most resistance. Marked changes were also observed in cell walls that appeared striated when stressed. These effects are illustrated in Plate I. Similar general observations were made with white clover (Engin, 1974). In this species, which has the elongate type of nodule, stress effects also varied with cell age and differentiation. Cells of the meristematic tip were most resistant. Those of the nitrogen-fixing region showed changes consistent with stress accelerating senescence. An interesting observation, which may have more general significance, was that stress affected the staining properties of membranes, particularly in the meristematic region, where negative staining was frequently observed. Severe stress resulted in changes of bacteroid structure, the surface becoming wavy in outline. In stressed nodules, free-living (nonnitrogen-fixing) *Rhizobium* cells invaded the tissues more easily than in unstressed nodules, where they

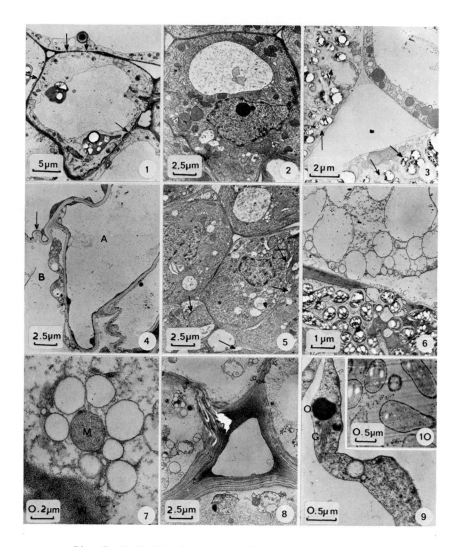

Plate I. (1–3) Fine Structure of Normal Soybean Nodules.

1. Cortical cell ×2500. Note large vacuole and numerous plasmodesmata (arrows). Starch granules and organelles such as mitochondria are also visible.

2. Pericycle cell from vascular trace in cortex ×4000. Note nucleus with prominent nucleolus, mitochondria, and endoplasmic reticulum. Pericycle cells always give the appearance of great metabolic activity.

3. Parts of 3 cells from the central infected region, surrounding an intercellular air space. ×6300. The left hand and bottom cells contain the nitrogen-fixing bacteroid form of *Rhizobium japonicum*. Groups of bacteroids are surrounded by membrane envelopes of host origin (arrows) and the infected cells also contain the usual eukaryotic organelles. They are not vacuolate. The cell on the right is a typical noninfected cell from the infected region. It has a large central vacuole and the cytoplasm contains numerous organelles. Noninfected cells are connected by plasmodesmata to infected cells.

were confined to the most senescent regions. Presumably, these bacteria utilized the contents of disrupted host cells as a source of nutrient; structurally they appeared normal. This accords with a suggestion put forward by Wilson (1942) that beans (*Vicia faba*) may shed nodules during drought and that such nodules may act as a source of soil nutrient. No evidence of nodule shedding in soybean has been found, but some nodule shedding may occur in white clover (Engin, 1974).

III. PHYSIOLOGICAL EFFECTS ON NODULES

A. WATER CONTENT OF NODULES AND NITROGEN-FIXING ACTIVITY

There is a general relationship between water content of nodules and acetylene-reducing activity as shown in Fig. 4. The following species have been shown to conform to this pattern, *Glycine max, Vicia faba, Phaseolus vulgaris, Trifolium repens,* and *Lupinus arboreus* (Sprent, 1971b, 1972c, 1973, 1975; Engin and Sprent, 1973). That the responses genuinely represent effects on nitrogen fixation has been confirmed with ^{15}N uptake in *G. max* (Sprent, 1971b) and by nitrogen analyses in *T. repens* (Engin and Sprent, 1973) and *Pisum sativum* (Minchin and Pate, 1975). Thus, it is reasonable to suggest that activity of nitrogen-fixing root nodules is

---

Plate I. (4–9) Effects of Water Stress on Soybean Nodule Cells.

4. Cells from the outer cortex, ×4000. The cytoplasm in cell A has largely broken down. Fragments remain near the wall. In cell B the cytoplasm is beginning to tear away from the wall. In places cell walls are becoming convoluted as the nodule shrinks (arrow).

5. Pericycle cells ×4000. Damage resulting from stress is far less obvious than in vacuolate cells. Some shrinkage of cell contents away from walls is visible (arrows).

6. Cells from the central infected region ×10,000. The upper, noninfected cell shows shrinkage of cytoplasm away from the wall and also extensive vacuolation. The lower, infected cell appears normal.

7. Detail of damage to cytoplasm of cortical cell ×50,000. Cytoplasmic membranes have formed into vacuoles, but the mitochondrion (M) still retains much structural integrity. Cellulose fibrils in the wall (W), cut obliquely, appear distinct.

8. Cells surrounding an intercellular space in the cortex ×4000. Note the typical striated appearance of the cell walls. The cytoplasm is in an advanced state of degeneration.

9. Cytoplasm that has completely separated from its associated wall (from a cortical cell, ×25,000). Note osmiophilic granule (O) and Golgi body (G).

10. Bacteriods from control cells that have been incubated in diaminobenzidine reagent. ×18,000. Note discrete regions of activity in cytomembrane.

Parts 1–9 are unpublished micrographs of Sprent and Part 10 of Marks and Sprent. For further details, see Sprent (1972a) and Marks and Sprent (1974.)

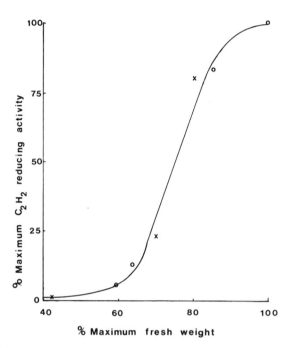

FIG. 4. Effect of water stress on acetylene reducing activity of *Lupinus arboreus* nodules grown at two light levels X, 20,000 lux; ○, 6,500 lux. Data of Sprent (1973).

generally sensitive to water stress. As in many other systems discussed in these volumes, there is difficulty in expressing the degree of water stress in nodules. Initially, percentage of full turgidity was used (Sprent, 1971b) as were fresh weight:dry weight ratios (Sprent, 1972c). These measures are fairly satisfactory for experiments conducted on one age of nodules, but they present problems when nodules are sampled at intervals during growth, since the fresh to dry weight ratio in control plants changes with age (Sprent, 1973). Pankhurst and Sprent (1975b) tried to measure the water potential of nodules (a) by the Shardakov dye method (Knipling, 1967) and (b) by measuring the water potential of sand equilibrated with stressed nodulated root systems using a dew-point psychrometer. The relationship between these estimates and the fresh to dry weight ratio for soybean nodules is given in Fig. 5. It is likely that under natural conditions a gradient of water potential exists in nodules since water is lost from the surface and supplied by the vascular system. This is confirmed by the gradient of effects of stress on fine structure (Section II,E). All estimates of average nodule water potential are oversimplified, but nevertheless useful in comparing levels of stress having different physiological effects. In general, loss of up to about 25% in fresh weight (equivalent to −8 to −10

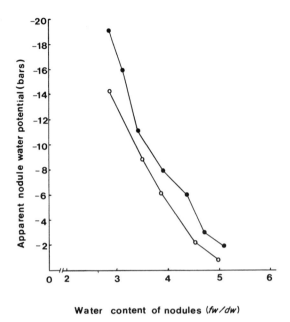

Fig. 5. Relationship between water potential Ψ of soybean (*Glycine max*) nodules measured by the Shardakov dye method (●) or by psychrometry on sand equilibrated with nodules (○) and fresh to dry weight ratio. Data of Pankhurst and Sprent (1975b).

bars) results in reversible effects on nitrogen fixation, but beyond this, the effects become progressively more severe (Sprent, 1971a, 1972c; Pankhurst and Sprent, 1975b).

B. Osmotic Stress

Solutes have effects on nitrogen fixation which are quite distinct from those of atmospheric water stress. Experiments have been conducted, using detached nodules, detached root systems, and whole plants (Sprent, 1972b). Detached nodules of soybean were immersed in salt or mannitol solutions and effects on acetylene-reducing activity followed. Because submerged nodules are oxygen deficient (Sprent, 1969), an atmosphere of 90% oxygen and 10% acetylene was used, together with shaking. Salt (sodium or potassium chlorides) had a very rapid inhibitory effect on acetylene reduction (Fig. 6). In experiments on detached root systems, direct contact between nodules and salt was necessary for the effects to be shown. Watering the root systems of growing soybean plants with salt

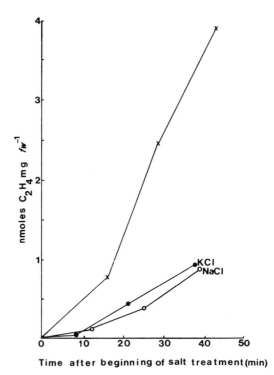

Fig. 6. Effect of NaCl (○) and KCl (●) each at 0.1 M and water (X) on acetylene reduction by detached soybean nodules. Nodules incubated under an atmosphere of 90% oxygen, 10% acetylene, and shaken. Each point is the mean of 5 replicates. Data of Sprent (1972c).

solutions also had a very rapid inhibitory effect (detectable in 5 min) which was related to the concentration of salt supplied. Flushing the salt out with water reversed the effects. The concentrations of salt required to give an effect (0.04 M upward) and the rapidity of the response precluded an explanation on the basis of osmotic withdrawal of water and a more direct effect on nodule metabolism was suggested (Sprent, 1972b). For this reason, the problem will not be considered further here.

Mannitol solutions had a much slower effect on nodules than salt solutions and the effective concentrations were greater, suggesting that the effects may have been osmotic. Water loss from nodules and wilting of plant leaves occurred at around 0.4 M. However, prolonged exposure (2 hr or more) to 0.2 M mannitol also depressed nitrogen-fixing activity; this coupled with the lack of reversibility of the effect of mannitol when nodules were returned to water suggested that the effects may be complex. Clearly, it is unwise to simulate the effects of atmospheric water stress on root nodules by using osmotica.

C. WATER STRESS, NITROGEN FIXATION, AND RESPIRATION

Both atmospheric water stress (Sprent, 1971b) and salt stress (Sprent, 1972b) reduce oxygen uptake by nodules as well as depress acetylene reducing activity. This reflects the very close link, which has been known for many years, between nitrogen fixing and respiratory activities. Various workers (e.g., Bergersen, 1962) have shown that soybean nodules fix nitrogen most actively at 50% oxygen, although there is not complete agreement on this point (Dart and Day, 1971). The nitrogenase enzyme is inactivated by oxygen (Mortensen, 1966) and nitrogen-fixing systems use a variety of methods to exclude oxygen from the enzyme (Jones et al., 1973). However, the conversion of nitrogen to ammonia requires considerable quantities of ATP and reductant. In legume root nodules, the current opinion is that bacteroids can carry out oxidative phosphorylation to provide their own ATP and, hence, need a source of oxygen, but the level must be strictly controlled to avoid inactivation of nitrogenase (Bergersen, 1971; Bergersen and Goodchild, 1973b). Leghemoglobin, which bathes the bacteroids within their membrane envelopes, probably both facilitates and regulates the oxygen supply to bacteroids (Bergersen and Goodchild, 1973b; Appleby et al., 1973; Wittenberg et al., 1974). Thus, in view of this delicate balance between oxygen and nitrogen metabolism, it is perhaps not surprising to find that both are affected by water stress. In a series of experiments planned to see whether the depressed oxygen uptake was a cause or a consequence of reduced nitrogen fixation. Pankhurst and Sprent (1975b) tested the effects of a range of oxygen concentrations on response of detached soybean nodules to water stress. Both acetylene reduction and oxygen uptake were measured. Some typical results are shown in Fig. 7. It is clear that the effects of moderate water stress can be overcome completely by increasing the oxygen concentration. Lesser, but significant, recovery can be achieved with severely stressed nodules. It was concluded that the prime effect of water stress on nodules was to depress oxygen uptake, which reduced the supply of available metabolites (e.g., ATP) essential for nitrogen fixation. Two ways in which oxygen uptake might be reduced were suggested (a) because oxygen diffusion barriers were increased under stress and (b) because oxygen requiring reactions were inhibited. A combination of these two is not excluded.

The pathway for oxygen diffusion in nodules has not entirely been elucidated, even for soybeans upon which most work has been done. Bergersen and Goodchild (1973a) consider that free diffusion occurs

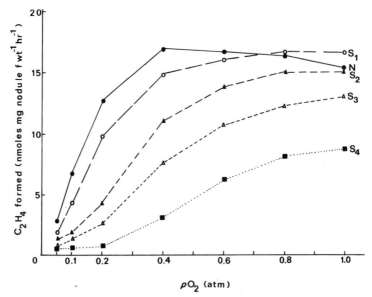

Fig. 7. Effect of increasing oxygen concentration on acetylene reducing activity of turgid (N) and water stressed (S) soybean root nodules. Four levels of water stress were used S1, $\Psi = -4.5$ bars; S2, $\Psi = -7.5$ bars; S3, $\Psi = -11.0$ bars; S4, $\Psi = -18.0$ bars. Each point is the mean of 3 replicates. Data of Pankhurst and Sprent (1975b).

from the soil atmosphere through an intercellular space system to the center of nodules. Tjepkema and Yocum (1973, 1974) postulated on the basis of temperature coefficients (Q_{10}) for oxygen uptake and on oxygen probe data that there is a barrier in the cortex, possibly located at the endodermis. Pankhurst and Sprent (1975a,b) favor a compromise viewpoint where diffusion into the nodules is restricted to the "lenticel-like" ridges on nodule surfaces. Calculations of rates of oxygen uptake in relation to nodule volume (see Table I) suggest that water stress decreases rates of diffusion, presumably by increasing the effectiveness of some barrier. Such a barrier could be anywhere between the nodule surface (where ridge cells collapse as described earlier) and the bacteroids. Possibly hemoglobin structure and hence oxygen-carrying properties are altered.

The second alternative, that oxygen-requiring processes are depressed, cannot be excluded. It is possible, for example, to visualize macromolecular changes resulting in increased K_m value for terminal oxidases which could account for the response of stressed nodules to increasing pO_2. This is consistent with data discussed in the next section and also with the fact that ethanol production by nodules varies with both water stress and oxygen tension (Sprent and Gallacher, 1975).

D. EFFECTS OF ENZYME LOCALIZATION IN NODULES

An alternative approach to the problem was attempted by trying to locate key respiratory enzymes histochemically in nodules and then see how this localization was affected by water stress. Detailed localization required electron microscope techniques and Marks and Sprent (1974) demonstrated the presence of ATPases in the host plasmalemma (Mg^{2+} stimulated) and the cytomembrane of bacteroids (Ca^{2+} stimulated), which also contained a diaminobenzidine oxidase, presumed to be some type of terminal oxidase of the general cytochrome oxidase type (see Plate I). When nodules were water stressed to the point where N_2-fixing activity was stopped, this oxidase could no longer be demonstrated (J. I. Sprent and I. Marks, unpublished). A similar situation in roots was found by Nir *et al.* (1970) for a mitochondrial oxidase and was thought to represent a change in association of membrane-bound enzymes rather than an actual loss of enzyme. Miller *et al.* (1971) and Bell *et al.* (1971) have found that water stress affects both membrane characteristics and respiratory oxidations in isolated corn mitochondria. Changes in membrane structure and hence function of membrane-bound enzymes occur in many chilling sensitive plants (see Raison, 1973, for review) and it seems possible that analogous changes could occur as a result of water stress. These, like the temperature-induced changes, might be reversible if the stress is not too severe or prolonged. The water potential at which changes in oxidase activity in bacteroids were detectable was around -9 bars, which corresponds to the lowest potential at which depression of acetylene-reducing activity was reversed by increasing the oxygen tension (Pankhurst and Sprent, 1975b). The cytomembrane ATPase in bacteroids is also stress sensitive. This could be linked to the oxygen sensitivity and result in impaired oxidative phosphorylation. It is interesting to note that Keck and Boyer (1974) concluded that in chloroplasts stress levels up to -11 bars probably had their major effect on electron transport, but, beyond that, the limiting process was probably photophosphorylation.

E. EFFECTS IN RELATION TO PHYSIOLOGY OF THE WHOLE PLANT

In the preceding section, the detailed effect of water stress at the site of nitrogen fixation, the nodule, was considered. This section will now examine the relationship between the stressed nodule and the rest of the plant. Pate *et al.* (1969) suggested that limitations in water supply to the nodule might affect nodule activity by restricting export of fixation products which would then accumulate in inhibitory concentrations. At one time, many investigators thought that ammonia or simple amino acids might act as end-product inhibitors of nitrogenase, but it is now known that this is not the case (Carnahan *et al.,* 1960). Pate and co-workers (1969) showed that exported amino acids and amides are near the limits of their solubility. Minchin and Pate (1974) found that export of nitrogenous products from pea nodules was closely linked to transpiration rate, and Sprent (1972c) showed that soybean root nodules are efficient at extracting tritiated water en route from the root to the shoot. Thus, even if accumulation of products does not have a direct effect upon nitrogenase, lack of water and impaired water flow must interfere with vital transport processes to and from nodules. Minchin and Pate (1975) concluded from nitrogen contents that water stress reduced flow of nitrogen from roots to shoots of peas.

For pea nodules, it has been estimated that export of 1 mg fixed nitrogen requires at least 0.35 ml water (Minchin and Pate, 1973) and that half this requirement might be met by mass flow into nodules from the phloem. Water stress is known to affect phloem translocation (see, for example, Plaut and Reinhold, 1965). This could have the obvious effect of reducing the carbon supply to nodules and the less obvious one of inducing a water deficit in them.

The relationship between carbon and nitrogen fixation has been investigated both directly and indirectly. Kuo and Boersma (1971), working with soybean, found that the ratio of nitrogen fixed (measured by analysis of total nitrogen) to carbon dioxide absorbed (measured using an infrared gas analyzer) decreased as soil water suction was increased. Engin and Sprent (1973) calculated nitrogen content per unit dry weight of white clover plants and found that this was reduced by water stress. Minchin and Pate (1975) found similar effects in pea, but the effects of water stress were considerably smaller than those of waterlogging. All these observations are consistent with the view that the nitrogen-fixing process is more sensitive than carbon fixation to stress, even though nodules are nearer to the normal water supply (roots) than leaves. However, from the results of an elegant series of experiments with intact soybeans, Huang *et al.* (1975a,b) concluded that reduced supplies of photosynthate were responsible for the depression of acetylene reduction under stress (induced by withholding water from pots over a perid of several days). By increasing the carbon dioxide supply to the shoots, they were able partially to overcome the effects of stress. It is well known (Mague and Burris, 1972) that detached nodules and nodulated roots show lower acetylene reducing activity than whole plants. Thus it is not surprising to find that inhibition of shoot photosynthesis by water stress will in turn effect nodule activity. Any effects of this nature would be additional to, and could interact with the direct effects of water stress on nodule physiology (Section III,C).

F. RECOVERY FROM STRESS

It was mentioned briefly in Section III,A that the effects of water stress on nitrogen fixation are, within limits, reversible. The limits in soybean approximate to the degree of stress that can be overcome by increasing the oxygen supply (Section III,C). Sprent (1971b) reported that she could not get detached water-stressed soybean nodules to recover, although attached nodules could recover quickly following rewatering (Sprent, 1971a). The fact that tritiated water passes quickly from roots to nodules (Sprent, 1972c) suggests that this is the major pathway for water movement into stressed soybean nodules. This view is supported by the fact that Pankhurst and Sprent (1975b) obtained partial recovery in detached soybean nodules only after recutting the severed vascular connection with the root. Pate *et al.* (1969) reported that various detached legume nodules, including soybean, could be induced to bleed by placing them on moist filter paper. It is thus possible that unstressed soybean nodules can take up water at their surfaces.

Engin and Sprent (1973) carried out a series of recovery experiments with water-stressed white clover plants and concluded that recovery in this species is a two-stage process. The first stage involves rehydration of existing nitrogen-fixing tissue and the second involves production of new nitrogen-fixing tissue as a result of renewed growth of the nodule meristem. These conclusions, based on measurements of nodule specific activity and growth in fresh weight, are summarized in Fig. 8. *Vicia faba* (Sprent, 1972c) behaves similarly. Thus, meristematic nodules have a greater recovery potential than spherical nodules, in which rehydration of existing tissue is the only possibility, unless new nodules are produced.

IV. FIELD STUDIES

A. NODULE ACTIVITY AND SOIL MOISTURE CONTENT

Sprent (1972c) measured specific activity of *V. faba* nodules grown on a well-drained loam from July through September. The activities were compared with a variety of environmental parameters (soil temperature at various depths, radiation, rainfall) and the factor most closely related with activity was soil-moisture content. Maximum activity occurred at around field capacity: above and below this acetylene reduction was depressed. This very strong indication of the significance of water stress in the field (even in Scotland!) was confirmed by watering plots during dry periods and measuring recovery. Subsequently, in a planting density trial

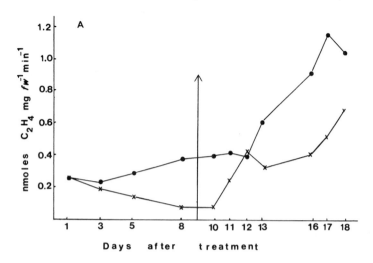

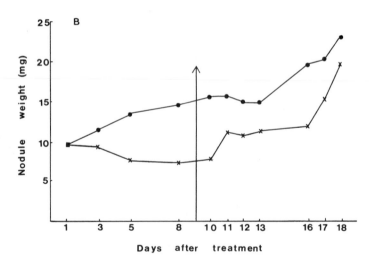

Fig. 8. Effect of water stress followed by rewatering on specific activity (A) and nodule fresh weight (B) of *Trifolium repens* nodules. Each point is the mean of 9 replicates. ●, control; X, stressed plants. Data of Engin and Sprent (1973).

with *V. faba* (Sprent, 1975), water content of soil was a major factor affecting specific activity. In a preliminary trial with a grass/clover sward, irrigation during a dry period (water potential of soil —2.5 bars) produced a tenfold increase in nodule specific activity (Sprent, 1975), which is similar in magnitude to the effect reported for nonlegumes by Paul *et al.* (1971). Although no detailed measurements were made, soil moisture was considered to be a factor affecting nitrogen fixation by *Lupinus arboreus* grown on reclaimed sand dunes in the North Island of New Zealand (Sprent and Silvester, 1973). Sau (1970), working on white clover in Russia, concluded that fixed nitrogen was important for yield at rainfalls above 300 mm in the vegetative period, but below this, nitrogen fertilizers were more important, which implies that nitrogen fixation is more sensitive to water stress than uptake and utilization of combined nitrogen. More field studies in different regions are urgently needed.

B. Possible Economic Significance

Legumes are notorious for their profligate use of water. Ludlow and Wilson (1972) showed for the tropical grasses *Pennisetum purpureum* and *Sorghum album* a transpiration ratio (gm water transpired per gm CO_2 fixed) of just under 80 compared with 180 for the legumes *Calopogonium mucunoides* and *Glycine wightii*. Thus, it is not surprising that the literature contains numerous reports of the depressing effect of moisture stress on yield of forage (e.g., Bourget and Carson, 1962) and grain (e.g., Mack, 1973) legumes. The responses vary among varieties, for example, in soybean (Mederski and Jeffers, 1973) and at least in grain legumes stress is particularly harmful during flowering (Singh and Tripathi, 1972). In some species, this may be related to cessation of root growth at this stage of growth (Salter and Drew, 1965), but not all legumes show this effect (El Nadi, Brouwer and Locher, 1969). At flowering and into pod-filling stages nitrogen fixation decreases in many species (e.g., Pate, 1958): this may reflect competition for carbohydrate between nodules and developing fruits (Lawn and Brun, 1974). This competition, compounded with cessation of root growth and thus impaired water uptake, would make the symbiotic plant especially sensitive to water stress at this time.

Recent transfer of nitrogen-fixing genes between prokaryotic genera (Dixon and Postgate, 1972) has encouraged research into the possibility of synthesizing nitrogen-fixing higher plants, especially cereals. For various reasons (see Postgate, 1974, for summary) this possibility is still a long way off. In the interim, much can be done by selection of drought-tolerant cultivars and improved management practices to make the best use of the legume species which we now grow.

REFERENCES

Appleby, C. A., Wittenberg, B. A., and Wittenberg, J. B. (1973). Nicotinic acid as a ligand affecting leghemoglobin structure and oxygen reactivity. *Proc. Nat. Acad. Sci. U. S.* **70**, 564–568.

Bell, D. T., Koeppe, D. E., and Miller, R. J. (1971). The effects of drought stress on respiration of isolated corn mitochondria. *Plant Physiol.* **48,** 413–415.

Bergersen, F. J. (1962). The effects of partial pressure of oxygen upon respiration and nitrogen fixation by soybean root nodules. *J. Gen. Microbiol.* **29,** 113–125.

Bergersen, F. J. (1970). The quantitative relationship between nitrogen fixation and the acetylene reduction assay. *Aust. J. Biol. Sci.* **23,** 1015–1025.

Bergersen, F. J. (1971). Biochemistry of nitrogen fixation in legumes. *Annu. Rev. Plant Physiol.* **22,** 121–140.

Bergersen, F. J., and Goodchild, D. J. (1973a). Aeration pathways in soybean root nodules. *Aust. J. Biol. Sci.* **26,** 729–740.

Bergersen, F. J., and Goodchild, D. J. (1973b). Cellular location and concentration of leghaemoglobin in soybean root nodules. *Aust. J. Biol. Sci.* **26,** 741–756.

Bieberdorf, F. W. (1938). The cytology and histology of the root nodules of some Leguminosae. *J. Amer. Soc. Agron.* **30,** 375–389.

Bourget, S. J., and Carson, R. B. (1962). Effect of soil moisture stress on yield, water-use efficiency and mineral composition of oats and alfalfa grown at two fertility levels. *Can. J. Soil Sci.* **40,** 7–12.

Brockwell, J., and Whalley, R. D. B. (1970). Studies in seed pelleting as an aid to legume seed inoculation. 2. Survival of *Rhizobium meliloti* applied to medic seed sown into clay soil. *Aust. J. Exp. Agr. Anim. Husb.* **10,** 455–459.

Burris, R. G., and Wilson, P. W. (1957). Methods for measurement of nitrogen fixation. *In* "Methods in Enzymology" (S. P. Colowick and N. O. Kaplan, eds.), Vol. 4, pp. 3–55. Academic Press, New York.

Cameron, D. G., and Mullaly, J. D. (1972). The performance of lucerne (*Medicago sativa*) lines in pure stands under irrigated and raingrown conditions in subcoastal Queensland. *Aust. J. Exp. Agr. Anim. Husb.* **12,** 646–652.

Carnahan, J. E., Mortensen, L. E., Mowar, H. F., and Castle, J. E. (1960). Nitrogen fixation in cell free extracts of *Clostridium pasteurianum*. *Biochim. Biophys. Acta* **44,** 520–535.

Chatel, D. L., and Parker, C. A. (1973) Survival of field-grown rhizobia over the dry summer period in Western Australia. *Soil/Biol. & Biochem.* **5,** 415–423.

Dart, P. J., and Day, J. M. (1971). Effects of incubation temperature and oxygen tension on nitrogenase activity of legume root nodules. *Plant Soil, Spec. Vol.* pp. 167–184.

Dilworth, M. J. (1966). Acetylene reduction by nitrogen-fixing preparations from *Clostridium pasteurianum*. *Biochim. Biophys. Acta* **127,** 285–294.

Dixon, R. A., and Postgate, J. R. (1972). Genetic transfer of N_2 fixation from *Klebsiella pneumoniae* to *Escherichia coli*. *Nature (London)* **237,** 102–103.

Doku, E. V. (1970). Effect of daylength and water on nodulation of cowpea (*Vigna unguiculata* (L.) Walp) in Ghana. *Exp. Agr.* **6,** 13–18.

El Nadi, A. H., Brouwer, R., and Locher, J. Th. (1969). Some responses of the root and shoot of *Vicia faba* plants to water stress. *Neth. J. Agric. Sci.* **17,** 133–142.

Engin, M. (1974). The effects of water stress on the physiology and fine structure of nitrogen fixing *Trifolium repens* L. Ph.D. Thesis, University of Dundee.

Engin, M., and Sprent, J. I. (1973). Effects of water stress on growth and nitrogen-fixing activity of *Trifolium repens*. *New Phytol.* **72,** 117–126.

Evans, N. T. S., and Ebert, M. (1960). Radioactive oxygen in the study of gas transport down the root of *Vicia faba*. *J. Exp. Bot.* **11,** 246–257.

Foulds, W. (1971). Effect of drought on three species of *Rhizobium*. *Plant Soil* **35,** 665–667.

Frazer, H. L. (1942). The occurrence of endodermis in leguminous root nodules and its effect upon nodule function. *Proc. Roy. Soc. Edinburgh, Sect. B* **61**, 328–343.

Furuhata, I., and Monsi, M. (1973). An analytical study on the ecophysiological adaptation of soybean plants to limited water supply. *J. Fac. Sci., Univ. Tokyo, Sect. 3* **9**, 243–262.

Greenwood, D. J. (1967). Studies on the transport of oxygen through the stems and roots of vegetable seedlings. *New Phytol.* **66**, 337–347.

Hamdi, Y. A. (1970). Soil water tension and the movement of rhizobia. *Soil/Biol. & Biochem.* **3**, 121–126.

Hardy, R. W. F., Burns, R. C., Hebert, R. R., Holsten, R. D., and Jackson, E. K. (1971). Biological nitrogen fixation: A key to world protein. *Plant Soil, Spec. Vol.* pp. 561–590.

Hardy, R. W. F., Burns, R. C., and Holsten, R. D. (1973). Applications of the acetylene-ethylene assay for measurement of nitrogen fixation. *Soil/Biol. & Biochem.* **5**, 47–81.

Hitch, C., and Stewart, W. D. P. (1973). Nitrogen fixation by lichens in Scotland. *New Phytol.* **72**, 509–524.

Huang, C-Y., Boyer, J. S. and Vanderhoef, L. N. (1975a). Acetylene reduction (nitrogen fixation) and metabolic activities of soybean having various leaf and nodule water potentials. *Plant Physiol.* **56**, 222–227.

Huang, C-Y., Boyer, J. S. and Vanderhoef, L. N. (1975b). Limitation of acetylene reduction (nitrogen fixation) by photosynthesis in soybean having low water potentials. *Plant Physiol.* **56**, 228–232.

Jensen, C. R., Luxmoore, R. J., van Gundy, S. D., and Stolzy, L. H. (1969). Root air space measurements by a pycnometer method. *Agron. J.* **61**, 474–475.

Jones, C. W., Bruce, J. M., Wright, V., and Ackrell, B. A. C. (1973). Respiratory protection of nitrogenase in *Azotobacter vinelandii*. *FEBS (Fed. Eur. Biochem. Soc.) Lett.* **29**, 77–81.

Keck, R. W., and Boyer, J. S. (1974). Chloroplast response to low leaf water potentials. III. Differing inhibition of electron transport and photophosphorylation. *Plant Physiol.* **53**, 474–479.

Kershaw, K. A. (1974). Dependence of the level of nitrogenase activity on the water content of the thallus in *Peltigera canina, P. evansiana, P. polydactyla* and *P. praetextata*. *Can. J. Bot.* **52**, 1423–1427.

Knipling, E. B. (1967). Measurement of leaf water potential by the dye method. *Ecology* **48**, 1038–1040.

Kuo, T., and Boersma, L. (1971). Soil water suction and root temperature effects on nitrogen fixation in soybeans. *Agron. J.* **63**, 901–904.

Lawn, R. J., and Brun, W. A. (1974). Symbiotic nitrogen fixation in soybeans I. Effects of photosynthetic source-sink manipulations. *Crop Sci.* **14**, 11–16.

Ludlow, M. M., and Wilson, G. L. (1972). Photosynthesis of tropical pasture plants. IV. Basis and consequences of differences between grasses and legumes. *Aust. J. Biol. Sci.* **25**, 1133–1145.

Mack, A. R. (1973). Soil temperature and moisture conditions in relation to the growth and quality of field peas. *Can. J. Soil Sci.* **53**, 59–72.

Mague, T. H. and Burvis, R. H. (1972). Reduction of acetylene and nitrogen by field-grown soybeans. *New Phytol.* **71**, 275–286.

McKee, G. W. (1961). Some effects of liming, fertilization, and soil moisture on seedling growth and nodulation in birds-foot trefoil. *Agron. J.* **53**, 237–240.

Marks, I., and Sprent, J. I. (1974). The localization of enzymes in fixed sections of soybean root nodules by electron microscopy. *J. Cell Sci.* **16**, 623–637.

Masefield, G. B. (1952). The nodulation of annual legumes in England and Nigeria: Preliminary observations. *Emp. J. Exp. Agr.* **20**, 175–186.

Masefield, G. B. (1961). The effect of irrigation on nodulation of some leguminous crops. *Emp. J. Exp. Agr.* **29**, 51–59.

Mederski, H. J., and Jeffers, D. L. (1973). Yield response of soybean varieties grown at two moisture stress levels. *Agron. J.* **65**, 410–412.

Miller, R. J., Bell, D. T., and Koeppe, D. E. (1971). The effects of water stress on some membrane characteristics of corn mitochondria. *Plant Physiol.* **48**, 229–231.

Minchin, F. R., and Pate, J. S. (1973). The carbon balance of a legume and the functional economy of its root nodules. *J. Exp. Bot.* **24**, 259–271.

Minchin, F. R., and Pate, J. S. (1974). Diurnal functioning of the legume root nodule. *J. Exp. Bot.* **25**, 295–308.

Minchin, F. R., and Pate, J. S. (1975). Effect of water, aeration and salt regime on nitrogen fixation in a nodulated legume—definition of an optimum root environment. *J. Exp. Bot.* **26**, 60–69.

Mortensen, L. E. (1966). Components of cell-free extracts of *Clostridium pasteurianum* required for ATP-dependent H_2 evolution from dithionite and for N_2 fixation. *Biochim. Biophys. Acta* **127**, 18–25.

Nir, I., Poljakoff-Mayber, A., and Klein, S. (1970). The effect of water stress on mitochondria of root cells. A biochemical and cytochemical study. *Plant Physiol.* **45**, 173–177.

Nutman, P. S., ed. (1975). "Symbiotic Nitrogen Fixation in Plants." Cambridge Univ. Press, London and New York.

Pankhurst, C. E., and Sprent, J. I. (1975a). Surface features of soybean root nodules. *Protoplasma* **85**, 85–98.

Pankhurst, C. E., and Sprent, J. I. (1975b). Effects of water stress on the respiratory and nitrogen-fixing activity of soybean root nodules. *J. Exp. Bot.* **26**, 287–304.

Pate, J. S. (1958). Nodulation studies in legumes. II. The influence of various environmental factors on symbiotic expression in the vetch (*Vicia sativa* L.) and other legumes. *Aust. J. Biol. Sci.* **11**, 496–515.

Pate, J. S., Gunning, B. E. S., and Briarty, L. G. (1969). Ultrastructure and functioning of the transport system of a leguminous root nodule. *Planta* **85**, 11–34.

Paul, E. A., Myers, R. J. K., and Rice, W. A. (1971). Nitrogen fixation in grassland and associated cultivated ecosystems. *Plant Soil, Spec. Vol.* pp. 495–507.

Plaut, Z., and Reinhold, L. (1965). The effect of water stress on ^{14}C sucrose transport in bean plants. *Aust. J. Biol. Sci.* **18**, 1143–1155.

Postgate, J. R. (1974). New advances and future potential in biological nitrogen fixation. *J. Appl. Bacteriol.* **37**, 185–202.

Quispel, A., ed. (1974). "The Biology of Nitrogen Fixation." North-Holland Publ., Amsterdam.

Raison, J. K. (1973). Temperature induced phase changes in membrane lipids and their influence on metabolic regulation. *Symp. Soc. Exp. Biol.* **27**, 485–512.

Roughley, R. J., and Vincent, J. M. (1967). Growth and survival of *Rhizobium* spp. in peat cultures. *J. Appl. Bacteriol.* **30**, 362–376.

Salter, P. J., and Drew, D. H. (1965). Root growth as a factor in the response of *Pisum sativum* L. to irrigation. *Nature (London)* **206**, 1063–1064.

Sau, A. (1970). Legumes and fertilizers as sources of grassland nitrogen in temperate regions of the USSR. *Proc. Int. Grassland Conf. Surfers Paradise, 11th, 1969* pp. 416–420.

Schöllhorn, R., and Burris. R. H. (1966). Studies of intermediates in nitrogen fixation. *Fed. Proc., Fed. Amer. Soc. Exp. Biol.* **24,** 710.

Shimshi, D., Schiffmann, J., Kost, Y., Bielorai, H., and Alper, Y. (1967). Effect of soil moisture regime on nodulation of inoculated peanuts. *Agron. J.* **59,** 397–400.

Singh, A., and Tripathi, N. C. (1972). Effect of moisture stress on soybean (*Glycine max* (L.) Merr.). *Indian J. Agr. Sci.* **42,** 582–585.

Sprent, J. I. (1969). Prolonged reduction of acetylene by detached soybean nodules. *Planta* **88,** 372–375.

Sprent, J. I. (1971a). Effects of water stress on nitrogen fixation in root nodules. *Plant Soil, Spec. Vol.* pp. 225–228.

Sprent, J. I. (1971b). The effects of water stress on nitrogen-fixing root nodules. I. Effects on the physiology of detached soybean nodules. *New Phytol.* **70,** 9–17.

Sprent, J. I. (1972a). The effects of water stress on nitrogen-fixing root nodules. II. Effects on the fine structure of detached soybean nodules. *New Phytol.* **71,** 443–450.

Sprent, J. I. (1972b). The effects of water stress on nitrogen-fixing root nodules. III. Effects of osmotically applied stress. *New Phytol.* **71,** 451–460.

Sprent, J. I. (1972c). The effects of water stress on nitrogen-fixing root nodules. IV. Effects on whole plants of *Vicia faba* and *Glycine max. New Phytol.* **71,** 603–611.

Sprent, J. I. (1973). Growth and nitrogen fixation in *Lupinus arboreus* as affected by shading and water supply. *New Phytol.* **72,** 1005–1022.

Sprent, J. I. (1975). Nitrogen fixation by legumes subjected to water and light stresses. *In* "Symbiotic Nitrogen Fixation in Plants" (P. S. Nutman ed.). Cambridge Univ. Press, London and New York 405–420.

Sprent, J. I., and Gallacher, A. (1975). Effects of water stress on respiratory pathways in nitrogen-fixing root nodules. Abstracts XII *Intern. Bot. Congr.* (*Leningrad*) Vol. II, p. 476.

Sprent, J. I., and Silvester, W. B. (1973). Nitrogen fixation by *Lupinus arboreus* grown in the open and under different aged stands of *Pinus radiata. New Phytol.* **72,** 991–1003.

Tjepkema, J. D., and Yocum, C. S. (1973). Respiration and oxygen transport in soybean nodules. *Planta* **115,** 59–72.

Tjepkema, J. D., and Yocum, C. S. (1974). Measurement of oxygen partial pressure within soybean nodules by oxygen microelectrodes. *Planta* **119,** 351–360.

Trinick, M. J. (1974). Symbiosis between *Rhizobium* and the non-legume, *Trema aspera. Nature* (*London*) **244,** 459–460.

Vincent, J. M. (1965). Environmental factors in the fixation of nitrogen by the legume. *In* "Soil Nitrogen" (W. V. Bartholomew and F. E. Clark, eds.), p. 384–435. Amer. Soc. Agron., Madison, Wisconsin.

Wilson, J. K. (1931). The shedding of nodules by beans. *J. Amer. Soc. Agron.* **23,** 670–674.

Wilson, J. K. (1942). The loss of nodules from legume roots and its significance. *J. Amer. Soc. Agron.* **34,** 460–471.

Wittenberg, J. B., Bergersen, F. J., Appleby, C. A., and Turner, G. L. (1974). Facilitated oxygen diffusion. The role of leghemoglobin in nitrogen fixation by bacteroids from soybean root nodules. *J. Biol. Chem.* **249,** 4057–4066.

CHAPTER 8

PLANT BREEDING FOR DROUGHT RESISTANCE

E. A. Hurd

RESEARCH BRANCH, CANADA DEPARTMENT OF AGRICULTURE,
SWIFT CURRENT, SASKATCHEWAN, CANADA

I. BREEDING AND TESTING METHODS

A. INTRODUCTION

1. Historical and New Emphasis

The importance of breeding for drought resistance in plants is emphasized by the fact that water is the main factor limiting growing crops in a large proportion of the world's land area (Burton, 1964; Kozlowski, 1968). According to Raheja (1966), 36% of the land area is classed as arid to semiarid, receiving only 5 to 30 in. of rainfall annually. Much of the other 64% undergoes temporary drought during the crop season.

Breeding for resistance in all crops is beyond the scope of this chapter because of the diversity of plant characters and methods used. This chapter will be confined to agricultural crops of arid and semiarid areas, with emphasis on cereal crops, especially wheat (*Triticum aestivum*). The fundamental principles of breeding for drought resistance in cereals apply to other self-pollinated crops. Some reference will be made to adaptation of the techniques discussed to cross-pollinated crops.

Breeding for drought resistance involves breeding for yield in environments dominated by water deficits. A large number of genes control yield by direct or indirect contributions. Frankel (1947) claimed that most increases in yield were achieved by overcoming limiting factors whose effects were simply inherited rather than by assembling productivity genes. Bell (1963) agreed. Frankel's statement is still valid, although the empirical methods of breeding are being reexamined in the light of increasing pressure for higher yield. The inability of breeders to grow large enough populations was nevertheless expressed as early as 1913 by Nilsson-Ehle (MacKay, 1953).

The Swedish and Minnesotan pedigree and mass-breeding methods of the early 1900's have been modified, but are still basic to methods presently used. Knowledge of gene action and heritability of quantitative characters, as well as useful breeding tools, have become available, but breeders have not commenced to use them (Grafius, 1963). The standard pedigree method is still suitable for transferring and combining simply inherited, qualitative characters, such as disease resistance (Hayes and Immer, 1942; Akermann and MacKay, 1948). The backcross method was introduced (Briggs, 1930) as an efficient means of maintenance type of breeding, so necessary in keeping ahead of the changing race, and strain, disease phenomenon.

The simply inherited genes, which control factors limiting full expression of yield potential have been called negative or minus genes, while those controlling quantitative characters have been referred to as plus genes (Hurd, 1971). Hence, breeding for yield involves accumulation of plus genes. Few breeders have exploited the full potential of their crosses because they used small plant populations and did not make yield comparisons until late in a breeding program.

El-Haddad (1974) suggested that even selection for yield in F_3 was expected to give response. He found almost continuous variation in segregating generations studied in F_4 of wheat crosses, and concluded that the inheritance was typically quantitative. Heritability estimates for total weight and weight of grain increased from F_3 to F_4 and El-Haddad believed this was at least partly due to decreased nonadditive effects. Fairly high heritability estimates within generations for yield were also reflected in parent-progeny relationships, especially between F_3 and F_4 means.

Yield tests in F_7 or later generations only test a few genetically fixed lines. Shebeski (1967) has shown the need for growing large populations. Lupton (1961) found that F_1 and F_2 trials were difficult to interpret, but were useful in discarding poor crosses. The F_3 and F_4 trials gave estimates of means and genetic variance of a cross and identified superior families. The variance provided a direct indication of the degree of transgressive segregation shown by the parents and indicated the likelihood of obtaining high-yielding segregates. Breeders should return to the identified high-yielding families and reselect in them because of their greater potential for having high-yielding combinations.

Diallel crosses identify good "combiners," but much time is required to make the many crosses and to grow out the generations. Thus, diallel crossing is too laborious to be popular.

Mass selection has limitations in a program where the objective is to combine many small-effect genes unless natural selection is strong. Several investigators have attempted to measure the effect of natural selection in cereals by growing a mixture of cultivars with different yielding capacity over a number of years without artificial selection (Harlan and Martini, 1938; Suneson, 1949; Sakai and Gotoh, 1955). The highest yielding cultivar was not the one that became dominant in these trials. The reason might be that root patterns differ; that one cultivar may have better competitive advantage than another. This competitive advantage for one cultivar over another is likely to be greater under moisture stress (Donald, 1963). By selecting in barley (*Hordeum vulgare*) crosses, after 29 years of selfing and only natural selection, Suneson (1949) obtained yields with most populations showing only slight increase in yield over controls. Donald (1963) and Suneson (1949) stated that some progress may be made by the mass method, but full potential of crosses will not be exploited.

Mutation breeding is a "long-shot" method resorted to when the required genes are not available in germ plasm. Hagberg and Akerberg (1962) reviewed mutations and polyploidy in breeding and pointed out difficulties in selecting superior types. Yield depends on a delicate balance of plant processes and induction of mutation upsets that balance, so if the desired character is found, it usually must be transferred to an otherwise suitable type to be useful.

Methods of breeding for quantitative characters (e.g., yield) were suggested by Grafius (1965) and others. There has been comparatively little breeding for drought resistance, probably because of lack of understanding of plant response to severe moisture stress (Burton, 1964). Breeders have increased yield, not because they consciously bred for it, but because some lines happened to yield more than others and they selected them (Donald, 1963). Suneson (1963) stated that "existing 'cookbooks' on breeding are obsolete."

Lack of support personnel has been a detriment to progress. Many plant breeders are supported by a technician and perhaps a summer student. Even with a highly mechanized and computer-oriented program as outlined later, a breeder with less than six man-years of support staff has little chance of combining an appreciable number of yield genes (Hurd, 1971). The nature of plant breeding for yield is such that a large support staff is necessary.

Hamilton (1959) made a strong case for improved methods of selecting for yield and suggested that selection be started in early generations using large populations. Hurd (1969, 1971) outlined a method that produced high-yielding cultivars. Its main features were few crosses, selection in early generations from large populations, growing yield trials in early generations, and reselecting from only few of the highest yielding lines.

2. Physiological Form and Function

Many investigators studied the association between yield and individual plant characters, i.e., number of tillers, kernels per spike, and size of kernel. As Donald (1963) pointed out, this is measuring different aspects of yield. Representative tissues are merely sinks and often are not limiting, especially under conditions of moisture stress (Kaul, 1974). Lupton (1961) stated that measuring such components of yield had little value and that the physiological factors underlying cultivarial differences should be studied.

The need for physiological studies was also emphasized by Eastin *et al.* (1969):

> Plant production processes must be better understood if maximum economic yields are to be realized and exploited. The competitive advantage of any biologic organism in the field, be it crop plant or pest, is dictated by its relative response to the prevailing environment. Environmental physiology research has scarcely touched on interdependencies amongst and control of the major physiologic processes dictating competitive advantage. Quantitating the environment and plant morphologic characters simultaneously with major physiologic process rates may provide much essential perspective concerning the order of these limiting factors. The subsequent exploitation of these yield-related factors will depend on their detailed characterization at cellular and molecular levels.

Jensen (1969) recommended two avenues for improvement in yield: (1) modifying breeding methods to increase predictability and odds of success, and (2) increasing our knowledge of form and function components of yield. In discussing manipulation of germ plasm in the future, Sprague (1969) advocated emphasis on physiological genetics, including aspects of physiology involving form or function which influenced economic worth or biological efficiency. Obviously, close cooperation is needed between physiologists and geneticists, so that there is a clearer understanding of crop response to environment. Researchers in the 1950's suggested that there was little opportunity of increasing yield through better net assimilation rate (Watson, 1952; Gregory et al., 1950). These views led to programs that tried to assemble populations from which to select cultivars that made best use of a limited environment rather than those that were resistant to it, i.e., heading in long days or filling during the rainy season. Corn (*Zea mays*), soybeans (*Glycine max*), and sorghum (*Sorghum bicolor*) were moved farther north in North America by this method of improvement; plants that grew with fewer heat units were selected.

This chapter will discuss a method of breeding for yield under moisture stress. It is not the only method available, but it is working with wheat in the Canadian Prairie (Hurd, et al., 1972a,b, 1973, Hurd and Spratt, 1975). Evans et al. (1972) increased yield by 24% using a similar method previously outlined by Shebeski (1972). The method depends on detailed study of the parents used in crosses. To this end, plant growth characteristics associated with drought resistance will be discussed. Methods of testing potential parents and some old and new screening techniques will be described.

3. Genetics of Yield

Yield increase depends on improving physiological processes in plants. Wallace et al. (1972) stated that genetic understanding and eventual control of physiological components of yield would make it possible to abandon trial and error methods of breeding for increase in yield. The identification of these components began about 20 years ago with beans (*Phaseolus vulgaris*) (Wallace et al. 1972). Such research is basic to adapting a scientific approach to breeding for yield in all crops and it requires the combined efforts of breeder-geneticists, plant physiologists, and biometricians.

Under favorable growing conditions, light interception, leaf-area index, and CO_2 exchange are of primary importance, while root weight and root pattern are less significant. Under drought, the reverse is true except that CO_2 may be important in either environment. Drought causes stomatal closure which limits CO_2 exchange. Variation in physiological components of yield within a species is genetically controlled, but because yield is affected by so many genes and is thus the product of many processes interacting with each other and with the particular environment, heritability often is not high. As Sprague (1966) emphasized, heritability estimates provide useful adjuncts to a selection program.

Gene action can be in the form of overdominance in which the F_1 exhibits a phenotype more extreme than the parents. In simple dominant action, the F_1 is like one or other parent. Anwar and Chowdry (1969) found that plant height, heading time, and grain yield were quantitatively inherited and both narrow-sense heritability (NSH) and broad-sense heritability (BSH) were computed, but NSH estimates were smaller in all four crosses studied. Wallace *et al.* (1972) discussed BSH which indicates the percentage of variation caused by genetic influence. High BSH indicates much genotypic variation. They discussed NSH also, which refers to predictability of progeny performance. Low NSH indicates that performance of selections will be ineffective in predicting progeny performance; thus, large populations in yield tests with extensive statistical procedures are required to identify superior genotypes. Sampson (1971a,b) studied additive (NSH) and nonadditive variance in oats (*Avena sativa*). He found that additive genetic variance was the most important component of phenotypic variance among progenies. Major differences occurred between F_1's and F_2's, but F_2's and F_3's agreed closely. The percentage of additive variance for F_2, F_3 analysis was high for height (91%), heading date (87%), and other simply inherited characters and lower (52%) for plot yield. Nonadditive variance for F_2, F_3 data gave plot yield estimates of 17%. These studies did not provide evidence of a shortcut to selection for yield. Syme (1970, 1971) found that high yield and harvest index were highly correlated. He also disproved Donald's (1968) suggestion that few leaves were advantageous to yield in wheat, at least in the cultivars studied. Smith (1966) found that many of the important improvements in wheat were bonuses obtained by transgressive segregation over and above the main objectives of making the cross. These often were of a quantitatively inherited character such as yield or quality. Such improvements are not recovered in a backcross program. Because of current food shortages, total yield of dry matter probably will be as important as grain yield. Wallace *et al.* (1972) recommended recording harvest indices as well as economic yield. One cultivar may have high total biological yield and another may have superior enzyme activity. By measuring grain yield only, the advantages of combining two cultivars may not be recognized. The contribution of specific physiological components can only be understood from running correlations between various characters and yield in large populations. Because of the work involved, Wallace *et al.* (1972) questioned whether expensive and sophisticated procedures would provide better results than selecting in the field. Real progress is made by identification and selection of superior combinations of parents. Wallace *et al.* (1972) stated that a large body of data showed genetic variability within all species for many different physiological components of yield. An example of a character that may have value in selection is nitrogen reductase activity. Hageman *et al.* (1967) stated that NR is genetically controlled and that corn hybrids can be developed with known NR. But NR is probably much more important in a favorable environment than under drought stress.

Grafius *et al.* (1952) explained that nonadditive genes gave high yield in F_1 which disappeared in subsequent generations. Selection for yield in F_3 and beyond is for additive genes that can be fixed. They recommended increasing F_3 to provide seed for replicated F_4 yield trials in which heritability could be measured.

Williams *et al.* (1969) studied inheritance of drought tolerance in sweet corn. From variance-covariance regression, they concluded that inheritance followed through patterns of partial to nearly complete dominance. Lines with high general combining ability produced a large number of drought-tolerant hybrids. Drought tolerance was controlled by no less than three gene pairs. Simple screening seemed to work in corn. Highly heritable genes, or genes not greatly affected by the environment, are readily concentrated in parents, but yield genes have low heritability and are much more difficult to manipulate. The nearer the environment of selection is to typical field conditions, the better are chances for success. Allard (1960) stated that linkage tends to combine favorable combinations of genes automatically, and this can be increased by backcrossing. Whenever many genes interact, or when heritability is low, normal distribution of population occurs. When selection for yield is taken in a large F_3 population (1000 lines) and repeated in an equally large F_5, the range of yield narrows and the average of the selections increases each cycle of selection (Hurd, 1971).

Sears (1954), Kuspira and Unrau (1957, 1959), and Person (1956) discussed ways of using aneuploids and substitution lines in wheat breeding. Substitution in breeding is useful following location of genes for desired characters on specific chromosomes. While aneuploidy is possible in breeding, it has been used infrequently because of the work and time required to locate the gene(s) and transfer it (them). Usually the same result can be achieved by backcrossing and in much less time. Larson and colleagues isolated genes for resistance to wheat stem sawfly (*Cephus cinctus*) (Larson and MacDonald, 1959a,b, 1964, 1966); for resistance to common root rot, a complex disease caused by *Cochliobolus sativum* and *Fusarium* sp. (Larson and Atkinson, 1970b) and resistance to wheat-streak mosaic (Larson and Atkinson, 1970a). Monosomic analyses are very useful in identifying genes conditioning disease and insect resistance and other qualitatively inherited characters, but are of less value in genetics of quantitative characters and in direct breeding for yield. If several major genes condition a desired character, aneuploid studies could be very useful in identifying location of genes and in transferring them to a desirable cultivar. Many of the more simply inherited characters, such as rust resistance in wheat, have been quickly transferred by backcrossing. L. P. Reitz (personal communication, 1973) believes that breeding for most types of resistance leads to reduced yields because the size of population only allows combining of a few genes. Breeders are thus fortunate if any resistant line equals the best parent in yield. Baker (1971), studying effects of rust on yield of wheat, suggested that genotype-environment interaction in quantitative traits may be due to simply inherited traits. Baker (1968) demonstrated that some simply inherited characters can have a major effect on yield. As previously stated, they prevent plants from expressing their yield potential. Thus, in breeding for high yield under drought stress, the breeder should not ignore major limiting factors.

4. Statistical Applications

Quantitative genetics and the closely allied statistical and population genetics have much to offer in breeding for yield in dry areas. Where heritability is low and selection is for genotype-environment interaction, the breeder has several objectives. The primary objective is to maximize chances of identifying substantially superior yielding lines. Once the breeder establishes a population with the desired genetic variability, he is in a game of chance in which he wishes to eliminate in each generation the vast bulk of inferior combinations of additive genes. Population improvement may take many forms, but selection is primarily for additive gene effects.

Statistical measurement of genetic variation is a useful tool and statistical analysis, an integral part of plot design (Fisher, 1951; Goulden, 1952; LeClerg, 1966). Replication, randomization, and local control (blocking) are used to minimize uncontrollable variation and to increase genetic variation. Systematically distributed check plots often are used. Briggs and Shebeski (1968) concluded that frequent controls were essential for efficient selection for yield in hybrid nurseries. Baker and McKenzie (1967), however, questioned the value of repeated controls on both theoretical and experimental grounds. Knott (1972) compared the use of repeated checks with the moving mean system of adjusting yields and found no differences, although both methods increased efficiency. Townley-Smith and Hurd (1973) reported that the moving mean of adjacent hybrid plots gave superior control of experimental error to the use of repeated controls even when used as frequently as every third plot. One moving average method (Rickey, 1924) used frequent checks rather than the hybrids in test. The moving mean of hybrid plots used to adjust yields of large trials is helpful in sorting out superior or inferior genetic combinations from large populations whether grown in single or replicated plots. More detail on plot design is presented in Section I,B,2.

Recognizing the need to improve selection techniques for complex characters such as yield, Grafius (1965) proposed that geometry might be used as a vehicle to understand biology. Yield predictions may be useful information in deciding how to handle succeeding generations. Pesek and Baker (1971) found that observed response to selection for yield in five crosses of common wheat agreed with the response predicted by multiplying estimates of heritability by the selection differential.

Elliot (1958) suggested that increase in outcrossing in F_2 and F_3 would overcome the normal limitation in the number of combinations of plus genes for yield obtainable in classical breeding programs. T. F. Townley-Smith (personal communication) used ethrel to induce pollen sterility in every other row of "cocktails" of F_2's to increase combinations of genes from single and double crosses. Fertility was high in treated plants and selfing was very low. Other investigators recommended intercrossing selected plants in F_2, but this is time-consuming and leads to small populations. Selection for yield in F_2 is inefficient (Bell, 1963; Knott, 1972).

As mentioned, breeding programs usually are limited by lack of manpower. Use of statistics and computers can increase efficiency of tests and greatly reduce work load per plot. At Swift Current, Saskatchewan, preparation of field notebooks, labels, and cards before the field season, saves technician time during the growing season and also reduces error. Together with other methods of accelerating operations, it is possible for a two-breeder wheat program to carry 30,000 plots in yield trials at four locations with seven man-years of support staff. Additional details on handling a large number of plots at seeding and harvest are given in Section III,A.

B. A METHOD OF BREEDING FOR YIELD IN SELF-POLLINATED CROPS

1. Selection of Parents and Handling of F_1 and F_2 Generations

The importance of careful selection of parents for crossing cannot be overemphasized. Many breeders seem to make crosses almost indiscriminately. All characters desired in the new cultivar must be possessed by the parents used. Occasionally, transgressive segregation or a combination of recessives carried by both parents has given unexpected results. To hope that yield genes of one parent will enhance those of another by chance is naive. Plant breeding is a game of chance in which every step possible must be taken to assure high probability of, first, having the desired combination of genes present in a population and, second, of being able to identify them. Ways of accomplishing the first objective will be discussed.

As pointed out in Section I,A,2, we know little about what constitutes yield especially under drought, but much research is now underway to help understand the physicochemical processes of plants. The functioning of plant systems in an environment determines yield. The efficiency of the photosynthetic mechanism and of the moisture use of one cultivar in a particular environment may mean little to their efficiency in another environment. And, yet, some breeders even conduct yield tests under irrigation when their cultivars are intended for growing under water-stress conditions. Such programs may result in adding simply inherited characters that are enhanced under improved growing conditions, but increasing the combination of yield genes for a dry regime, per se, is unlikely. Hurd (1969) demonstrated that the highest yielding cultivar of spring wheat under adequate moisture was the lowest yielding of many under drought stress. This was in accord with findings of Burton (1964). Reitz (1975) implied that to select for wide adaptability is to select for mediocrity or even low yield. To cite exceptions to these reasonable approaches is not to disprove them. Like mutation breeding, the "long shot" occasionally succeeds, but breeders should not strive for the near impossible unless there is no more logical or easier way of achieving their goal.

Plant adaptibility is a fine balance of many plant processes. To make a cross between two widely divergent cultivars adapted to different environments upsets the balance and makes the chances remote of combining genes that will meet the delicate balance in a specific environment. "Exotic" germ plasm can be introduced with a reasonable chance of success by using a double or three-way cross where the other parents used are all well adapted to the environment (e.g., drought resistant). Even then, the desired character from the introduced cultivar will only be found in a small number of well-adapted lines. This necessitates growing large populations.

Some criteria for identifying characteristics of plant processes that are adapted to moisture stress environment follow in Section II. As much as possible should be known about parents before they are considered for crossing. Then cultivars with all of the essential simply inherited characters such as disease resistance in at least one of the parents should be chosen for crossing. Also, cultivars should be chosen that to the best of available knowledge are resistant to drought for different reasons, e.g., tolerance and avoidance. Whereas one cultivar may have an extensive root system another may be resistant primarily because it exhibits desiccation tolerance. In this example, if neither of the parents had resistance to an essential disease, a third parent could be introduced by crossing it with either of the others and making a three-way cross. The disease resistance, if in a poorly adapted cultivar, should be put into one of the parents by backcrossing before a cross is made.

The use of parents with characters such as extensive root systems for which there is no screening technique can be successful if selection is made for yield in the moisture stress environment. This has been demonstrated by Raper and Barber (1970a,b) in soybeans and Hurd *et al.* (1972a,b, 1973) in durum wheat (*Triticum turgidum*).

The presence of awns in wheat and barley appears to be associated with high yield under drought stress (Grundbacher, 1963). Awns increased yield of wheat by 7% on the average and were especially beneficial when the flag leaf was inactivated (Suneson and Ramage, 1962). Saghir *et al.* (1968) found that clipping awns reduced yield by 20.8% and seed size by 13.4%. According to Holmes (1974), the best assumption is that awns play a multiple role in determining growth of grain and size of seed. When drought causes leaves to senesce, the plant cannot produce photosynthate rapidly enough to fill the kernels before they ripen, so they shrivel. Awns increase the photosynthetic surface of the spike and up to 90% of the photosynthate goes to the kernel. Thus, most of the increase in yield resulting from awns is in increased seed size.

Perhaps more important than the direct contribution of awns to yield is their mechanical value in western Canadian harvest systems. Crops are windrowed or swathed to accelerate drying and harvesting. If crops are thin because of drought, some spikes will fall through the standing stubble. Often those on the ground are not picked up by the combine. Awns on the spikes cause them to stick together. They prevent the formation of a dense swath and facilitate drying. If awns or other specific characters have advantage, they should be in one or more of the parents used in crosses.

Techniques for handling crosses will vary with available facilities. If growth chambers are available, crossed seed can be produced with a minimum of effort and greater assurance of obtaining the required number of seeds. The F_1 can be produced in growth chambers, greenhouses, or out-of-main season in field plots (winter nurseries). The F_2 should be very large and subject to typical stresses of the area. Where disease resistance is important, the nursery should be infected naturally or artificially if necessary. Seeds should be spaced so that individual plants can be eliminated from time to time during the growing season as undesirable features become evident. Since selection for yield in F_2 has little or no relationship to F_3 performance (McGinnis and Shebeski, 1968; Bell, 1963), competition is not important. Many of the negative yield genes are eliminated in the F_2

generation. Thus the size of the population will depend on the number
of negative (simply inherited) genes and their mode of inheritance. If
double recessive genes are required, only one in sixteen will be left after
roguing for plants not possessing that character. From the following, Section I,B,2, it is shown that a minimum of 1000 plants should be retained
for yield testing. On the Canadian Prairies, common root rot is a serious
limiting factor to yield under moisture stress. Some cultivars are more resistant than others to root rot, but all are far from immune. For this reason,
all plants are pulled and the subcrown internode examined (Sallans and
Tinline, 1965). Often two-thirds of a population have root rot and are
discarded at harvest. Because of the heavy discards for several diseases,
straw strength and length, shattering, etc., populations of 20,000 to 50,000
are required to provide the 1000 acceptable plants for advancement. A
similar discard pattern will be found in most other crops.

2. Number of Crosses and Handling Procedure for Segregating Generations

The number of crosses is likely to be small if the breeder follows
the requirements recommended in Section I,B,1. A more important reason
for limiting the number of crosses is to yield test large populations in F_3
or F_4 generations and again in F_5 or F_6. Unless there is an effective way
of evaluating F_2's, and that means separating additive and nonadditive
effects, then there is little reason for making many crosses. Shebeski
(1967) and Knott (1972) believed it was impractical to try to select for
yield in F_2.

While F_3 may be the preferred generation to commence yield testing,
F_2 plants may not produce enough seeds in a semiarid climate. If the seed
of selected F_2's is increased out-of-season, the first yield test will be in
F_4 and the second in F_6. By out-of-season increase, F_4 yield tests can be
grown at two locations using at least two replicates of multirow plots. Frey
(1965) and many others have stated that hill plots are useful in early generation testing for small grains, but few breeders use them. Baker and
Leisle (1970) found increased variability with hills as compared to rod-row
plots. It is important to know if competition effects exist and if such competition in hills alters relative performance of cultivars or lines being tested.
Even if differential competition is insignificant, it is unlikely that hills will
be used until mechanization of planting and harvesting is perfected. Using
modern plot combines, a multiple row plot can be seeded or cut and
threshed more quickly than a hill.

The moving mean analysis is suited to improving yield comparisons
in large populations at this stage in the program. To minimize interplot
competition and to facilitate harvesting, an alternative crop should be

grown between plots. For this purpose, a strong competitor for moisture is preferred. It should be a crop that does not head or grow tall enough to interfere with harvesting. If such a crop is not available, the between-plot rows can be mowed before seed set. Winter wheat that remains vegetative is suitable for use with spring cereals. It has little tendency to cause intercultivarial interaction (A. L. D. Martin, A. B. Campbell, K. W. Buchannon, and J. N. Welsh, personal communication, 1960). The use of two locations increases chances of the test year giving good results. Test failure due to hail or extreme drought and the chances of having no moisture stress are reduced.

Frey (1964) found that selecting in small populations under stress conditions (e.g., eroded infertile gravelly outcrop on a knoll) was less successful than selection in fertile, high moisture-retaining soil with four crosses of oats. Because of their own successful selection techniques under drought conditions, Hurd (1968) and Burton (1964) disagreed with Frey. Hurd *et al.* (1972a, 1973) released two wheats selected in large populations under moisture-stress conditions that increased yield in the dry prairie area of western Canada. These cultivars, Wascana and Wakooma, had 15% higher yields than previously grown cultivars. They also are higher yielding in the wetter areas of Manitoba. Macoun, an even newer cultivar selected by the same procedure (Townley-Smith *et al.,* 1975), is equally high yielding in both dry and moist areas. Macoun comes from a cross made in Winnipeg where attempts to select a superior line failed. Success in the dry area and not in the moist one could be due to chance or to a different selection procedure. It does not mean that selecting under stress is more likely to be successful for moist areas but that selection under stress is not a handicap. Selection in large populations grown under drought stress is believed useful in selecting for extensive rooting (Hurd *et al.,* 1972a). Probably similar criteria for selection for other characters conditioning drought resistance or tolerance may facilitate success.

Briggs and Shebeski (1968) and Shebeski (1967) tested the efficiency of using frequent checks in yield trials. Shebeski reported a rank correlation of 0.847 ($P < 0.01$) between F_3 yields expressed as a percentage of an adjacent control and the means of their respective F_5's. DePauw and Shebeski (1973) evaluated an early generation yield testing procedure and concluded that by growing a large sample of the progeny of each F_2 genotype adjacent to a control it was possible to select F_3 lines for heritable quantitative differences. The value of control plots was discussed in Section I,A,4 where it was reported that the moving mean analysis gave superior control of experimental error.

Early generation yield testing to identify lines with a large number of plus genes for yield is recommended over the more common practice

of using visual selection in head rows. While visual selection requires much
less work and less land, it is ineffective (Briggs and Shebeski, 1968; Townley-Smith *et al.*, 1973). In general, visual selection for yield has varied
from modestly successful to useless. Simply inherited characters are amenable to visual selection, whereas yield is not.

Why use large populations? Shebeski (1967) demonstrated that where
parents differed by 25 important independent genes, only one plant containing all 25 was present in 1330 F_2 plants. If nearly all genes contribute
to yield directly or indirectly, it is conservative to consider that about 20
genes may have an important effect. Baker *et al.* (1971) stated that plant
breeders strive to increase heritability by testing in several seasons or at
several locations, but if the interaction of genotype and year is negligible,
the same increase in heritability can be realized by more extensive testing
in a single year. Based on these considerations, 1000 F_3 or F_4 lines derived
from F_2 plants and grown at two locations comprise the objective in the
South Saskatchewan Wheat Program (Hurd, 1971). Substantial increases
in yield have been made from populations of about 1500 lines in F_4 and
F_6 yield tests (Hurd, 1969, 1975). Shebeski's (1967) tables show that
yield tests of 200 lines in F_5 can only hope to combine seven or eight
desirable genes from two parents.

Handling of 30,000 plots in yield trials requires a systems approach
with certain characteristics. Automation, where possible, and use of computers reduce labor requirements. Two men seeding or three men combining at the rate of 15 sec per plot means that 1000 to 1500 plots can be
completed per day with one crew. Plots are desiccated to speed up harvest
and make it possible to cut and thresh in the order in which they were
sown. The grain from plots is placed in order on trays which are placed
on portable racks and wheeled into a dryer at the end of the day. Drying
all plot samples for 15 hr, or more if necessary, reduces differences in
moisture content before weights are recorded. Considerable time is saved
by not handling each plot sample except when filling and weighing and
by having samples in the same order as on the record sheet.

Since further selections are taken from F_4 and F_6 yield trials, it is
not necessary to keep grain samples particularly clean. The grain sample
of all those lines being advanced is used for quality tests, but not replanted.

Many breeders start yield testing in later generations with fewer lines.
Such programs at best can eliminate negative genes—genes that prevent
the plant from expressing its true yield potential—but do not allow for
accumulation of an appreciable number of plus genes. The best lines may
be as well balanced and productive as the best parent. Figure 1 illustrates
a probable distribution of the yield of 10, 100, and 1000 lines from a
cross. Even if yield tests are not as precise as they would be in later genera-

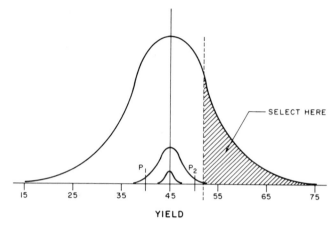

YIELD

FIG. 1. A normal distribution for yield as determined by additive genes for populations of 10, 100, and 1000 lines from a cross. P_1 and P_2 represent the yield of the two parents.

tions when there is more seed and genetic homogeneity, they will ensure that most lines selected will be in the highest yielding third of the population. Figure 1 shows what can happen in different sizes of populations of a cross. If 100 crosses were made and only ten lines from each cross were yield tested in advanced generations, it is unlikely that any would exceed the highest yielding parent by a significant amount. If 1000 lines are yield tested and reselection confined to the high-yielding lines (for additive genes), chances of obtaining a high-yielding line may be better in the one cross than in the 100. Visual selection for characters not associated with yield tends to keep pieces of chromosome and thus limits selection for yield.

The polyploid genetic makeup of wheat permits a type of chromosome engineering not possible in most crops. Because of several sets of chromosomes, wheat can tolerate whole or partial chromosome substitutions from even related, but different, species. It may also tolerate induced mutation better than diploid crops, so it is possibly more amenable to this method of creating variability. Genetic stocks are maintained in many countries and in wheat, hundreds of thousands of separate items provide a large germplasm bank. The preservation of germ plasm is important to the future even though man is a long way from exploiting the potential of presently available variations.

3. Advanced Testing and Multiplication of Seed

Heads should be taken from the lines in F_6 yield trials, increased out-of-season, and the seed produced used for replicated trials at three or four locations in F_8. Since advanced tests are also harvested with the plot combine and scrupulous clean out between plots is not desirable, increase plots of each line are grown separately. These can be rogued and the seed kept pure. Two-hundred heads are taken from the increase plot to form the breeder's lines for all cultivars placed in advanced tests. These lines are increased simultaneously with regional testing of the cultivars. When the superior strain from a cross has been identified and released for commercial production, part of the seed of the 200 lines is bulked and distributed to industry. Some seed of each line is retained to provide breeders with seed in subsequent years.

Where demand is great, the bulked seed may be grown in an out-of-season increase. Great Plains cultivars are frequently increased in southern California in winter. The North Dakota Agricultural Research Station developed procedures which can multiply a pound of seed to 150,000 bushels in 15 mo. Ebeltoft (1969) considered such a system necessary because of increased demands for food.

C. MODIFICATION OF BREEDING PROCEDURES WITH CROSS-POLLINATED CROPS

Because of differences in compatibility and diversity of breeding behavior, as well as deterioration of subsequent generations, breeding of cross-pollinated crops cannot be described simply (Allard, 1960). The general principles of combining genetically controlled characters are similar to those described for self-pollinated crops. The most commonly used methods are mass selection, backcross and hybridization of inbreds to form hybrid cultivars, recurrent selection, and formation of synthetic cultivars. In the mass method, selection is only partially effective because there is no control of pollination. The backcross method can be used with about equal success. The use of hybrid vigor, although used for both self- and cross-pollinated crops, is better suited to the latter in production of hybrid cultivars. However, Breese and Hayward (1972) concluded that breeding methods are largely conditioned by the stage of evolution and development of a particular species as a crop plant. Continued selection and isolation by man are likely to influence the amount of genetic variability, mating system, and type of gene action. All these are interdependent and together determine applicability of a particular breeding method. In herbage plants, the most common breeding method is by population improvement through synthetic varieties. The stated aims are to increase gene frequency with favorable expression in particular characters, while avoiding inbreeding depression. Forage crop breeders recognized a high degree of additive genetic variation for many physiological characters. Possibilities for exploiting such variation through phenotypic selection in improvement of synthetic varieties depend on ability (1) to translate nutritional requirements and agronomic performance into identifiable plant characteristics and (2) to measure plastic responses of a genotype to the environment.

Though methods of breeding vary considerably between cross- and self-pollinated crops and with cross-pollinated ones, some of the same general principles described in Section I,B,1 and I,B,2 are applicable in breeding for drought resistance in cross-pollinated crops.

Careful study of potential parents to enable precision breeding rather than haphazard mating will increase likelihood of success. The genetics of drought resistance in cross-pollinated crops is conditioned by quantitative inheritance. Once the drought resistant characters have been identified through physicochemical research, populations should be created that combine more than one such character. These should be grown under drought-stress conditions and selected for superior performance. The use of large populations in yield tests and at more than one location will assist in identification of superior lines. Lawrence and Townley-Smith (1975) found in breeding grasses that analysis of yield data by using the moving mean gave the same kind of control of soil variability that was found for wheat (Townley-Smith and Hurd, 1973).

II. PARAMETERS OF DROUGHT RESISTANCE IN A BREEDING PROGRAM

Physiological genetic studies may contribute greatly to breeding for drought resistance. But, before genetic aspects of drought resistance can be studied, an understanding is needed of physiological processes within the plant and their responses to environment. Parker (1968, 1972) discussed drought resistance in general terms and, specifically, mechanisms of protoplasmic resistance. In this chapter, discussion will be limited to some specific studies that may have application in breeding programs. Only differences between cultivars are useful to the plant breeder. Screening techniques developed from the concepts discussed in this section will be itemized in Section III,A.

A. DESICCATION TOLERANCE AND PHOTOSYNTHESIS

Kaul (1967, 1969, 1974) and Kaul and Crowle (1971, 1974) studied drought resistance in cereal crops. Kaul (1969) suggested that if one cultivar maintained lower stress levels than another in the same environment, both seed setting and plasmatic efficiency should be superior. His work did not show negative correlation between water status and productivity, although he reported that many other studies had. Kaul used a variety of cultivars that were thought to differ in many genetically controlled processes. It is difficult to distinguish between plants with plasmatic desiccation tolerance (ability to survive and/or grow under reduced water content), drought avoidance (ability to maintain a favorable internal water balance under stress), or those plants being economical of water use. Kaul's (1967) survey of water suction forces revealed that prairie wheat cultivars made their growth at relatively high suction tensions of 14 to 25 bars compared to German cultivars of up to 5 bars suction tension. Wheats selected for production under stress have narrow leaves and low shoot-root ratios. He speculated that wheats with this "cautious" growth habit have high suction tension and low yield.

Kaul and Crowle (1971) found that plant-water potentials dropped to −35 bars under severe drought. They then (1974) followed the water potential, stomatal opening, and net photosynthetic potential through the life cycle of six wheat cultivars. Water potentials declined in a similar pattern, but photosynthetic potential was specific for the cultivars, which indicates differences in photosynthetic desiccation tolerance. Stomatal apertures were specific for cultivars when grown under stress. Kaul (1974) found that, under more severe moisture stress, yield performance of five of six wheat cultivars was predictable from the photosynthetic potential of the flag leaves. Pelissier, a very drought-resistant spring durum wheat, was uniformly underestimated. Kaul concluded that Pelissier depended on longevity of its heavily awned spike for additional grain filling. Pelissier also has a very extensive root system (Hurd, 1964) and the cut flag leaf has great resistance to desiccation (Dedio, 1975). Kaul also found Pitic 62 to rank above the other four hexaploid wheat cultivars tested. Kaul and Crowle (1974) suggested that single photosynthetic potential measurements on slowly stressed plants could be used to select highest yielding cultivars. Lawes and Treharne (1971) compared the rate of photosynthesis of seedling leaves and flag leaves with yield and total dry matter. No correlations were obtained, although cultivars differed. They suggested that breeders might be able to transfer increased photosynthetic potential to an elite gene pool where selection may allow fuller expression in the new genetic background.

Dedio (1975) measured water saturation deficit (WSD) of wheat cultivars under moderate stress. Pitic 62 had the lowest WSD among the cultivars tested, but Durum cultivars possessed higher WSD than hexaploid wheats. Pitic 62 always had the highest water content (WC), suggesting that WC may be a more useful measurement than WSD. At 3 wk of age, excised leaves of Pelissier were much slower to dry than leaves of any other cultivar. Pitic 62 dried more slowly than others at later stages of growth and hybrid F_3's of Pitic 62/ACEF 125 were better than either parent at retaining moisture. Oleinikova (1969) found that retention of water was associated with protoplasmic permeability. Levitt (1956) found close correlation of hardiness of plants to frost, drought, and osmotic water removal. He found dehydration hardiness associated with small cell size, high concentration of cell sap, increased protoplasmic permeability, and high viscosity of cytoplasm. Parker (1972) concluded that the "cause of protoplasmic resistance of plants to desiccation continues to elude us even today." Gates (1964) proposed that drought resistance depended on the degree of resistance of protoplasmic proteins to hydrolytic activity of protein-splitting enzymes. He reported that enzymatic action changed during drought when enzymes promoted hydrolysis. Stoker (1961) stated that the anatomical, morphological, and physiological constitution of a resistant cultivar tended to maintain high photosynthesis under stress by restricting transpirational water loss. Levitt (1972) emphasized the importance of measuring both avoidance and tolerance to give overall resistance. The avoider retains high water potential, while the tolerant cultivar is less affected by low water potential.

Wallace *et al.* (1972) claimed that almost all biochemical and physiological processes in plants are relevant physiological components of yield. Thus, all genes that affect photosynthesis, even by diverting it from one metabolic pathway to another, influence yield. They cited research showing that narrow-sense heritability varies from 0% to as high as 80% and may increase with advancing generations. Broad-sense heritability was more stable at about 60%; thus it was under genetic control and provided opportunity to select for improvement. Vegetative growth and yield are not always positively correlated. Under moisture stress, increased vegetative growth depressed grain yield (Fischer and Kohn, 1966). High soil nitrogen enhanced this trend. Excessive vegetative growth wastes a limited moisture supply. Donald (1963) advocated single culm plants to provide control of vegetative growth. In some plants transpirational area is reduced through leaf senescence and shortening of stems (Asana *et al.*, 1958). Wardlaw (1967, 1969, 1971) found that several days of wilting did not affect grain yield of Gabo wheat, but assimilates were diverted from lower parts of the plant to the grain. Under slight wilting, lower leaves drooped one-third below normal, whereas in severe wilting, they hung lower and were slightly curled. Photosynthesis in corn was reduced by stress and translocation within leaves was reduced, although only slightly less, in conducting tissue. Temporary drought during 7 days after anthesis reduced final grain number and weight, although reduction in seed set was associated with increase in cell division in the endosperm. Water deficit resulted in a reduction in dry matter storage in stems, a temporary stop in tillering, and inhibition of root growth.

Frank *et al.* (1973) found that as the flag leaf of wheat matured, stomata closed at progressively lower leaf-water potentials. Recovery of photosynthesis was related to recovery of stomatal opening at tillering and heading stages, but the recovery was never quite to the prestress levels. They recommended that plant breeders look for plants whose flag leaf stomata are similar in reaction to stomata on the fifth leaf, which close at leaf water potentials that do not appear to reduce yield. According to Kaul (1974), wheat in semiarid areas is usually subject to a slowly decreasing water potential and thus has time to adapt and harden. As a result, wilting is rare even under severe stress. Drought damages chloroplasts and lowers output of the photosynthetic apparatus (Boyer and Bower, 1970; Wardlaw, 1971).

B. Stomatal Response and Transpiration

Control of stomatal aperture is important in drought avoidance because some species and cultivars close stomata early during developing drought and others do not. Examples of variation in stomatal behavior were cited by Kozlowski (1972). He discussed advantages of one potato (*Solanum tuberosum*) over another. Excessive wilting of tomato (*Lycopersicon esculentum*) mutants over normal plants was found to be due to behavior of stomata (Tal, 1966). Severe drought may cause permanent damage to plant processes and the relationship to stomatal behavior may be direct or indirect. After severe stress, stomata may open slowly or not at all (Iljin, 1957). In some species, the amount of recovery is related to the length of the drought. Kaul (1974) demonstrated that stomata of Pitic 62 leaves were more sensitive to drought, and Dedio (1975) showed that Pitic 62 also maintained a higher water potential than other wheat cultivars tested. This agreed with findings of Oppenheimer (1961) and Salim *et al.* (1969). The maintenance of a higher water potential may have been due to the more rapid closing of the stomata during development of drought. Since most transpiration occurs through open stomata, measuring stomatal apertures is important. Porometry is the usual means of measuring stomatal aperture (Slatyer, 1967, 1969; Barrs, 1968; Sullivan, 1971; Sullivan and Eastin, 1975; Kramer, 1969).

Ways of reducing transpiration, other than by stomatal manipulation, have been found. Lowering temperature of leaves by increased hairiness or albedo was suggested to plant breeders (Eslick and Hackett, 1975). Isogenic lines of barley cultivars have been developed with light-green, golden, and dark-green surfaces. Light-green and golden leaves reflect more light than dark-green leaves and thus remain cooler. Reduced leaf area and thus greater root/shoot ratio is also associated with lower transpiration

(Kramer, 1969). Leaf shape and orientation are related to transpiration and are amenable to genetic manipulation (Slatyer, 1967). Cuticular thickness or waxiness of leaf surfaces are also genetically controlled characters that appeared to affect transpiration (Kramer, 1969). For any such character, significant effects of intercultivarial differences should be identified before any of these characters are used as selection criteria.

Parker (1968) stated that some grasses of the Mediterranean region reduced transpiration as much as 46 to 63% by rolling their leaves. He found that leaf rolling did not occur in many plants until the water content of the leaves was below the lethal level. However, some durum cultivars grown in Saskatchewan roll their leaves, but do not lose their ability to photosynthesize. The most drought-resistant cultivars seemed to roll their leaves to protect the plant from severe stress (E. A. Hurd, unpublished observation). Such cultivars are considered drought resistant because they continue to grow and fill the kernels in the spike. These same cultivars have more extensive roots and only roll their leaves in certain years (under severe stress at a critical stage of growth). Poorer rooting cultivars showed no sign of rolled leaves. Leaf rolling and root pattern may be independent characters or plants with extensive root systems may suffer less in a drought because of a more effective moisture replacement mechanism than plants with less root. Levitt (1972) stated that plants with roots of low osmotic potential may continue to remove water from soil that is too dry for absorption of water by other plants with roots of high osmotic potential. Extensive rooting systems and lower water potential have the same effect on water uptake and plants under stress. Kramer (1969) stated that more plants are injured or killed by excessive transpiration under conditions of low water uptake than by any other factor.

C. Rooting Patterns

Root development and capacity of plants to absorb water are closely related. Generally, as width, depth, and branching of root systems increase, plant-water stress decreases (Singh, 1952; Gliemeroth, 1952; Donald, 1963; Bertrand, 1965; Weatherley, 1965; Wadleigh et al., 1965; Watson, 1968). Passouri (1972) found the reverse in water-culture experiments. According to Levitt (1972), deep-rooted plants show greater drought avoidance than shallow rooted ones, if ground water is available; but they may show lower avoidance, if it is not present.

Close relations have been found between growing root systems and grain production under drought conditions (Belzakov, 1968; Danilchuk et al., 1971). The most important soil depth for root development of wheat is below 60 cm (Weaver, 1926; Russell, 1957; Hurd and Spratt, 1973).

Most wheat roots are in the 0 to 60 cm layer. In semiarid regions, the available soil moisture of this layer is depleted by grain-filling time, but additional moisture usually is available below 60 cm even after harvest (Hurd, 1975). Conflicting data have been reported on root growth relative to moisture stress. Those that report more root growth occurring in dry than in wet soil were using about field capacity for the "wet soil." In such moist soil, poor aeration may have retarded root growth, whereas the "dry soil" was not dry enough to seriously limit root growth (Baldy, 1973; Pearson, 1965, 1974). With wheat, severe drought reduces root growth to a small percentage of that found under ideal moisture levels (Simmons and Sallans, 1933; Hurd, 1964, 1968). Cultivars that have the greatest root mass under drought are important in breeding for drought resistance, especially if that mass extends below 60 cm. The proportion of roots in the deeper layers is considerably less for grasses than for cereals (Goedewaagen and Schuurman, 1957).

In studies of semi-dwarf and taller winter wheats, Lupton *et al.* (1974) found that the relative growth rates of aerial parts followed a sigmoid curve, but those of the root showed little change between germination and anthesis. They found little evidence of cultivarial difference in root growth, though there was some indication that at depth the roots of semi-dwarf cultivars were more extensive and absorbed more phosphate than those of the taller cultivars.

Breeding for root patterns associated with drought resistance has been carried out for several crop plants. One soybean cultivar had twice as large a root system as another cultivar (Raper and Barber, 1970a,b) and one barley cultivar had deeper roots than another one (Engledow and Wardlaw, 1923). Garkavy *et al.* (1970), working with crosses of barley cultivars, selected lines that exceeded the parent in depth of rooting and amount of branching. These selections were higher yielding. Kirichenko (1963) reported that Russian researchers found a direct relationship between the length of primary rootlets and yield, but they later found exceptions because secondary root growth did not always correlate with primary growth. Increases in yield of 25% were found from F_3 selections based on root characters. He cautioned against visual selection based on aboveground parts only. Kirichenko reported success in selecting from root patterns in soft and hard wheat, barley, corn, and sunflower (*Helianthus annuus*).

Several investigators reported an association between early root growth and amount of roots at maturity (Pinthus and Eshell, 1962; Hurd, 1964, 1975). Pohjakallio (1945) crossed two oat cultivars and combined drought characters from each parent, one being deep roots. Roma (1962) studied heritability of root systems in 14 cultivars of wheat and found that parent systems were passed on in crosses. Danilchuk (1972), Danilchuk *et al.* (1971), and Danilchuk and Yatsenko (1973) studied the physiological and biochemical development of root and shoot growth of winter wheat cultivars in relation to productivity. Some cultivars (e.g., Odessbaya) grew rapidly in the spring and built up reserves that were useful when stress curtailed nutrient movement into the plant. Forest-Steppe cultivars (Besoskaya) grew slowly early in the season, but later, surpassed growth of the others and produced more grain. Kaul (1974) demonstrated similar patterns in Canada. The Thatcher-type wheat grows rapidly in spring and produces deep vigorous roots (Hurd and Spratt, 1973). Early growth provides reserve photosynthate to fill the grain if severe stress inhibits photosynthesis. Bunting *et al.* (1964) also found that cultivars which germinated rapidly and had high assimilation rate in the seedling stage often had considerable yield advantage. In western Canada, the Pitic 62 cultivar is slow growing, but continues to function under severe stress, which results in high-grain yield. In the USSR, Kirichenko and Urazaliev (1970) studied heterosis in F_1, F_2, and F_3 and were able to transfer desirable root characters to 8 of 100 lines of wheat. Using substitution lines, Monyo and Whittington (1970) concluded that variations in shoot and root characteristics were markedly influenced by single genes affecting duration of the vegetative growth period as well as by polygenic systems. Hurd (1964) identified a cultivar with a very extensive root system, especially at lower soil depths. In crosses with a high-quality durum wheat, he and his colleagues (Hurd *et al.*, 1972a,b, 1973) produced two cultivars, now the most commonly grown durum cultivars in the Canadian prairies. They have the root pattern of the drought resistant parent and the superior quality of the other parent, plus high yield. Hurd (1975) concluded that selection for yield made over several seasons in semiarid areas will produce the advantageous root characters of the better parent. Similar methods outlined in Sections I,B,1 and I,B,2 were used to develop these superior durums.

D. PROLINE CONTENT

Proline accumulates in leaves of plants subjected to drought. Barnett and Naylor (1966) and Singh *et al.* (1972) suggested that proline content was a measure of drought resistance. They recorded cultivarial differences in accumulation of proline in Burmuda grass (*Cynodon dactylon*). Kaufmann (1972) recorded increase in proline in oranges associated with low water potentials. Parker (1968) suggested that proline accumulation may not be related to a protective mechanism, but may result from some protein breakdown, or it may simply be a storage compound for nitrogen. Naylor (1972) suggested that water stress leads to blockage of synthesis of some amino acids at one or more points in the metabolic pathway. W. Dedio (personal communication, 1974) found that proline levels in 16 wheat cultivars were related more closely to water stress of leaves than to cultivars. He suggested that proline level could be used as an indicator of degree of water stress, but was of little use in identification of drought resistance.

Too little is known about the relationship of proline or protein synthesis to drought resistant mechanisms to make these useful in screening methods. Further research is needed along these lines.

E. DROUGHT HARDENING

Hardening of cereal crops by presowing treatments has been attempted since 1883 with varying success (Salim and Todd, 1968). Mild drought stress in early stages of growth increases resistance to water stress under subsequent and more severe drought. Todd and Webster (1965) found that four cultivars of wheat and Arkwin oats had higher photosynthetic rates at lower turgor after the plants had been previously subjected to a single drought. They concluded that cultivars differed in hardening capacity and the most and least drought-resistant ones could be identified by measuring survival after several cycles of stress.

Seed conditioning treatments to induce drought hardiness were proposed by Henkel (1960) and Badanova (1963). The first hardening treatment imposed was a single soaking in water for 24 hr followed by drying. Various modifications have been added by Henkel and others. Henkel claimed that hardened plants had higher yields under drought, higher water contents, increased viscosity of protoplasm, more bound water, higher metabolic rate, and stronger roots. Badanova (1963) reported that "Henkel-treated" sunflowers had more stable nitrogen and phosphorus metabolism than untreated plants. Treated plants also were more productive under dry conditions. May *et al.* (1962) and Salim and Todd (1968) tested the Henkel method on cultivars of cereal crops and found little or no effect of treatment. Their work suggested possible reasons for the discrepancy in results obtained in the two areas. The genetic background of cultivars grown in the two areas differs widely. Salim and Todd found that some treated barley cultivars showed no differences in water-holding capacity from untreated ones, while others showed pronounced increases.

Salim and Todd (1968) also found that Ponca wheat, which was the most responsive cultivar in water-holding capacity, also had higher stomatal frequency when treated for hardening. They concluded that treatment did not influence changes of some characters in the same direction as they did in other cultivars. Thus, it seemed that each species and cultivar required independent investigation. Woodruff (1969) studied presowing drought-hardening of wheat and found that the rate of drop to the critical level of water content increased over untreated plants.

Cultivarial differences in presowing and field drought-hardening exist, but more research and more consistent positive results are needed before any practical use can be made of these influences.

F. Crop Pests

Plant disease and insect damage usually are considered to be problems of higher producing areas rather than areas undergoing drought. However, if disease or insects cause losses in marginal crops, the damage is serious.

In cereal crops on the Great Plains, root rot is particularly serious in dry years or in dry areas. Plants seem to be able to grow away from root rots if growing conditions are favorable. Ledingham *et al.* (1973) reported that common root rot of wheat and barley was aggravated by drought in the Canadian prairies. The Brown soil zone, which had lowest amounts of available water, had the highest losses (4.6 to 7.3%) compared to the Dark Brown (3.8 to 6.0%), and the Black soil zones (5.0 to 6.7%). Neal *et al.* (1970) showed how cultivars and even substitution lines can affect the rhizosphere microflora of spring wheat and how this can lead to development of resistant strains.

Also in the Great Plains area, leaf rust, when combined with severe drought at filling time, caused up to 58% loss in susceptible wheat cultivars and 28% in resistant ones (Samborski and Peturson, 1960). Leaf rust destroys the photosynthetic capability of the leaves and causes filling to depend on photosynthesis in the head and stem. Drought that hastens ripening for largely denuded plants thus combines with the disease to reduce yields.

On these same crops, wheat stem sawfly and grasshoppers (*Melanoplus sp.*) in the semiarid prairies reduce yield. When combined with drought, they may make crop production in some years uneconomical.

Since these crop pests do severe damage in some years, breeders should build in resistance where possible. This will involve breeding for negative genes and is practical in most crops grown in semiarid areas.

III. APPLICATION OF DROUGHT-RESISTANCE MEASUREMENTS IN BREEDING PROGRAMS

A. Screening Segregating Populations

Some parameters of drought resistance are sufficiently understood and testing methods simple enough for them to be useful in breeding for drought resistance. Others are promising, but untried. Sullivan (1971) summarized techniques that he considered useful in practical research on drought resistance. By comparison, Kaul (1974) stated that a fast screening test for selecting drought-resistant wheat lines was not available. Ashton (1948) described a number of elaborate procedures that measured drought resistance in plants. Recently, several investigators have made progress in understanding drought resistance and in developing simple tests that can be used on large segregating populations in breeding programs. Some of the most promising screening techniques will be mentioned briefly. The order is not indicative of their potential use.

1. Water-Stress Parameters

Barrs (1968) used the term water deficit (WD) for both water content (WC) and water potential (Ψ) of plant tissue. Water-saturation deficit (WSD) is similar to relative water content (RWC) and should be related to full turgidity and not dry or fresh weight (Sullivan, 1971). Barrs (1968) discussed techniques of measuring WC and limitations in use of this parameter.

Dedio (1975) tested various indices of water stress in screening methods. These were water potential, water saturation deficit, water content, and rate of drying of cut leaves at room temperature and humidity. He found that WSD varied from 14.7 to 33.9% among wheat cultivars subjected to moderate soil-water stress. Pitic 62 had the lowest deficit and durum cultivars had consistently higher ones than red spring wheats. Under severe stress, Neepawa had the lowest WSD. Pitic 62, with superior field resistance to drought, always had the highest WC, but not WSD. He concluded that WC was more useful in screening wheats than WSD, but emphasized the importance of using young plants of the same physiological age. Dedio was surprised to obtain wide differences among cultivars, since Kaul (1969) reported no consistent differences in some of the same cultivars. Kaul studied more mature plants in the field at three locations. Dedio found that Pelissier, one of the most drought-resistant wheats, had high WSD. He considered that its massive root system was not an influencing factor, since all four plants tested were grown in the same pot and, thus, subject to the same stress. He suggested that resistance may have been a physicochemical process. Oleinikova (1969) concluded that water-retention capacity was associated with protoplasmic permeability.

Oleinikova and Kozhushko (1970), Sandhu and Laude (1958), and Salim et al. (1969) favored water retention of excised leaves on whole plants as a test for drought resistance even though differences among cultivars were small. Dedio had better results. In 3-wk-old plants, Pelissier was the best water retainer among six cultivars tested. At anthesis, both Pitic

62 and Pelissier were significantly better water retainers than were Koga, Glenlea, or Manitou.

Pitic 62 and ACEF 125, which differed in water-retaining capacity, were crossed and the progeny from F_2 plants were compared with the parents. The hybrid gave data which when plotted showed two peaks: one major peak similar to that of Pitic 62 and one minor peak corresponding to ACEF 125. Mean water contents of excised leaves after 24-hr drying were 7.5% for ACEF 125, 54.2% for Pitic 62, and 56.7% for the hybrid.

Salim *et al.* (1969) tested water retention of 16 cultivars of wheat, oats, and barley by the whole-plant or cut leaf method. They found that hastening drying with $CaCl_2$ proved too rapid, but slower drying gave reliable and reproducible measurements.

Several investigators used tissue water potential (Ψ) as an index of water stress under drought conditions. The most common way of determining Ψ is with a thermocouple psychrometer. Other methods were discussed by Barrs (1968) and Brown (1970).

Psychrometers are expensive and considerable time and skill are involved in their use. Using a portable field psychrometer, Kaul (1967) found differences in Ψ among cultivars, but considered his technique too time-consuming for screening large numbers of lines in a breeding program. Nevertheless, Ψ is widely recognized as a useful index of plant-water stress (Kramer, 1969) and measurements of this parameter may play an increasingly important role in screening for drought resistance.

As mentioned in Section II,B, stomatal aperture often is an index of capacity of plants to resist drought. For references on methods of determining stomatal aperture, the reader is referred to Section II,B.

2. Photosynthetic Rates

The capacity of plants to photosynthesize during and following moisture stress is an important index of drought resistance. Pitic 62 has scored high in several tests. Perhaps, the combined effect of these can be determined by measuring Pitic's capacity to photosynthesize under stress or to withstand irreversible damage under severe desiccation. Dedio (1975) tested several methods of measuring photosynthesis of wheat cultivars under stress. Exposing leaves to $^{14}CO_2$ followed by determination of ^{14}C-photosynthate in plant tissue by scintillation spectrometry was considered most useful. The method showed that photosynthetic potential was higher in Pitic 62 than in other cultivars tested (Dedio *et al.*, 1975).

3. Yield Tests

The ultimate test of drought resistance is a yield comparison in the field conducted under typical drought conditions (Aufhammer *et al.*, 1959;

Hageman *et al.*, 1967; Hurd, 1971). Screening techniques will help the breeder eliminate poor yielding lines as well as many average lines. Such upgrading of populations will permit larger numbers of promising lines to be grown in yield tests. The larger the number of lines grown, the better the chances of having a really superior line present.

4. Other Simple Screening Techniques

Williams *et al.* (1967) subjected plants or seeds to three stresses: (1) 20-day-old seedlings were exposed to 52°C temperature for 6 hr and those recovering after 1 wk of normal growth conditions were counted, (2) seeds were germinated in mannitol solution at 15 atm osmotic pressure and selections made for highest percent germination, and (3) seedlings were subjected to a 14-day permanent wilting period and wilting ratings were taken. All three methods gave highly significant correlations with each other and with field ratings and all were considered useful for screening a large number of corn genotypes for drought tolerance. Kaul, as cited by Hurd (1975) using Williams *et al.* method (2) with mannitol (20 atm) on wheat seeds, obtained germinations of 49% for Pitic 62, 27% for Manitou, 5% for Giza, and 3% for Carazinho. Sorghum cultivars were divided into two separate groups by testing their germination capacity in osmotic solutions (Vasudevan and Balasubramaniam, 1965). Rain-fed cultivars were superior to irrigated strains. Sullivan and Eastin (1975) stated that when two sorghum cultivars were subjected to a water potential of −33 bars, one demonstrated 90% recovery and the other did not recover at all.

Root lengths of seedlings grown for 5 to 7 days in sand were measured by Townley-Smith and McBean (cited by Hurd, 1975). They found that cultivars known to have extensive roots at maturity consistently had the longest roots in the early seedling stage. The selected plants could be replanted following root measurement and grown to maturity. Several thousand plants can be screened for seedling root length in the greenhouse during a winter season.

B. Testing Potential Parental Material

The importance of careful selection of parents for crosses was emphasized in Section I,B,1. One or other of the parental cultivars used in each cross must have all of the characters desired in the new cultivar. When selecting parents for crosses, a combination of tests should be made to ensure that the parents have desiccation tolerance, capacity to maintain low water deficit under stress, extensive root systems, and high yield potential.

Breeders in semiarid areas should build up their own genetic stocks of cultivars which possess various combinations of drought-resistant characters. With recent emphasis on drought research in many crops, genetic stocks are becoming available at an increasing rate. Plant breeders should transfer particular, desirable characters associated with drought resistance into locally adapted cultivars to provide genetic stocks for their area. Wide crosses used to transfer a quantitatively inherited character usually carry many genes for undesirable characters and this greatly reduces the chances of obtaining the required combination of favorable genes. Screening techniques can be readily applied to all potential parents as a routine procedure. In addition, more sophisticated and time-consuming methods of testing can also be applied to check drought resistance of all parents for crosses prior to their use. Such procedures will remove much trial and error from plant breeding. To know what characters the parent has for drought resistance will assist the breeder in selection of the best screening procedures to use for a cross involving that parent.

ACKNOWLEDGMENTS

I wish to acknowledge the assistance of Dr. T. Lawrence in preparing the section on cross-pollinated crops, Dr. J. D. McElgunn in checking the plant physiology sections, Dr. T. F. Townley-Smith for advice with the statistical discussion, and Mr. D. S. McBean for reading the manuscript.

REFERENCES

Akerman, A., and MacKay, J. (1948). The breeding of self-fertilized plants by crossing. *In* "Svalof 1886–1946; History and Present Problems" (A. Akerman, *et al.*, ed.), pp. 46–71. Carl Bloms Boktryckeri A-B., Lund.

Allard, R. W. (1960). "Principles of Plant Breeding." Wiley, New York.

Anwar, A. R., and Chowdry, A. R. (1969). Heritability and inheritance of plant height, heading date and grain yield in four spring wheat crosses. *Crop Sci.* **9**, 760–761.

Asana, R. D., Saine, A. D., and Ray, D. (1958). Studies in physiological analysis of yield. 3. The rate of grain development in wheat in relation to photosynthetic surface and soil moisture. *Physiol. Plant.* **11**, 655–665.

Ashton, T. (1948). Techniques of breeding for drought resistance in crops. *Commonw. Bur. Plant Breed. Genet., Tech. Commun. No. 14.*

Aufhammer, G., Fischbeck, G., and Grebner, H. (1959). Testing drought resistance in spring cereals. *Z. Acker- Pflanzenbau* **110**, 117–134.

Badanova, K. A. (1963). The effect of drought and dry wind on the metabolism of drought resistant plants. In "The Water Regime of Plants as Related to Metabolism and Productivity," pp. 230–234. Timiryazen Inst. Plant Physiol. U.S.S.R. Acad. Sci.

Baker, R. J. (1968). Genotype-environment interaction variances in cereal yields in Western Canada. *Can. J. Plant Sci.* **48**, 293–298.

Baker, R. J. (1971). Effect of stem rust and leaf rust of wheat on genotype-environment interaction for yield. *Can. J. Plant Sci.* **51**, 457–461.

Baker, R. J., and Leisle, D. (1970). Comparison of hill and rod row plots in common and durum wheats. *Crop Sci.* **10**, 581–583.

Baker, R. J., and McKenzie, R. H. (1967). Use of control plots in yield trials. *Crop Sci.* **7**, 335–337.

Baker, R. J., Tipples, K. H., and Campbell, A. B. (1971). Heritabilities of and correlations among quality traits in wheat. *Can. J. Plant Sci.* **51**, 441–448.

Baldy. C. (1973). Progrès recents concernant l'étude due systeme racinaire du ble (*Triticum sp.*) *Ann. Agron.* **149**, 250–276.

Barnett, N. M., and Naylor, A. W. (1966). Amino acid and protein metabolism in Burmuda grass during water stress. *Plant Physiol.* **41**, 1222–1230.

Barrs, H. D. (1968). Determination of water deficits in plant tissue. *In* "Water Deficits and Plant Growth" (T. T. Kozlowski, ed.), Vol. 1, pp. 235–368. Academic Press, New York.

Bell, G. P. H. (1963). Breeding techniques—general techniques. Barley genetics. I. *Cent. Agr. Publ. Doc., H. Veenman en yonen N.V. Wageningen* pp. 285–302.

Belzakov, I. (1968). The growth and development of wheat and barley roots in the semi-desert zone. *Vestn. Sel'skokhoz. Nauki* **13**, 31–33.

Bertrand, A. R. (1965). Water conservation through improved practices. *In* "Plant Environment and Efficient Water Use" (W. H. Pierre *et al.*, eds.), pp. 207–235. Amer. Soc. Agron.—Soil Sci. Soc. Amer, Madison, Wisconsin

Boyer, J. S., and Bowen, B. L. (1970). Inhibition of oxygen evaluation in chloroplasts isolated from leaves with low water potentials. *Plant Physiol.* **45**, 612–615.

Breese, E. L., and Hayward, M. D. (1972). The genetic basis of present breeding methods in forage crops. *Euphytica* **21**, 324–336.

Briggs, F. N. (1930). Breeding wheat resistant to bunt by the backcross method. *J. Amer. Soc. Agron.* **22**, 239–244.

Briggs, K. C., and Shebeski, L. H. (1968). Implications concerning the frequency of control plots in wheat breeding nurseries. *Can. J. Plant Sci.* **48**, 49–153.

Brown, R. W. (1970). Measurement of water potential with thermocouple psychrometers: Construction and application. *U.S., Forest Serv., Pap. Int.* **80**, 27.

Bunting, A. H., Drennan, D. S. H., de Silva, W. H., and Krishnamurthy, K. (1964). The structure of yield in wheat varieties in England. *Proc. Int. Bot. Congr., 10th, 1964.*

Burton, G. W. (1964). The geneticists role in improving water use efficiency by crops. "Research on Water," Spec. Publ., Ser. 4. Amer. Soc. Agron., Madison, Wisconsin.

Danilchuk, P. V. (1972). The development of the shoot system and root and their physiological activity in winter wheat plants of various ecotypes. *Dokl. All-Union Akad. Agr. Sci. Kolos* **8**, 10–12.

Danilchuk, P. V., and Yatsenko, G. K. (1973). Physiological and biochemical peculiarities of the development of the shoot system and roots as related to the productivity of winter wheat grown in southern Ukraine. *Vop., Genet., Sel'ksii, Seminovodstva* pp. 152–171.

Danilchuk, P. V., Yatsenko, G. K., and Shlifasovsky, V. A. (1971). The development of roots and ground mass in winter wheat. *Vestn. Sel'skokhoz. Naukt* **10**, 50–55.

Dedio, W. (1975). Water relations in wheat leaves as screening tests for drought resistance. *Can. J. Plant Sci.* **55**, 369–378.

Dedio, W., Stewart, D. W., and Green, D. G. 1975.. Evaluation of photosynthesis measuring methods as possible screening techniques for drought resistance in wheat. *Can. J. Plant. Sci.* (In press).

DePauw, R. M., and Shebeski, L. H. (1973). An evaluation of an early generation yield testing procedure in *Triticum aestivum*. *Can. J. Plant Sci.* **53**, 465–470.

Donald, C. M. (1963). Competition among crop and pasture plants. *Advan. Agron.* **15**, 1–118.

Donald, C. M. (1968). The design of a wheat ideotype. *Proc. Int. Wheat Genet. Symp., 3rd, 1968* pp. 377–387.

Eastin, J. D., Haskins, F. A., Sullivan, C. Y., and van Bavel, C. H. M., eds. (1969). "Physiological Aspects of Crop Yield." Amer. Soc. Agron.—Crop Sci. Soc. Amer., Madison, Wisconsin.

Ebeltoft, D. C. (1969). Rapid increase of barley, spring wheat. *N. Dak. Farm Res.* **26**, No. 4, pp. 3–6.

El-Haddad, M. M. (1974). Continuous variation in generations derived from crosses between Scandinavian and Mexican wheat varieties. *Meld. Norg. Landbrukshoegsk.* **53**, 1–27.

Elliot, F. C. (1968). "Plant Breeding and Cytogenetics." McGraw-Hill, New York.

Engledow, F. L., and Wardlaw, S. (1923). Investigation on yield in cereals. *J. Agr. Sci.* **13**, 390–439.

Eslick, F. R., and Hackett, E. A. (1975). Genetic engineering as a key to water use efficiency. *Agr. Meteorol.* **14** Elsevier Sci. Publ. Co. pp. 13–22.

Evans, L. E., Shebeski, L. H., McGinnis, R. C., Briggs, K. G., and Zuzens, D. (1972). Glenlea red spring wheat. *Can. J. Plant Sci.* **52**, 1081–1082.

Fisher, R. A. (1951). "The Design of Experiments," 6th ed. Hafner, New York.

Fischer, R. A., and Kohn, G. D. (1966). The relationship of grain yield to vegetative growth and post flowering leaf area in the wheat crop under conditions of limited soil moisture. *Aust. J. Agr. Res.* **17**, 255–280.

Frank, A. B., Power, J. R., and Willis, W. O. (1973). Effect of temperature and water potential in spring wheat. *Agron. J.* **65**, 777–780.

Frankel, O. H. (1947). The theory of plant breeding for yield. *Heredity* **1**, 109–120.

Frey, K. J. (1964). Adaptation reaction of oat strains selected under stress and non-stress environmental conditions. *Crop Sci.* **4**, 55–58.

Frey, K. J. (1965). The utility of hill plots in oat research. *Euphytica* **14**, 196–208.

Gardner, W. R. (1960). Soil-water relationships in arid and semiarid conditions. *In* "Plant Water Relationships in Arid and Semi-Arid Conditions," pp. 37–62. UNESCO, Paris.

Garkavy, P. F., Danilchuk, P. V., and Linchevsky, A. A. (1970). The selection of malting varieties of spring barley according to the vigor of root development. *Vop., Genet. Sel'ksii, Semenovodstva* pp. 53–65.

Gates, C. T. (1964). The effect of water stress on plant growth. *J. Aust. Inst. Agr. Sci.* **30**, 3–22.

Gliemeroth, G. (1952). Water content of the soil in relation to the root development of some cultivated plants. *Z. Acker-Pflanzenbau* **95**, 21–46.

Goedewaagen, M. A. J., and Schuurman, J. J. (1957). Root development of grassland with special reference to water conditions of the soil. *Proc. Int. Grassland Congr., 7th, 1956* pp. 45–55.

Goulden, C. H. (1952). "Methods of Statistical Analysis," 2nd ed. Wiley, New York.

Grafius, J. E. (1963). Breeding for yield. Barley genet. I. *Cent. Agr. Publ. Doc., H. Veenan en yonen N.V. Wageningen* p. 267–277.

Grafius, J. E. (1965). A geometry of plant breeding. *Mich.* Agr. Exp. Sta. *Res. Bull.* **7**.

Grafius, J. E., Nelson, W. L., and Dirks, D. A. (1952). The heritability of yield in barley as measured by early generation bulked progeny. *Agron. J.* **44**, 253–257.

Gregory, W. C., Yarbrough, J. A., and Smith, B. W. (1950). The morphology, genetics, variation and breeding of peanuts. *Proc. 46th Annu. Conv. Ass. 5th Agr. Workers* pp. 160–188.

Grundbacher, F. J. (1963). The physiological function of the cereal awn. *Biol. Rev. Cambridge Phil. Soc.* **29**, 366–381.

Hagberg, A., and Akerberg, E. (1962). "Mutations and Polyploidy in Plant Breeding." Heinermann, Educational Books Ltd., London.

Hageman, R. H., Leng, E. R., and Dudley, J. W. (1967). A biochemical approach to corn breeding. *Advan. Agron.* **19**, 45–86.

Hamilton, D. G. (1959). Improving Canada's wheat. *Agr. Inst. Rev.* **14**(6), 18–22.

Harlan, H. V., and Martini, M. L. (1938). The effect of natural selection in a mixture of barley varieties. *J. Agr. Res.* **57**, 189–199.

Hayes, H. K., and Immer, F. R. (1942). "Methods of Plant Breeding." McGraw-Hill, New York.

Henkel, P. A. (1960). Drought resistance in plants; Methods of recognition and intensification. *In* "Plant-Water Relationships in Arid and Semi-Arid Conditions," Vol. 16, pp. 167–174. UNESCO, Paris.

Holmes, D. P. (1974). Physiology of grain filling in barley. *Nature (London)* **247**, 297–298.

Hurd, E. A. (1964). Root study of three wheat varieties and their resistance to drought and damage by soil cracking. *Can. J. Plant Sci.* **44**, 240–248.

Hurd, E. A. (1968). Growth of roots of seven varieties of spring wheat at high and low moisture levels. *Agron. J.* **60**, 201–205.

Hurd, E. A. (1969). A method of breeding for yield of wheat in semi-arid climates. *Euphytica* **18**, 217–226.

Hurd, E. A. (1971). Can we breed for drought resistance? *In* "Drought Injury and Resistance in crops" (K. L. Larson and J. D. Eastin, eds.), Publ. No. 2, pp. 77–88. Crop. Sci. Soc. Amer., Madison, Wisconsin.

Hurd, E. A. (1975). Phenotype and drought tolerance in wheat. *In* "Modification for More Efficient Water Use" (J. E. Stone, ed.) pp. 37–55. *Agr. Meteorol.* **14**, Elsevier Sci. Pub. Co., Amsterdam.

Hurd, E. A., and Spratt, E. D. (1975). Root patterns in crops as related to water and nutrient uptake. *In* "The Physiological Aspects of Dryland Farming" (U. S. Gupta, ed.) Oxford and IBH Publ. Co., New Delhi. 166–235.

Hurd, E. A., Townley-Smith, T. F., Patterson, L. A., and Owen, C. H. (1972a). Wascana, a new durum wheat. *Can. J. Plant Sci.* **52**, 687–688.

Hurd, E. A., Townley-Smith, T. F., Patterson, L. A., and Owen, C. H. (1972b). Techniques used in producing Wascana wheat. *Can. J. Plant Sci.* **52**, 689–691.

Hurd, E. A., Townley-Smith, T. F., Mallough, D., and Patterson, L. A. (1973). Wakooma durum wheat. *Can. J. Plant Sci.* **53**, 261–262.

Iljin, W. C. (1957). Drought resistance in plants and physiological processes. *Annu. Rev. Plant Physiol.* **8**, 257–274.

Jensen, N. F. (1969). Discussion of Sprague—germ plasm manipulation of the future. *In* "Physiological Aspects of Crop Yield" (J. D. Eastin *et al.*, eds.), pp. 387–388. Amer. Soc. Agron.—Crop Sci. Soc. Amer., Madison, Wisconsin.

Kaufmann, M. R. (1972). Water deficits and reproductive growth. *In* "Water deficits and Plant Growth" (T. T. Kozlowski, Ed.) Vol. 3, pp. 91–124. Academic Press, New York.

Kaul, R. (1967). A survey of water suction forces in some prairie wheat varieties. *Can. J. Plant Sci.* **47**, 323–326.

Kaul, R. (1969). Relations between water status and yield of some wheat varieties. *Z. Pflanzenzuech.* **62,** 145–154.

Kaul, R. (1974). Potential net photosynthesis in flag leaves of severely drought-stressed wheat cultivars and its relationship to grain yield. *Can. J. Plant Sci.* **54,** 811–815.

Kaul, R., and Crowle, W. L. (1971). Relation between water status, leaf temperature, stomatal aperture, and productivity of some wheat varieties. *Z. Pflanzenzuech.* **65,** 233–243.

Kaul, R., and Crowle, W. L. (1974). An index derived from photosynthetic parameters for predicting grain yields of drought stressed wheat cultivars. *Z. Pflanzenzuech.* **71,** 42–51.

Kirichenko, F. G. (1963). The effect of plant selection according to root strength on increasing the yield and improving its quality in the progeny. *Vestn. Sel'skokhoz. Nauki* No. 4, pp. 3–20.

Kirichenko, F. G., and Urazaleiv, R. A. (1970). Degree of inheritance of strength of root system and the plant mass in interspecies hybrids of wheat. *Dokl. All-Un. Acad. Agr. Sci.* **4,** 9–13.

Knott, D. R. (1972). Effects of selection for F_2 plant yield on subsequent generations in wheat. *Can. J. Plant Sci.* **52,** 721–726.

Kozlowski, T. T. (1968). Introduction. *In* "Water Deficits and Plant Growth" (T. T. Kozlowski, ed.), Vol. I, pp. 1–21. Academic Press, New York.

Kozlowski, T. T. (1972). Shrinking and swelling of plant tissues. *In* "Water Deficits and Plant Growth" (T. T. Kozlowski, ed.), Vol. III, pp. 1–64. Academic Press, New York.

Kramer, P. J. (1969). "Plant and Soil Water Relationships: A Modern Synthesis." McGraw-Hill, New York.

Kuspira, J., and Unrau, J. (1957). Genetic analysis of certain characters in common wheat using chromosome substation lines. *Can. J. Plant Sci.* **37,** 300–326.

Kuspira, J., and Unrau, J. (1959). Theoretical ratios and tables to facilitate genetic studies with aneuploids. I. F_1 and F_2 analysis. *Can. J. Genet. Cytol.* **1,** 267–312.

Larson, R. I., and Atkinson, T. G. (1970a). Identity of the wheat chromosomes replaced by Agropyron chromosomes in a triple alien chromosome substation line immune to wheat streak mosaic. *Can. J. Genet. Cytol.* **12,** 145–150.

Larson, R. I., and Atkinson, T. G. (1970b). A cytogenetic analysis of reaction to common root rot in some hard red spring wheats. *Can. J. Bot.* **48,** 2059–2067.

Larson, R. I., and MacDonald, M. D. (1959a). Cytogenetics of solid stem in common wheat. II. Stem solidness of monosomic lines of the variety S615. *Can. J. Bot.* **37,** 365–378.

Larson, R. I., and MacDonald, M. D. (1959b). Cytogenetics of solid stem in common wheat. III. Culm measurements and their relation to stem solidness in monosomic lines of the variety S615. *Can. J. Bot.* **37,** 379–391.

Larson, R. I., and MacDonald, M. D. (1964). Cytogenetics of sawfly resistance in wheat. *Can. Entomol.* **94,** 124 (abstr.)

Larson, R. I., and MacDonald, M. D. (1966). Cytogenetics of solid stem in common wheat. V. Lines of S615 with whole chromosome substitutions from apex. *Can. J. Genet. Cytol.* **8,** 64–70.

Lawes, D. A., and Treharne, K. J. (1971). Variation in photosynthetic activity in cereals and its implications in a plant breeding programme. I. Variation in seedling leaves and flag leaves. *Euphytica* **20,** 86–92.

Lawrence, T., and Townley-Smith, T. F. (1975). Use of moving mean in grass yield trials. *Can. J. Plant Sci.* **55,** 587–592.

Le Clerg, E. L. (1966). Significance of experimental design in plant breeding. *In* "Plant Breeding" (K. J. Frey, ed.), pp. 243–313. Iowa State Univ. Press, Ames.

Ledingham, R. J., Atkinson, T. G., Horricks, J. S., Mills, J. T., Piening, L. J., and Tinline, R. D. (1973). Wheat losses due to common root rot in the prairie provinces of Canada, 1969–71. *Can. Plant Dis. Surv.* (W. L. Seaman, Ed.) **53,** 113–122.

Levitt, J. (1956). "The Hardiness of Plants." Academic Press, New York.

Levitt, J. (1972). "Responses of Plants to Environmental Stresses." Academic Press, New York.

Lupton, F. G. H. (1961). Studies in the breeding of self pollinating cereals. 3 further studies in cross prediction. *Euphytica* **10,** 209–224.

Lupton, F. G. H., Oliver, R. H., Ellis, E. R., Barnes, B. T., Howes, K. R., Welbank, P. J., and Taylor, P. J. (1974). Root and shoot growth of semi-dwarf and taller winter wheats. *Ann. Appl. Biol.* **77,** 129–144.

McGinnis, R. C., and Shebeski, L. H. (1968). The reliability of single plant selection for yield in F_2. *Proc. Int. Wheat Genet. Symp., 3rd, 1968* pp. 410–415.

Mackay, J. (1953). Autogamous plant breeding based on already hybrid material. *In* "Recent Plant Breeding Research," (A. Akerberg and A. Hagberg, Eds). pp. 73–88. Wiley, New York.

May, L. H., Milthorpe, E. J., and Milthorpe, E. L. (1962). Pre-sowing hardening of plants and drought. An appraisal of the contribution by P. A. Henkel. *Field Crop Abstr.* **15,** 193–198.

Monyo, J. H., and Whittington, W. J. (1970). Genetic analysis of root growth in wheat. *J. Agr. Sci.* **74,** 329–338.

Naylor, A. W. (1972). Water deficit and nitrogen metabolism. *In* "Water Deficits and Plant Growth" (T. T. Kozlowski, ed.), Vol. 3, pp. 241–254. Academic Press, New York.

Neal, J. L., Jr., Atkinson, T. G., and Larson, R. I. (1970). Changes in the rhizosphere microflora of spring wheat reduced by disomic substitution of a chromosome. *Can. J. Microbiol.* **16,** 153–158.

Oleinikova, T. V. (1969). The effect of drought on protoplasm permeability in cells of spring wheat leaves. *Byull. Vses. Inst. Rasteniev* No. 14, pp. 25–30.

Oleinikova, T. V., and Kozhushko, N. N. (1970). Laboratory methods of drought resistance evaluation of some cereal crops. *Tr. Prikl. Bot., Genet. Selek.* **43,** 100–111.

Oppenheimer, H. R. (1961). Adaptation to drought. Xerophytism. *In* "Plant-Water Relationships in Arid and Semi-Arid Conditions," pp. 105–138. UNESCO.

Parker, J. (1968). Drought-resistance mechanisms. *In* "Water Deficits and Plant Growth" (T. T. Kozlowski, ed.), Vol. 1, pp. 195–235. Academic Press, New York.

Parker, J. (1972). Protoplasmic resistance in water deficit. *In* "Water Deficits and Plant Growth" (T. T. Kozlowski, ed.), Vol. 3, pp. 125–176. Academic Press, New York.

Passouri, J. B. (1972). The effect of root geometry on the yield of wheat growing on stored water. *Aust. J. Agr. Res.* **23,** 745–752.

Pearson, R. W. (1965). Soil environment and root development. *In* "Plant Environment and Efficient Water Use" (W. H. Pierre *et al.,* eds.), pp. 95–126. Amer. Soc. Agron.—Soil Sci. Soc. Amer., Madison, Wisconsin.

Pearson, R. W. (1974). Significance of rooting pattern to crop production and some problems of root research. *In* "The Plant Root and its Environment" (E. W. Carson, ed.), pp. 247–270. Univ. of Virginia Press, Charlottesville.

Person, C. O. (1956). Some aspects of monosomic wheat breeding. *Can. J. Bot.* **34,** 60–70.

Pesek, J., and Baker, R. J. (1971). Comparison of predicted and observed responses to selection for yield in wheat. *Can. J. Plant Sci.* **51,** 187–192.

Pinthus, M. J., and Eshel, Y. (1962). Observations on the development of the root system of some wheat varieties. *Isr. J. Agr. Res.* **12,** 13–20.

Pohjakallio, O. (1945). The question of the resistance of plants to drought periods in Finland. *Nord. Jordbrugs forsk.* **5–6,** 206–226.

Raheja, P. C. (1966). Aridity and salinity. *In* "Salinity and Aridity: New Approahces to Old Problems" (H. Boyko, ed.), pp. 10–42. Junk, The Hague.

Raper, C. D., and Barber, S. A. (1970a). Rooting systems of soybeans. I. Differences in root morphology among varieties. *Agron. J.* **12,** 581–584.

Raper, C. D., and Barber, S. A. (1970b). Rooting systems of soybeans. II. Physiological effectiveness of nutrient absorption surfaces. *Agron. J.* **62,** 585–588.

Reitz, L. P. (1975). Breeding for more effective water use. Is it real or a mirage? *Agr. Meteorol.* **14,** Elsevier Sci. Publ. Co. pp. 3–11.

Rickey, F. D. (1924). Adjusting yields to their regression on a moving average, as a means of correcting for soil heterogeneity. *J. Agr. Res.* **27,** 79–90.

Roma, A. (1962). Heritability in the root system characters of wheat in correlation with lodging resistance. *Lucr. Stiinti. (Cluj)* **18,** 81–90.

Russell, M. B. (1957). "Water and its Relation to Soils and Crops," Dept. of Agronomy, University of Illinois, Urbana.

Saghir, A. R., Khan, A. R., and Worzella, W. W. (1968). The effect of plant parts on grain yield, kernel weight and plant height of wheat and barley. *Agron. J.* **60,** 95–97.

Sakai, K., and Gotoh, K. (1955). Studies on competition in plants. IV. Competition ability of F_1 hybrids in barley. *J. Hered.* **46,** 139–143.

Salim, M. H., and Todd, G. W. (1968). Seed soaking as a pre-sowing, drought-hardening treatment in wheat and barley seedlings. *Agron. J.* **60,** 179–182.

Salim, M. H., Todd, G. W., and Stutte, C. A. (1969). Evaluation of techniques for measuring drought avoidance in cereal seedlings. *Agron. J.* **61,** 182–185.

Sallans, B. J., and Tinline, L. D. (1965). Resistance in wheat to *Cochleobolis sativus.* A cause of common root rot. *Can. J. Plant Sci.* **45,** 343–351.

Samborski, D. J., and Peturson, B. (1960). Effect of leaf rust on yield of resistant wheats. *Can. J. Plant Sci.* **40,** 620–622.

Sampson, D. R. (1971a). Evaluation of nine oat varieties as parents in breeding for short stout straw with high grain yield using F_1, F_2 and F_3 bulked progenies. *Can. J. Plant Sci.* **52,** 21–28.

Sampson, D. R. (1971b). Additive and nonadditive genetic variances and genotypic correlations for yield and other traits in oats. *Can. J. Genet. Cytol.* **13,** 864–872.

Sandhu, A. S., and Laude, H. H. (1958). Tests of drought and heat hardiness in winter wheat. *Agron. J.* **50,** 78–81.

Sears. E. R. (1954). The aneuploids of common wheat. *Mo., Agr. Exp. Sta., Res. Bull.* **572,** 1–58.

Shebeski, L. H. (1967). "Wheat and Breeding," Can. Centennial Wheat Symp., pp. 253–272. Modern Press, Saskatoon.

Shebeski, L. H. (1972). Wheat and breeding. *FAO Inform. Bull.* **7,** (2).

Simmons, P. M., and Sallans, B. J. (1933). Some observations on the growth of Marquis wheat with special reference to root development. *Proc. World Grain Exhib. Conf., 1933* vol. 2, pp. 163–177.

Singh, K. (1952). Effect of soil cultivation on the growth and yield of winter wheat. IV. Effect of cultivation on root development. V. *Sci. Food Agr.* **3**, 514–525.

Singh, T. N., Aspinall, D., and Paleg, L. G. (1972). Proline accumulation and varietal adaptability to drought in barley: A potential metabolic measure of drought resistance. *Nature (London), New Biol.* **236**, 188–190.

Slatyer, R. O. (1967). "Plant-water Relationships." Academic Press, New York.

Slayter, R. O. (1969). Physiological significance of internal water relations in crop yield. *In* "Physiological Aspects of Crop Yield" (J. D. Eastin *et al.*, eds.), pp. 53–83. Amer. Soc. Agron.—Crop Sci. Soc. Amer., Madison, Wisc.

Smith, G. S. (1966). Transgression segregation in spring wheats. *Crop Sci.* **6**, 310–312.

Sprague, G. E. (1966). Quantitative genetics in plant improvement. *In* "Plant Breeding" (K. J. Frey, ed.), pp. 315–354. Iowa State Univ. Press, Ames.

Sprague, G. E. (1969). Germ plasm manipulation of the future. *In* "Physiological Aspects of Crop Yield" (J. D. Eastin *et al.*, eds.), pp. 53–83. Amer. Soc. Agron.—Crop Sci. Soc. Amer., Madison, Wisconsin.

Stoker, O. (1961). Contributions to the problem of drought resistance of plants. *Indian J. Plant Physiol.* **4**, 87–102.

Sullivan, C. Y. (1971). Techniques for measuring plant drought stress. *In* "Drought Injury and Resistance in Crops" (K. L. Larson and J. D. Eastin, eds.), Publ. No. 2, pp. 1–18. Crop Sci. Soc. Amer. Madison, Wisconsin.

Sullivan, C. Y., and Eastin, J. D. (1975). Plant physiology response to water stress. *Agr. Meteorol.* **14**, Elsevier Sci. Publ. Co. pp. 113–127.

Suneson, C. A. (1949). Survival of four barley varieties in a mixture. *Agron. J.* **41**, 459–461.

Suneson, C. A. (1963). Breeding techniques-composite crosses and hybrid barley. Barley Genetics I. *Agr. Publ. Doc. H. Veenman in Zonen N.V. Wageningen* pp. 303–309.

Suneson, C. A., and Ramage, R. T. (1962). Competition between near-isogenic genotypes. *Crop Sci.* **2**, 249–250.

Syme, J. R. (1970). A high yielding Mexican semi-dwarf wheat and the relationship of yield to harvest index and other varietal characteristics. *Aust. J. Exp. Agr. Anim. Husb.* **10**, 350–353.

Syme, J. R. (1971). Features of high-yielding wheats grown at two seed rates and two nitrogen levels. *Aust. J. Exp. Agr. Anim. Husb.* **11**, 165–170.

Tal, M. (1966). Abnormal stomatal behavior in wilty mutants of tomato. *Plant Physiol.* **41**, 1387–1391.

Todd, G. W., and Webster, D. L. (1965). Effects of repeated drought periods on photosynthesis and survival of cereal seedlings. *Agron. J.* **57**, 399–404.

Townley-Smith, T. F., and Hurd, E. A. (1973). Use of moving mean in wheat yield trials. *Can. J. Plant Sci.* **53**, 447–450.

Townley-Smith, T. F., Hurd, E. A., and Leisle, D. (1975). Macoun durum wheat. *Can. J. Plant Sci.* **55**, 317–318.

Vasudevan, V., and Balasubramaniam, V. (1965). Germination in osmotic solutions as an index of drought resistance in sorghum. *Madras Agr. J.* **52**, 386–390.

Wadleigh, C. H., Raney, W. A., and Herschfield, D. M. (1965). The moisture problem. *In* "Plant Environment and Efficient Water Use" (W. H. Pierre *et al.*, ed.), pp. 1–19. Amer. Soc. Agron.—Soil Sci. Soc. Amer., Madison, Wisconsin.

Wallace, D. H., Ozbun, J. L., and Munger, H. M. (1972). Physiological genetics of crop yield. *Advan. Agron.* **24,** 97–146.

Wardlaw, I. F. (1967). The effect of water stress on translocation in relation to photosynthesis and growth. I. Effect during grain development in wheat. *Aust. J. Biol. Sci.* **20,** 25–39.

Wardlaw, I. F. (1969). The effect of water stress on translocation in relation to photosynthesis and growth. II. Effect during leaf development in *Lolium temulentum* L. *Aust. J. Biol. Sci.* **22,** 1–16.

Wardlaw, I. F. (1971). The early stages of grain development in wheat: Response to water stress in a single variety. *Aust. J. Biol. Sci.* **24,** 1047–1055.

Watson, D. J. (1952). The physiological basis of variation on yield. *Advan. Agron.* **4,** 101–145.

Watson, D. J. (1968). A prospect of crop physiology. *Ann. Appl. Biol.* **62,** 1–9.

Weatherley, P. E. (1965). Some investigations on water deficit and translocation under controlled conditions. In "Water Stress in Plants," (Bohdan Slavik, Ed.) pp. 63–69. Junk Publ. The Hague.

Weaver, J. E. (1926). Root habits of wheat. *In* "Root Development of Field Crops," (J. E. Weaver, Ed.) pp. 133–161. McGraw-Hill, New York.

Williams, T. V., Snell, R. S., and Ellis, J. F. (1967). Methods of measuring drought tolerance in corn. *Crop Sci.* **7,** 179–182.

Williams, T. V., Snell, R. S., and Cress, C. E. (1969). Inheritance of drought tolerance in sweet corn. *Crop Sci.* **9,** 19–23.

Woodruff, D. R. (1969). Studies of presowing drought hardening of wheat. *Aust. J. Agr. Res.* **20,** 13–24.

AUTHOR INDEX

Numbers in italics refer to the pages on which the complete references are listed.

SUBJECT INDEX

A

ABA, 63, 90, 91, 93, 122, 130, 131, 170, 192
Abscisic acid, *see* ABA
Abscission, 191–222, 277, *see also* Abscission layer, Abscission zone
Abscission layer, 193–197
Abscission zone, 192
Absorption
  energy, 15
  gases, 153, 308, *see also* Photosynthesis
  minerals, 208, 212, 338
  radiant heat, 138
  radiation, 7, 141
  water, 38, 40, 111, 204, 205, 208, 209, 212, 213, 258, 267, 337
Absorption hygrometer, 23, 24
Accessory cells, 165, 168, *see also* Subsidiary cells
Acetate, 244, 262
Acetyl CoA, 243, 244
Acetylene, 277, 281, 292, 301–306, 308, 309
Acid phosphatase, 69, 273
Acidity, *see* pH
Actinomycetes, 292
Active transport, 92
Adaptation, 212, 245, *see also* Drought avoidance, Drought hardening, Drought resistance
Additive genes, 324, 328, 331
Adsorption, 3, 6, 32, 242
Adventitous roots, 206, 208, 210, 212
Aeration, 206, 208, 209, 295, 338, *see also* Flooding, Soil atmosphere
Aging, 130, 132–134, *see also* Senescence
Air pollution, 131

Albedo, 336
Alkaloids, 238, 243
Alvim porometer, 114
^{241}Am, 15, 16, 18
Amides, 308
Amino acids, 240, 242, 307, 308, 339, *see also* individual amino acids
Ammonia, 222, 307
Ammonium nitrate, 221, 222
Amphistomatous species, 109, 119, 120, 141, 142
Anaerobiosis, 211, 212
Aneuploidy, 323, 324
Angiosperms, 65–72, 82, 83, 206, 291
Anions, 83–86, 168, *see also* individual anions
Anthercerotae, 61
Anthesis, 122, 135, 335, 338, 342
Antitranspirants, 145
Apical meristem, 132, 294
Arsenic acid, 221
Ascorbic acid, 240
Aspartate, 87, 88
Assimilation, 142, 144, 145, 155, 321, 339, *see also* Photosynthesis
Atmometer, 247, 248
ATP, 90, 92, 162, 164, 243, 244, 305
ATPase, 91, 92, 307
Attenuation method, 11–16
Auxin, 192, 193, 210, 278–280, *see also* 2,4-D, 2,4,5-T, IAA, NAA

B

Backcross, 318, 322–324, 332
Backscattering, 12
Bacteria, 215, 281, 292, 293
Bacterial wilt, 215
Bacteroids, 300, 305–307

370